The Earth
Problems and Perspectives

The Earth
Problems and Perspectives

Barbara C. Atkin

Department of Geology and Geophysics
University of California, Berkeley

Jeffrey A. Johnson

Department of Geology and Geophysics
University of California, Berkeley

with contributions by

Garniss Curtis
Luna Leopold
Lionel Weiss

Blackwell Scientific Publications

Palo Alto   Oxford   London   Edinburgh   Boston   Melbourne

Sponsoring Editor: John H. Staples
Development: Andrew Alden
Interior and Cover Design: Gary Head Design
Production Editor: Larry Olsen
Artists: Robin Mouat, Colleen Donovan, Pauline Phung
Color Separation: Colorprep
Composition: Graphic Typesetting Service
Printer and Binder: Arcata Graphics, Kingsport
Cover: Marble Canyon, Grand Canyon
      National Park. Photograph by Barbara Atkin.

Editorial Offices

667 Lytton Avenue, Palo Alto, California 94301
Osney Mead, Oxford, OX2 0EL, UK
8 John Street, London WC1N 2ES, UK
23 Ainslie Place, Edinburgh, EH3 6AJ, UK
52 Beacon Street, Boston, Massachusetts 02108
107 Barry Street, Carlton, Victoria 3053, Australia

Library of Congress Cataloging-in-Publication Data

Atkin, Barbara C., 1938-
   The earth: problems and perspectives.

   Bibliography: p.
   1. Geology—Laboratory manuals.   I. Johnson, Jeffrey
A., 1956-   . II. Title.
QE44.A85   1988   551   87-31985
ISBN 0-86542-325-3

Distributors

USA and Canada
Blackwell Scientific Publications
P.O. Box 50009
Palo Alto, California 94303
(415) 965-4081

UK
Blackwell Scientific Publications
Osney Mead
Oxford OX2 0EL
011 44 865-240201

Australia
Blackwell Scientific Publications (Australia) Pty Ltd
107 Barry Street, Carlton
Victoria 3053

Preface

*The end of all our exploring will be to arrive where we started
and know the place for the first time.*

T. S. ELIOT

The Earth: Problems and Perspectives is intended for use as a practical text and laboratory manual in a first course in geology. In this manual, students are introduced to the basic concepts of geology and the working methods and philosophy of the geologist. Each exercise discusses an important topic in geology and illustrates the topic with photographs, maps, charts, or tables that are to be used to solve geologic problems. It is expected that the instructor also will provide some materials, such as rock and mineral samples and some local maps, for some of the problems. This book is perforated and three-hole punched, allowing students to remove and hand in answer sheets for grading. When returned, the pages can be stored in a three-ring binder. The Glossary, Appendices, and the color photographs and maps at the end of the book are not to be removed from the book and are intended to be used for reference throughout the course.

This manual is divided into five parts. Part A deals with the materials that make up the earth—minerals and rocks. The first exercise discusses order in nature, the crystalline state, and minerals and their properties. In contrast to other approaches, we discuss minerals in the context of the rocks in which they most often occur. For example, minerals that are formed in igneous processes are studied along with plutonic and volcanic rocks. Our study of rocks emphasizes their settings and formative processes, from local features to field relationships to the global picture of blueschist metamorphism or basaltic volcanism. Rock samples and hand specimens in the laboratory must be understood in terms of their large-scale settings and the processes that produced them.

Geology is a historical as well as an analytical science. Part B begins with an exercise on geologic time that features a correlation of stratigraphic sequences across the Colorado Plateau. The remaining exercises in this part are concerned with two important tools for the geologist, topographic and geologic maps.

Part C discusses the geologic processes that are powered by the earth's external source of energy, the sun. The exercises focus on weathering, slope processes, groundwater, rivers and streams, open bodies of water and shorelines, glaciers, and wind. The discussion emphasizes the geologic processes involved, rather than simply the landforms produced.

Processes caused by the earth's internal energy are the subject of Part D. The earth's topography, heat, gravity, and magnetism are explored in these exercises. Earthquakes and plate tectonics are a main focus in this part. Exercise 18, Plate Tectonics, brings together concepts and data from several of the previous exercises in order to provide a comprehensive model of the earth's structure. Students can construct a simplified model of plate movements along the western margin of North America at the end of this exercise. A plate-tectonic "movie" showing the evolution of the plates appears in the lower right pages of Exercises 21 and 22.

Part E is concerned with planetary geology and the way geology has influenced civilization and conversely how civilization has affected the earth. These exercises may serve as a comprehensive review of the preceding exercises, or sections of these exercises may be assigned as additional instruction for the preceding exercises on that topic. For example, the first section in Exercise 22 may be added to Exercise 2, Igneous Minerals and Rocks. These exercises deal with some unusual and interesting historical situations as well as with the traditional topics of geologic hazards and environmental geology.

At the end of each part there is a References section for further reading. At the end of the book are Appendices containing key illustrations and tables, a Glossary, color photographs of rock and mineral samples, and color maps.

Throughout this book we have emphasized the idea that the earth is a dynamic system that is constantly changing due to internal and external processes. Our physical environment is a beautiful, complex natural system in delicate equilibrium. We hope that students will come to see that knowledge and understanding of that natural system can be a source of pleasure and satisfaction. Another goal of this manual is to familiarize students with the scientific method. Whenever it is appropriate, we ask students to synthesize a general law from specific observations because this process, inductive reasoning, is at the heart of the scientific method. We also ask them to test hypotheses and provide alternative explanations, for this is the essence of how science is done.

Wherever possible, we encourage instructors to use local maps and materials in these exercises so that students may become familiar with their local surroundings. For example, the soil profile materials used in Exercise 9 should be collected locally, not purchased. We hope to stimulate students to understand and perhaps really see for the first time the local landscape they inhabit.

The Field Study sections in the exercises are intended as a comprehensive synthesis of the main ideas in each exercise. Field Studies focus mainly on national parks, national monuments, and national recreation areas. If students have not already visited these areas, we hope that reading about them and understanding how their features came into being will stimulate students to visit and enjoy the special scenic areas of our country.

ACKNOWLEDGMENTS

We would like to acknowledge the help of many people who contributed both moral and tangible support to our project.

- Luna Leopold got us started.

- Garniss Curtis read the manuscript and kept us honest.

- Lane Johnson read and commented on the earthquake exercise.

- Peggy Gennaro, photographer par excellence, printed the black and white photographs that begin each part and supplied rock and mineral specimens for the color photographs from the Museum collection of the Department of Geology and Geophysics, University of California at Berkeley.

- Janet M. Sowers made valuable contributions to Part C, Surface Processes.

- Mike Wopat and Kevin Stewart kept rocks and minerals in line and provided many helpful suggestions on the use of this manual.

- The staff of the Earth Sciences Library at the University of California at Berkeley allowed us to rummage undisturbed through their maps and books.

- Hallam Noltimeier of Ohio State University and Robert Twiss of the University of California at Davis read the manuscript and made helpful suggestions.

- Barbara Tewksbury of Hamilton College furnished some material for Exercise 19.

- Our teaching assistants ferreted out glitches, typos, and muddy waters. Our students struggled through these exercises and let us know where we went wrong and occasionally where we went right.

- Andrew Alden quibbled over details and made major contributions while doing a fine job of editing the manuscript; Robin Mouat did the beautiful illustrations; Gary Head was the designer; and Larry Olsen coordinated the publication with unfailing good humor and patience.

- And, most of all, Lionel Weiss was a godsend! He provided facts, source materials, comments, criticism, and unfailing interest in the project. He read the entire manuscript and made valuable comments on everything, especially rocks, maps, geophysics, and structural geology.

A few great people taught us what geology and geologists are all about:

- Charles Meyer and Clyde Wahrhaftig, University of California, Berkeley

- Donald B. Potter, Hamilton College

- Gertrude Spremulli, University of North Carolina at Chapel Hill

- Richard Hay, University of Illinois

- John Breyer and Art Ehlman, Texas Christian University

Finally, many thanks to the late John Staples of Blackwell Scientific Publications. Without his unfailing support, encouragement, advice, and occasional polite prodding, this book would not have happened.

BARBARA ATKIN

JEFFREY JOHNSON

Credits

Several of the illustrations in this book were adapted from a variety of published and unpublished sources. Credits for recently published figures appear below. We have attempted to acknowledge the sources of older, unpublished figures as well, but in some cases the source was unknown or unavailable to us. We would like to thank all who have provided data and illustrations that appear in this book.

Frontispiece and part opening photographs courtesy of the late Vaughn Culler.

Plate-tectonic movie courtesy of Christopher Scotese, Shell Oil Company, Houston, Texas.

Plate-tectonic model in Exercise 18 courtesy of Lionel Weiss.

3.2   Adapted from M.C. Powers, 1953, *Journal of Sedimentary Petrology*, v. 23, figure 1, copyright by the Society of Economic Paleontologists and Mineralogists.

3.3   Adapted from Davies, D.K., and W.R. Moore, 1970, *Journal of Sedimentary Petrology*, v. 40, figure 1, copyright by the Society of Economic Paleontologists and Mineralogists.

4.5   Adapted from A.L. Albee in *Studies of Appalachian Geology* by E. Zen, W.S. White, J.B. Hadley, and J.B. Thompson. New York: Wiley Interscience, 1968.

4.7   Adapted from USGS.

6.3   Adapted from USGS.

6.5   USGS, Tau Rho Alpha.

6.7   Aerial photograph, USGS.

6.8   Adapted from USGS, Tau Rho Alpha.

6.10   Adapted from *Elements of Geography—Physical and Cultural* by Vernor C. Finch, Glenn T. Trewartha, Arthur H. Robinson, and Edwin H. Hammond, 4th edition. New York: McGraw-Hill, 1957, p. 216, figure 11.10.

7.12   Adapted from USGS.

7.16   Adapted from USGS.

8.4   Aerial photo, USGS.

8.6   USGS.

9.5   Adapted from L. Peltier, "The Geographical Cycle in Periglacial Regions," *AAAG* 40(1950), pp. 214–236.

9.7   Adapted from Janet Sowers.

10.2   Adapted from USGS.

10.6   Aerial photo, USGS.

10.7   Aerial photo, USGS.

10.8   USGS.

10.9   USGS.

10.10   USGS.

10.11   USGS.

10.12   Aerial photo, USGS.

11.1   Adapted from USGS.

11.3   *U.S. Department of Agriculture Yearbook*, 1955, H.E. Thomas, ed.

11.4   Adapted from Janet Sowers.

11.5   K.J. Hsu, "Studies of the Ventura Field, California, II: Lithology, Compaction, and Permeability of Sands." *American Association of Petroleum Geology Bulletin*, v. 61, figure 6B.

11.6   L.B. Leopold, *Water, A Primer*, W.H. Freeman and Company.

11.8   Adapted from USGS.

11.9   USGS.

11.10   Aerial photo, USGS.

11.11   Aerial photo, USGS.

11.12   USGS.

11.14   USGS.

12.4   Adapted from USGS.

12.6   From *Fluvial Processes in Geomorphology* by Luna B. Leopold, M. Gordon Wolman, and John P. Miller. Copyright 1964, W.H. Freeman and Company, figure 7.5, p. 205.

12.9   Adapted from N.D. Smith, *Geological Society of America Bulletin*, v. 81, p. 2996, figure 6, 1970.

12.11   USGS.

12.13   Aerial photo, USGS.

12.14   USGS.

12.15   USGS.

12.18   Aerial photo, USGS.

12.20   USGS.

12.21   Aerial photo, USGS.

12.22   USGS.

12.23   USGS.

12.24   Adapted from *Fluvial Processes in Geomorphology* by Luna B. Leopold, M. Gordon Wolman, and John P. Miller. Copyright 1964, W.H. Freeman and Company, figure 7–13, p. 228.

13.7   Aerial photo, USGS.

13.11   Aerial photo, USGS.

13.13   Aerial photo, USGS.

13.14   Adapted from Gawne, C.E., 1966, "Shore Changes on Fenwick and Assateague Islands, Maryland and Virginia." University of Illinois Bachelors thesis in Geology.

13.15   Courtesy of Blackwell Scientific Publications.

13.16   USGS.

13.17   Courtesy of Blackwell Scientific Publications.

13.20   From Dolan, Robert, and P. Godfrey, 1973, *Geological Society of America Bulletin*, v. 84, p. 1331, figure 3.

13.21   Adapted from USGS.

13.22   USGS.

14.5   USGS.

14.8   Aerial photo, Canadian Geological Survey.

14.10 USGS.

14.13 Aerial photo, USGS.

14.14 From K. Graetz and F.T. Thaites. University of Wisconsin, 1933.

14.15 USGS.

14.16 Aerial photo, USGS.

15.10 Aerial photo, USGS.

15.11 Courtesy E.D. McKee, *Sedimentology*, v. 7, 1966, figure 4, p. 13.

18.4 Reprinted with permission from Heirtzler, J.R., LePichon, X., and Baron, J.G., "Magnetic Anomalies over the Rekjanes Ridge," *Deep-Sea Research*, v. 13. 1966, Pergamon Journals, Ltd.

18.6, 18.8 Sykes, L.R., *Journal of Geophysical Research*, v. 72, pp. 2131–2153, 1967, copyright by the American Geophysical Union.

19.1 Viking Lander 2 image courtesy of National Space Science Data Center, Michael H. Carr, Team Leader.

19.2 Voyager 1 image courtesy of National Space Science Data Center, Bradford A. Smith, Team Leader.

19.3 Mariner 9 image courtesy of National Space Science Data Center, Harold Masursky, Team Leader.

19.4, 19.5 Lunar Orbiter image courtesy of National Space Science Data Center, Leon J. Kosofsky, Principal Investigator.

19.6, 19.7, 19.8, 19.9, 19.10 Viking Orbiter 1 image courtesy of National Space Science Data Center, Michael H. Carr, Team Leader.

19.11, 19.12 Voyager 1 and 2 images courtesy of National Space Science Data Center, Bradford A. Smith, Team Leader.

21.2 Adapted from Theodore C. Smith and Earl W. Hart, California Division of Mines and Geology.

21.3 Aerial photo, USGS.

21.4 Courtesy of California Division of Mines and Geology.

21.7 USGS.

21.8 Adapted from USGS.

21.12 USGS.

21.14 U.S. Department of Commerce.

12.15 Aerial photo, USGS.

22.1 Adapted from G. MacDonald, *Catalogue of the Active Volcanoes of the World.* Naples: International Volcanological Association, 1955.

22.2 USGS.

22.3 USGS.

22.4 USGS.

22.5 USGS.

22.10 Aerial photo, USGS.

22.11 USGS.

22.13 Courtesy of California Division of Mines and Geology.

Contents

To the Student

This manual has been specially designed to provide you with the materials you need to understand key geologic concepts and solve geologic problems. Each exercise contains photographs, maps, charts, tables, and illustrations for your use in solving the problems. At the end of the book are color photographs, color maps, Appendices, and a Glossary for your reference. Within the text pages, we have provided space for you to write your answers to the problems. The book is perforated and three-hole punched so that you can tear out the answers and hand them in to your instructor for grading. When returned, these pages should be kept in a three-ring binder for reference in solving problems that appear later in the book.

Several problems in this book require students to view stereo photographs and refer to a topographic map of that same area. Students should tear out the pages of the stereo photos and arrange them under the stereo viewer in the lab to get the three-dimensional effect. To avoid placing the corresponding topographic map under the photos, we have placed the maps a page or two pages away from the photos so that the map can also be removed from the book and viewed face up along with the photos. Page cross-references and figure numbers in the text will direct you to the appropriate maps.

Some of the exercises in this book require use of a hand lens, a calculator, and a metric ruler. Students will be asked to draw on photos and maps with different colored pencils. To use this book effectively, we suggest that you bring the following materials to your first lab:

• a three-ring binder

• a number 2 pencil with an eraser

• a 10× hand lens (not a magnifying glass)

• a ruler with metric and English units

• a calculator

• at least 6 colored pencils (red, yellow, green, blue, brown, and purple)

In Exercises 7, 8, and 18 you will construct models using forms printed on the pages. We suggest that you buy some thin card stock or construction paper and glue the model page to the heavy paper in order to make a sturdier model.

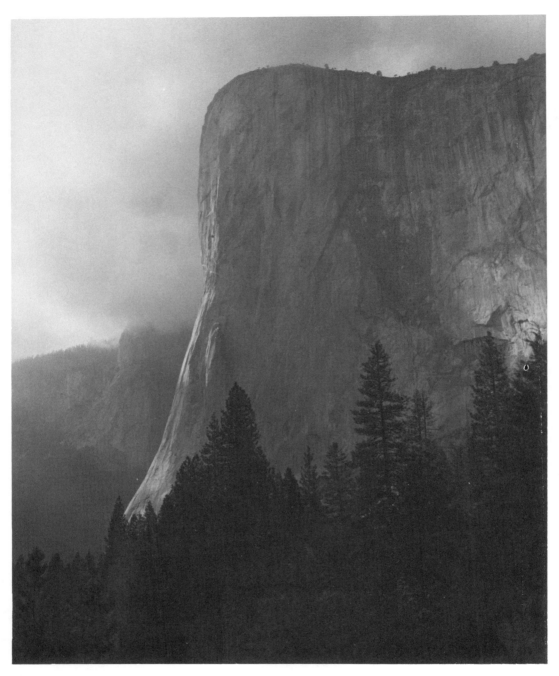

El Capitan, Yosemite National Park, California. This sheer cliff was carved from granite of the Sierra Nevada batholith by glaciers that covered the area 10,000 years ago.

To Our Parents
To Steve and Tim

The Earth

Problems and Perspectives

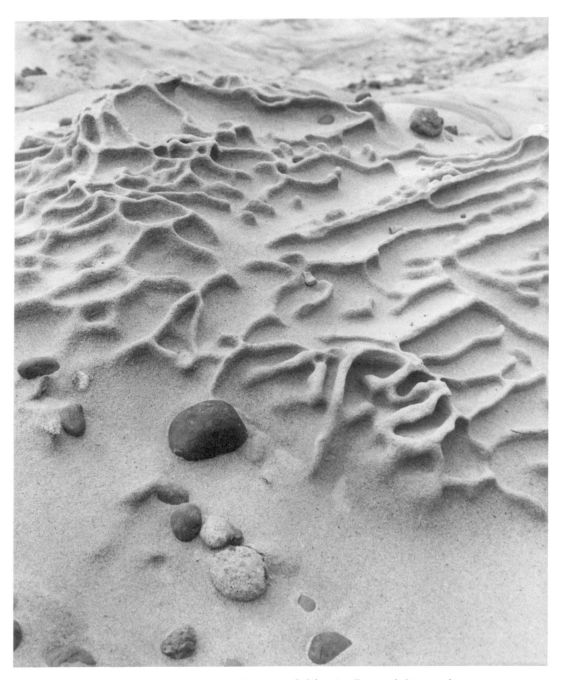

Weathered sandstone, Point Lobos State Reserve, California. Parts of this sandstone were preferentially eroded, leaving abstract patterns standing in relief.

Materials That Make Up the Earth

In the first four laboratory exercises, you will study the various minerals and rocks that make up the earth's crust. Minerals are important, not only because they are the building blocks of the earth's crust but also because they embody the concept of order in nature. By studying the physical and chemical characteristics of minerals, we can identify minerals in rocks. The study and identification of rocks is of great importance to geologists because much of the geologic history of an area may be inferred from a study of its rocks. Rocks are the history books of the earth's crust.

The principle of uniformitarianism, a rule of geological reasoning first clearly stated by geologist James Hutton in the late 1700s, holds that the physical processes operating on the earth at present are not too different from the ancient processes that produced the rocks we see around us. For example, a geologist watching a volcanic eruption might recognize that a certain type of rock, an olivine basalt with a ropy, wrinkled surface, forms when the lava flow cools. When this particular kind of rock occurs among ancient rocks elsewhere, the geologist assumes that it was produced in the same way as its modern analog—in a volcanic flow. Similarly, one can dig down into a modern sand dune, in the desert or on the beach, and see well-rounded quartz grains with a frosted appearance arranged in intersecting tilted layers several inches to several feet thick. The presence of rocks composed of quartz grains of similar appearance, arranged in the same kind of layering, implies that the area in which these rocks occur was once a beach or a dune-covered desert.

The uniformitarian concept is often stated as "The present is the key to the past." This concept implies that the laws of physics and chemistry do not change over time and that no unique events are necessary to explain what we see in the rock record.

To identify a rock and assess its environment of formation (and hence its geological history) requires close study of two properties, the *mineral composition* and the *texture* of the rock. Since rocks are aggregates of minerals, an understanding of what a mineral is and how it may be identified is fundamental to the study of rocks. A *mineral* is a naturally occurring crystalline solid whose chemical composition is fixed or varies only within certain limits. (Noncrystalline substances, such as opal or mercury, that satisfy all the other criteria of a mineral are called *mineraloids*.) You will explore some of these attributes of minerals in Exercise 1, and you will use some simple physical properties of minerals to identify them. In Exercises 2, 3, and 4, you will study the common minerals of the three major rock types, *igneous, sedimentary,* and *metamorphic,* and learn to identify important rock *textures*. The texture of a rock is the geologist's first diagnostic tool for evaluating how the rock formed and for inferring the geologic history of the area where it occurs.

States of Matter, Crystal Forms, and Minerals

Structure and pattern in the natural world are Nature's way of conserving energy.

NOVA

The study of geology in the field or laboratory is based on accurate and perceptive observation. The art of observation and, even more important, the ability to ask yourself relevant questions are skills that can be learned with practice. The first two parts of this laboratory exercise are designed to enhance those skills. The third part deals more specifically with the substances called minerals and the ways in which minerals reflect a basic order in nature. The final section concerns the identification of minerals.

OBSERVATION AND CLASSIFICATION

The physical world around you contains many things. To describe or remember each one individually would be impossibly cumbersome. For this reason people have developed methods of grouping constituents of the physical world into categories that emphasize their common properties. Many such classification systems are used. One of the simplest and most universal groupings is the division into the three physical states—solid, liquid, and gas.

Consider the properties of a single substance, water, in these three states—ice, liquid water, and water vapor. We have no trouble distinguishing among these three states, but it is not quite as easy to define them in such a way that one who had never seen them could distinguish among them. What physical properties may be used to characterize each of the three states of matter?

1. *Color:* Although color is an obvious property of matter, it is not relevant in distinguishing solids from liquids and gases.

2. *Density:* In general gases are less dense than liquids, and liquids are less dense than solids, but there are exceptions.

3. *Strength or hardness:* Here is a more diagnostic property. Liquids and gases have little of the cohesiveness that makes a substance hard or rigid.

4. *Shape:* Shape is probably the most definitive characteristic for distinguishing among the three states of matter. Gases have no shape and are not confined by an open vessel. Liquids assume the shape of their container and will flow relatively easily into another shape when poured. Only a solid possesses a true shape. A cube of ice will remain a cube regardless of the shape of its container.

Problem 1

What simple physical properties distinguish a piece of paraffin wax from a piece of window glass?

We shall begin our study of geology by examining rocks and minerals in the solid state.

Problem 2

Examine the rock furnished by your instructor. Clearly it is made up of a number of constituent grains. Your first commonsense step is to group all the grains with similar properties together and then describe the group ("the white, dull substance" or "the black flakes"). We need to decide what properties are useful to distinguish one substance from another. Reexamine carefully the discussion of properties that might be used to classify states of matter and your own answer to Problem 1. List five physical properties that will help you to distinguish one group of substances from another. Check these properties with your instructor before proceeding.

1.

2.

3.

4.

5.

Problem 3

Make a chart that describes the simple physical properties you find most helpful in distinguishing one group of grains (one substance) in the rock sample from another; list properties across the top, then fill in the spaces for that substance. There may be fewer than five different substances in your rock; it is up to you to decide.

Simple Physical Properties

Substance	1	2	3	4	5
1.					
2.					
3.					
4.					
5.					

MINERALS

Crystal Shape

The substances in the rock that you studied are called *minerals.* A *rock* is an aggregate of *minerals.* What exactly is a mineral? As we have seen, minerals are solid substances. Geologists make the further distinction that *minerals are naturally occurring* to distinguish them from manmade objects.

One of the two most important characteristics of a mineral is that, in addition to being a solid, *its atoms and molecules are arranged in an orderly fashion.* This orderly internal arrangement of atoms and molecules is expressed in the external *crystal shape* of the mineral when it forms in an unrestricted space. Your instructor will furnish well-formed crystals of several minerals, including quartz, halite, and pyrite (or galena). Use these, as well as Color Plate 1, to answer the following questions and problems.

Problem 4

Sketch the various mineral specimens. Draw both perspective and cross-sectional (end-on) views, as illustrated for quartz in Figure 1.1. Place a scale by each drawing to indicate the size of the sample.

	Perspective View	Cross-Sectional View
1. Quartz		
2. Halite		
3. Pyrite (or Galena)		

A.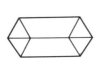

B.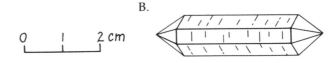

Figure 1.1
Quartz crystals of various shapes. A. Cross sections.
B. Perspective view of quartz crystal.

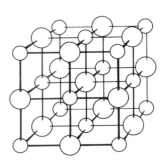 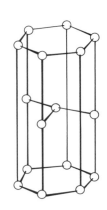 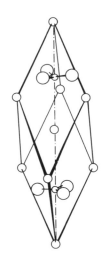

Figure 1.2
Some mineral structures.

The crystal shape of a mineral reflects the way its atoms are arranged. This atomic arrangement is called the *crystal lattice.* The atomic structures of most minerals are quite complex, however, and the relation of the shape of the crystal to the positions of its constituent atoms is not obvious in most cases. Only in some minerals with simple chemical formulas, such as halite (NaCl), does the atomic structure clearly correspond to the external shape of the well-formed crystal. Figure 1.2 shows the atomic structures of several types of crystals; the small circles represent atoms, and the lines represent chemical bonds.

Problem 5

From your sketches of the crystal shape of halite, mark which of the structures in Figure 1.2 you believe is closest to the structure of halite.

Problem 6

Using a contact goniometer, as shown in Figure 1.3, measure the angles between adjacent faces of the various minerals in a plane perpendicular to the two faces (in the plane of the page in the cross-sectional sketch of quartz). Label these angles on your cross-sectional sketches. These angles are called the *interfacial angles.*

Problem 7

Exchange quartz crystals with a neighbor, or examine other samples that may look different from your first specimen; for example, two or three sides may be elongated, as shown in Figure 1.1. Measure the inter-facial angles of several of these quartz crystals and record the results.

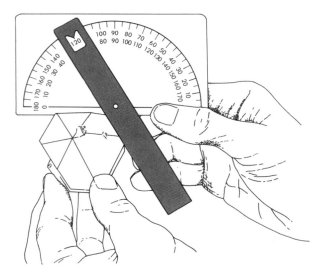

Figure 1.3
Use of the contact goniometer.

Problem 8

Crystals of quartz may look different, but their crystal structures are identical. Which property is the same for all the quartz crystals you examined: The number of crystal faces? The size of the faces? The shapes of the faces? The angle between faces?

The property you have just discovered is called the *law of constancy of interfacial angles.* The angles between corresponding faces of crystals of a particular mineral are constant, regardless of the shape or size of the crystal.

Problem 9

Compare crystals of quartz, halite, and pyrite. Which two minerals would you place in the same group, using the criterion of crystal shape?

Problem 10

If you grouped them according to color, which two would be in the same group?

Problem 11

Examine the specimens of various types of quartz provided by your instructor. What colors do you see? Would you say that color is a reliable and constant property of quartz?

Chemical Composition

Halite and pyrite have the same interfacial angles and are members of the same crystal system; that is, they have a very similar internal arrangement of atoms. They are not the same mineral, however, because their chemical compositions are very different. Several simple physical observations suggest that this is so. Properties most indicative of a different chemical composition are the following.

Problem 12

1. *Density:* Which mineral is more dense?

2. *Luster:* Do the two minerals differ in the way in which their surfaces reflect light?

3. *Color and transparency:* What is the difference between the two minerals?

The chemical formula of halite is NaCl (sodium chloride); that of pyrite is FeS_2 (iron sulfide). The chemical composition is the second of the two most important distinguishing characteristics of a mineral. *The chemical composition of a mineral is fixed or varies within fixed limits.*

Characteristics of Minerals

Problem 13

We have explored in some detail the characteristics that define the group of substances geologists call minerals. Write your own definition of a mineral. Be sure you include all three criteria (emphasized in the text above) that are essential to distinguish minerals from other classes of objects.

CRYSTALLINE AGGREGATES

Crystals with the ideal crystal shape, such as those you have seen in this exercise, occur only rarely in nature. More commonly, mineral crystals growing from liquids interfere with each other's shape at some point during their growth; the result is an aggregate of many single crystals, each crystal having an irregular outline.

Problem 14

Single crystals: Thymol is an organic compound that melts at 51.5°C. (*Note:* It is not poisonous, but you should avoid getting it on your skin because it may cause a burning sensation. Wash your hands when you finish this section of the exercise, and be sure not to rub your eyes.) Melt the thymol in a covered petri dish over a hot plate at low heat. Make certain it is entirely melted, but avoid overheating. Remove it from the hot plate and allow it to cool. As the melt reaches room temperature, small crystals of thymol will form. (You may have to add a small grain of thymol—a *seed crystal*—to the cooled melt to start crystallization.) These are *single crystals* of thymol, which grow by adding material to their faces. Sketch the typical shape of a thymol crystal, including any internal growth lines you see. Include a size scale.

Problem 15

Crystal aggregates: When the single crystals have grown in size to the point where they touch each other, growth cannot take place at the points of contact but only can take place where liquid is still present between the crystals. When the entire mass is solid, sketch a portion of it showing several crystals, their external boundaries, and any visible internal structure such as growth lines within the crystals.

Problem 16

How are the final external boundaries of most crystals related to the initial shape of the crystal?

The final solid product of the thymol melt is called a *crystalline aggregate;* it is crystalline in nature, that is, the atoms of each crystal are arranged in an orderly fashion, but it is not a single crystal. The rock you examined earlier also displays the typical *interlocking texture* of a crystalline aggregate. Compare the crystalline aggregate of thymol with the rock, and note the random orientation of the crystals of each. Crystalline aggregates of minerals such as these are far more common in nature than well-formed single crystals of minerals.

IDENTIFICATION OF MINERALS

A mineral is identified by its chemical composition and the physical arrangement of its atoms and molecules, that is, its crystals. Although some very simple chemical tests are helpful in identifying certain minerals, over 90 percent of the earth's crust is composed of silicate minerals (chemical compounds made up primarily of oxygen and silicon), and these minerals do not lend themselves to simple chemical analysis. The physical properties of minerals are much more useful for rapid identification. You have already learned some of the important physical properties of minerals for yourself. Now you will identify minerals using these and other properties. You may find a hand lens useful in mineral identification to enlarge what you see. Appendix 3 shows the proper use of a hand lens.

Crystal Shape and Habit

The *habit*, or characteristic appearance, of a mineral often reflects its basic internal structure—how the atoms are arranged to form that mineral. Silicate minerals whose atoms are arranged in chains or bands, for example, form needlelike shapes. Micas are silicates with two-dimensional sheetlike arrangements of silicon and oxygen atoms, and they have a characteristic sheetlike habit.

Only minerals that crystallize from the molten state earliest have the space to develop well-formed crystals, so crystal habit is diagnostic primarily for early forming minerals. Minerals crystallizing at lower temperatures are forced to grow in the spaces left between

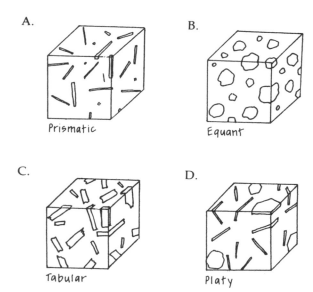

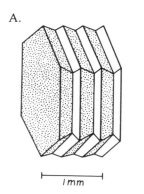

Figure 1.5
Albite twinning in plagioclase. A. Sketch of plagioclase crystal; one set of albite twins is stippled, the other plain. B. Albite twinning as it typically appears on a flat crystal face.

Figure 1.4
Some common crystal habits.
A. Prismatic (1 long dimension, 2 short dimensions)
B. Equant (all 3 dimensions equal)
C. Tabular (1 long dimension, 1 short dimension, 1 intermediate dimension)
D. Platy (2 long dimensions, 1 short dimension)

early formed crystals and are characteristically irregular in shape. Figure 1.4 shows sketches of four common crystal habits. These habits are also shown in the color photographs. Color Plate 16 shows equant habit in olivine; Color Plates 15 and 17 show prismatic habit in hornblende and pyroxene; Color Plates 7, 9, and 14 show tabular habit in feldspars; Color Plate 12 shows platy habit in biotite.

Certain crystal systems are prone to reflections or rotations of their crystal structure during growth, a phenomenon known as *twinning*. *Albite twinning*, a common example, manifests itself as a series of straight and parallel striations on the face of a single crystal. If you orient a twinned crystal so as to reflect the light to your eye from a broken face, twinning appears as straight, dark (nonreflecting) lines across the shiny reflecting face, similar to the grooves on a phonograph record. Figure 1.5 shows diagrams of albite twinning in the mineral plagioclase. Color Plate 2 shows albite twinning in a large crystal of plagioclase (the white mineral). Notice that albite twinning appears on only four of the six faces of a well-formed crystal of plagioclase.

Albite twinning is particularly important in distinguishing plagioclase feldspar, which is often twinned, from potassium feldspar, which does not exhibit albite twinning. Potassium feldspar commonly exhibits perthitic structure, which can be mistaken by the beginning student for albite twinning. Perthitic structure is the result of the intergrowth of two chemically different feldspar phases. Whereas the lines of albite twinning are straight and parallel, perthitic intergrowths are wavy subparallel bands of slightly different color or transparency. Color Plate 3 shows perthitic structure in potassium feldspar.

Problem 17

Examine the feldspar samples provided by your instructor. By the presence or absence of albite twinning, state which sample is plagioclase and which is potassium feldspar.

Cleavage

When some minerals fracture, they break along certain planes in the crystal structure. The bonding between atoms along these planes is weaker than bonding in other crystallographic directions. The preferred direction of breakage in a crystal is called *cleavage*. Like crystal form, cleavage reflects the internal structure of the mineral. In the sheet silicates called micas, the bonding between separate silicate sheets is much weaker than the covalent bonds within each sheet, resulting in the characteristic cleavage of micas into thin sheets. Micas cleave repeatedly along parallel planes in the crystal and therefore have one *cleavage direction*.

Some minerals have more than one cleavage direction. In many cases the angle between cleavage planes is used to distinguish one mineral from a similar-looking mineral—pyroxenes from amphiboles or calcite from halite, for example. Figure 1.6 illustrates several different types of cleavages.

Cleavage in some minerals, such as micas, calcite, and halite, is excellent, and the cleavage surface is smooth and shiny. In other minerals cleavage is poor to good, and cleavage surfaces are not as reflective.

The molecular structure of some minerals is such that the bonds are of similar strength in all directions. Such minerals do not have cleavage; they *fracture* in an irregular manner. Common types of fracture are conchoidal, as in quartz, or fibrous, as in asbestos. Figure 1.7 shows the difference between cleavage and fracture. Color Plate 4 shows excellent cleavage in three directions in calcite; Color Plate 20 shows conchoidal fracture in volcanic glass.

One final caution in identifying cleavage surfaces concerns the distinction between growth faces of a crystal and cleavage planes, which may resemble each other. Both reflect the geometry of the crystal structure, but crystal faces are growth phenomena of crystals in an open space, whereas cleavage planes result from crystal breakage. Every cleavage plane is a possible growth face of a crystal, but some minerals that develop good growth faces have no cleavage planes at all—quartz, for example. Most of the rock samples you collect in the field or examine in the laboratory have been broken; the shiny surfaces you see in these cases are cleavages or fracture surfaces.

Problem 18

Examine the mineral fragments provided by your instructor and use them to complete the chart below.

Sample Number	Number of Cleavage Directions	Angle Between Cleavages
1.		
2.		
3.		
4.		

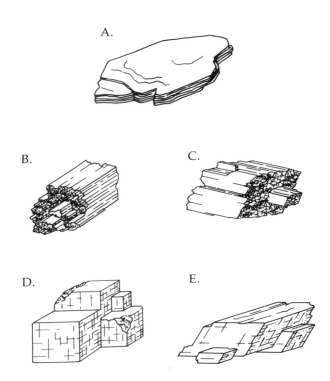

Figure 1.6
Cleavage in minerals.
A. Cleavage in one direction (micas).
B. Cleavage in two directions, at right angles (feldspars, pyroxenes).
C. Cleavage in two directions, not at right angles (amphiboles).
D. Cleavage in three directions, at right angles (halite, galena).
E. Cleavage in three directions, not at right angles (calcite, dolomite).

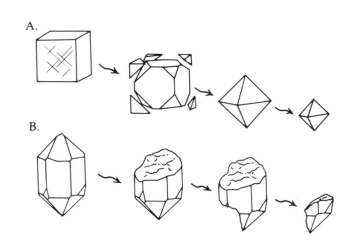

Figure 1.7
Cleavage and fracture. A. Cleavage (fluorite).
B. Fracture (quartz).

Hardness

The *hardness* of a mineral is a measure of its resistance to scratching and is related to the interatomic bond strength—the stronger the binding force between the atoms, the harder the mineral. Friedrich Mohs, a German mineralogist of the nineteenth century, devised a scale on which minerals are ranked by hardness. Talc (or graphite) is the softest mineral, with a hardness of 1; diamond, at 10, is the hardest mineral. It is interesting to note that diamond and graphite are chemically identical, consisting of pure carbon; the great difference in hardness is due to the difference in chemical bonding and crystal form. The *Mohs hardness scale* is shown at right. Some common articles are included in the scale as references. For example, a knife blade may be used to distinguish quartz from calcite. The hardness of calcite is 3 and that of quartz is 7; hence, a knife blade with hardness $5\frac{1}{2}$ will scratch calcite but not quartz.

Care should be exercised in determining the relative hardness of two minerals. Sometimes when one is softer than another, parts of the first may rub off on the second and leave a mark that may be mistaken for a scratch. Such a mark can be easily removed by a damp finger, whereas a true scratch is permanent. It is always advisable to confirm the result of a scratch test by reversing the procedure—see whether the second mineral will scratch the first.

Mohs Hardness Scale

Mineral	Hardness	Hardness Standard
Diamond	10	
Corundum	9	
Topaz	8	
Quartz	7	$6\frac{1}{2}$—steel file
Orthoclase	6	$5\frac{1}{2}$—window glass or knife blade
Apatite	5	
Fluorite	4	
Calcite	3	3—copper
Gypsum	2	$2\frac{1}{2}$—fingernail
Talc	1	

Luster

Luster refers to the way light is reflected from the surface of a mineral. The terms *metallic* and *nonmetallic* are used to describe luster. The luster of pyrite is a typical metallic luster. Nonmetallic lusters can be further described by such terms as *earthy, waxy, resinous, silky,* and *vitreous (glassy)*. The metallic luster of pyrite is shown in Color Plate 1. Vitreous luster in quartz is shown in Color Plates 2 and 6, earthy luster in Color Plate 39, waxy luster in Color Plate 42, and silky luster in Color Plate 60.

Problem 19

Determine the hardness of the samples provided by your instructor.

Sample Number	Hardness
1.	
2.	
3.	
4.	

Problem 20

Describe the luster of the samples provided by your instructor.

Sample Number	Luster
1.	
2.	
3.	
4.	

Color

The *color* of a mineral is one of its most obvious properties, and for some minerals such as olivine (green) and galena (silver), it can be diagnostic. For most metallic minerals, color is constant. Other minerals such as quartz occur in wide varieties of color, so that color is without diagnostic significance. Color is sometimes helpful in distinguishing between potassium feldspar and plagioclase, which have very similar physical properties. Potassium feldspar is commonly white but may range from pale pink to salmon red. Plagioclase ranges from white to dark gray or shades of blue.

Streak

Streak is the color of the powdered mineral, which is often very diagnostic for minerals with a metallic luster. Although the color of a given mineral may vary, its streak is usually constant. The streak of a mineral is determined by rubbing it across an unglazed porcelain streak plate. Many nonmetallic minerals have a white or pastel streak, hence streak is not helpful in identifying them. Minerals that are harder than the streak plate (about 7) will scratch the plate and not produce streak at all.

Problem 21

Determine the streak of the samples provided by your instructor.

Sample Number	Streak
1.	
2.	
3.	
4.	

Specific Gravity

The *specific gravity* of a mineral is the ratio of its weight to the weight of an equal volume of water, that is, the mineral's density relative to water. This quantity may be measured very accurately in the laboratory but may be roughly estimated simply by hefting the specimen in the hand. Common rock-forming silicates have specific gravities from 2.6 to 2.8, but sulfide and oxide ore minerals have specific gravities ranging upward to 8.

Other Properties

Other physical properties, such as magnetism, fluorescence, radioactivity, and taste, are either obvious or of limited value and will not be discussed. You will use the seven major physical properties described above to examine minerals and rocks of igneous, metamorphic, and sedimentary origins in the next three exercises. This discussion, as well as the determinative mineral tables in Appendix 1, will be very useful during the course of rock and mineral study.

Problem 22

Identify the unknown minerals provided by your instructor by determining the relevant physical properties and consulting Appendix 1. Fill in the chart on page 11.

Sample Number	Crystal Habit	Cleavage/Fracture (Number of cleavages and angle between them)	Hardness	Luster	Color	Streak	Name of Mineral

Exercise 2

Igneous Minerals and Plutonic and Volcanic Rocks

One who values stones is surrounded by treasures wherever he goes.

PAR LAGERKVIST

Igneous rocks are formed by the cooling and hardening of *magma*, complex molten material that originates deep within the earth's crust and upper mantle as a result of the earth's internal heat. (The word *igneous* is derived from a Latin word for fire.) The chemical elements making up the magma do not solidify in a random fashion but take the form of specific minerals. Although several thousand minerals are known, those important in the formation of igneous rocks are relatively few in number, partly because a typical magma is rich only in certain elements: oxygen (O), silicon (Si), aluminum (Al), iron (Fe), calcium (Ca), sodium (Na), potassium (K), and magnesium (Mg). These eight elements constitute 99 percent of the earth's crust. The important rock-forming minerals found in igneous rocks are all *silicates*, that is, various metal ions linked by ionic bonds to silicon-oxygen structures of greater or lesser complexity that are held together internally by covalent bonds. Seven of these silicate groups— the olivines, pyroxenes, amphiboles, micas, plagioclase feldspars, potassium feldspars, and quartz— constitute over 95 percent of the volume of the common igneous rocks, a fact that greatly simplifies the task of rock identification.

Most of the minerals within each silicate group are very similar in composition, structure, and appearance, so we usually refer to a group of minerals by one name—plagioclase, olivine, potassium feldspar, or pyroxene, for example. When one mineral is easily distinguished from others in the group, it is referred to by its mineral name. Hornblende is the common black amphibole of igneous rocks; biotite is the black mica; quartz is the common igneous form of SiO_2.

MINERALOGY OF IGNEOUS ROCKS

Knowledge of silicate mineral structures is essential to an understanding of the mineralogy of igneous rocks. The silicate ion, SiO_4^{-4}, is tetrahedral in shape, with the silicon atom in the center bonded to four oxygens at the corners of the tetrahedron. Figure 2.1A shows a model of the silicate ion. The two drawings are from slightly different positions in order to show the structure in three dimensions when the two drawings are viewed with a stereographic viewer. The viewing technique is the same as that used for aerial photographs and is described on page 75. Ionic bonds between the tetrahedra and various metal ions hold isolated silicate tetrahedra together in the structure called an *island silicate*. Olivine and garnet are island silicates.

When conditions in the magma are appropriate, silicate tetrahedra polymerize into more complex structures by sharing oxygens. When adjacent tetrahedra share a common oxygen, a *single-chain* polymer forms. The structure of single-chain silicates is shown in Figure 2.1B. The single chains are held to one another by ionic bonds between the chains and various metal ions. Pyroxenes are important single-chain silicates. Further polymerization results in *double-chain* (or *band*) silicate polymers, such as the amphibole group. If three oxygens of each tetrahedron are shared, a two-dimensional *sheet silicate* forms (Figure 2.1C). Micas are sheet silicates. Sharing of all four oxygens results in a three-dimensional *framework silicate* structure, the most common silicate structure. Feldspars and quartz are framework silicates.

12

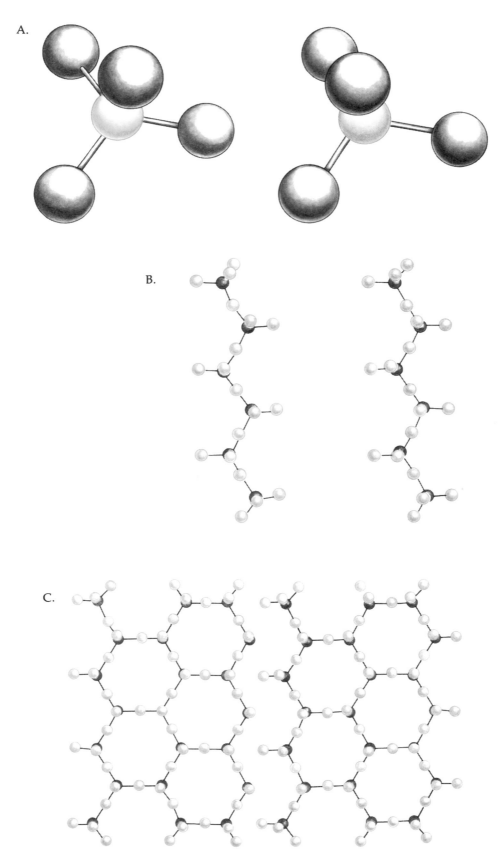

Figure 2.1
Silicate structures. A. Island structure. B. Single chain structure. C. Sheet structure.

Crystallization of silicate minerals from a magma occurs between 1700°C and 600°C. Experiments with silicate melts rich in iron and magnesium have established the general order in which the seven most common rock-forming igneous minerals crystallize. As a general rule, minerals with the highest melting points under natural conditions (in the presence of water, for example) crystallize first; then they react with the remaining magma to form other minerals. Figure 2.2 summarizes the important work in the 1920s of Norman L. Bowen in the field of silicate melt chemistry. You should not expect all the minerals listed in the chart to appear in the crystallization sequence of any one magma because the initial chemical composition of each different magma dictates which minerals will appear. The chart is useful, however, as a conceptual tool by which to remember silicate mineral structures and the minerals commonly found together in igneous rocks. In general, minerals in the same horizontal position in the chart are associated in igneous rocks. Gabbro, for instance, is a rock containing olivine, pyroxene, and calcium-rich plagioclase; granite contains quartz, biotite, and sodic plagioclase. Olivine and pyroxene are rarely seen in combination with biotite; conversely, rocks that contain significant amounts of quartz do not contain olivine. Crystallization and reaction of minerals in both the left and right sides of the chart proceed simultaneously in most magmas, so that minerals from each series are normally present in igneous rocks.

The minerals on the left side of the chart all contain iron or magnesium or both. These are the ferromagnesian minerals, also called *mafic* minerals (from *ma*gnesium and *fer*ro, meaning iron). The reaction series of mafic minerals is a *discontinuous reaction series;* that is, as the magma cools, minerals at the top of the chart crystallize first, then react with the remaining melt to form minerals lower in the chart, which have different mineral structures and compositions. Olivine is the first mineral in the series to crystallize from a mafic magma, and if crystallization proceeds slowly enough and the melt contains enough Si and O, it reacts with the magma at around 1550°C until all the olivine is converted to pyroxene. In theory any excess free silica crystallizes out last as quartz (SiO_2), and the rock contains pyroxene and quartz, plus plagioclase of some intermediate composition (from the right-hand series). If the initial composition of the melt is different, the crystallization and reaction series will be different. An amphibole may be the first mineral to crystallize from a less mafic melt, reacting with the melt at a lower temperature to form biotite.

The right-hand minerals in Figure 2.2 constitute the *continuous reaction series* of plagioclase feldspars, whose composition varies continuously between $CaAl_2Si_2O_8$ and $NaAlSi_3O_8$. (Calcium plagioclase tends to be darker and to exhibit more obviously the thin parallel striations of albite twinning than sodic plagioclase.) The continuous reaction series starts with a calcic plagioclase, which precipitates first from a melt. As the melt cools, sodium ions diffuse into the crystal lattice, replacing calcium ions. The final composition of the plagioclase depends on the initial chemical composition of the melt. There is no change in structure throughout the reaction series, only a gradual change in composition. The darker, more calcic plagioclase is most likely to be found in association with olivine and pyroxenes; lighter, more sodic plagioclase is found with

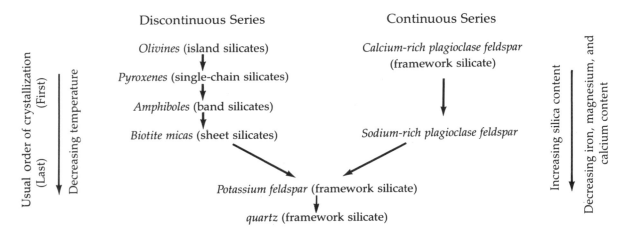

Figure 2.2
Order of crystallization of common igneous rock-forming silicate minerals according to the Bowen reaction series.

biotite, potassium feldspar (orthoclase), and quartz.

One common natural process shows the danger in assuming that crystallization of a melt, no matter what its composition, always begins with olivine. This process is *magmatic differentiation,* the separation of the liquid phase of a magma from the earliest formed crystals. In a mafic magma, olivine and calcium-rich plagioclase often do crystallize first, and by virtue of their greater density, they may settle to the bottom of the magma chamber. Alternatively the molten phase may be expelled from the magma chamber into another part of the crust, leaving the crystals behind. In either case the remaining magma is now depleted in iron, magnesium, and calcium (elements abundant in olivine and plagioclase) and is no longer mafic but is more *felsic* (from *fel*dspar plus *si*lica) in composition. This melt undergoes a different set of reactions, resulting in a different final mineral assemblage from the mafic melt that was the starting point. Quartz and potassium feldspar are commonly found as the early forming solids in some felsic rocks.

Notice the increasing complexity of silicate structure in the left-hand series as crystallization proceeds down the chart in the direction of lower temperatures. The first mineral to crystallize from a mafic silicate melt is olivine, an island silicate with the simplest silicate structure of any mineral in the chart. The pryoxenes (augite, enstatite, etc.) contain silicate groups linked into single chains; amphiboles (hornblende, etc.) are double-chain (band) silicates; biotite has a sheet silicate structure; feldspars and quartz have three-dimensional framework silicate structures.

To identify the minerals in igneous rocks, you should use the simple physical properties of minerals you learned in Exercise 1—crystal shape and habit, cleavage and fracture, hardness, luster, color, and streak.

Crystal Shape and Habit

Problem 1

Recall the six-sided form of the large single quartz crystal you sketched in Exercise 1. Examine the quartz in the plutonic igneous rocks provided by your instructor. Does any of the quartz occur in this well-shaped six-sided crystal form? Would you expect it to be in a well-shaped crystal? How does the Bowen reaction series explain your answers?

Problem 2

Examine the rock provided by your instructor, or see Color Plate 15. It is a volcanic rock containing small crystals of hornblende suspended in a fine-grained matrix. Is the typical crystal habit of hornblende equant? Tabular? Prismatic? Platy? (See Exercise 1 for a review of crystal habits.) Sketch the typical crystal habit of this mineral.

Problem 3

Examine another volcanic rock, this one containing small crystals of olivine. See also Color Plate 16. What is the crystal habit of olivine?

Problem 4

The third volcanic rock contains crystals of plagioclase. See also Color Plates 9 and 14. What is its crystal habit? Sketch a crystal.

Cleavage and Fracture

Problem 5

How might you recognize the difference between cleavage and fracture in a hand specimen of a rock with a freshly broken surface? Refer to Color Plates 20 and 4, which illustrate conchoidal fracture in volcanic glass and cleavage in calcite. This difference is important in discriminating between quartz and feldspar in some rocks.

Problem 6

Which of the mineral samples provided by your instructor display cleavage? Fracture? Both?

Sample	Cleavage	Fracture
1.		
2.		
3.		
4.		
5.		

Hardness

Problem 7

Hornblende (an amphibole) and biotite often resemble each other in igneous rocks. Will hardness distinguish between hornblende and biotite?

Problem 8

Will hardness distinguish easily between amphibole and pyroxene?

Luster

Problem 9

Luster can be a useful property to distinguish between quartz and feldspar. Examine the rock provided by your instructor and note the difference between them:

• Luster of quartz

• Luster of feldspar

Color

Problem 10

Compare the samples of biotite and muscovite. Which is the black mica?

Problem 11

The chemical formula for biotite is known to be $K(Mg,Fe)_3(AlSi_3O_{10})(OH)_2$; that of muscovite is $KAl_2(AlSi_3O_{10})(OH)_2$. What chemical elements seem to be responsible for the dark color?

Problem 12

Examine the rock provided, which contains both plagioclase and orthoclase. How can you distinguish between them in this rock?

Streak

Problem 13

Magnetite is a magnetic mineral sometimes found as small crystals in iron-rich igneous rocks. What is its streak?

Problem 14

What is the streak of quartz? Explain your answer.

Problem 15

There are several pairs of minerals that may look very much alike in igneous rocks. Usually one specific property is the most useful to tell the difference between the two. For the following combinations of similar-appearing minerals, determine the physical properties that may be used to distinguish readily between the two. Circle or star the single most valuable property in each case. Fill in only those properties that will distinguish readily between the two minerals of each pair, and leave the others blank.

Mineral Pairs	Crystal Form	Cleavage/ Fracture	Hardness	Luster	Color	Streak
• Quartz						
• Feldspar						
• Orthoclase						
• Plagioclase						
• Olivine						
• Pyroxene						
• Pyroxene						
• Amphibole						
• Amphibole						
• Biotite						
• Olivine						
• Quartz						

TEXTURES OF IGNEOUS ROCKS

The texture of a rock refers to the physical nature of its constituents, their size and shape, and the way they are put together to form the rock. Texture is one of the most important properties in determining the environment in which any rock formed. Recall the interlocking texture of the thymol melt in Exercise 1, referring to your sketches if necessary. Since the minerals of igneous rocks have crystallized from a molten state, the primary characteristics of almost all igneous textures is a network of interlocking crystals similar to that of thymol. See Figure 2.3.

The size of these interlocking crystals defines the two major types of igneous rocks. *Plutonic* rocks, also called *intrusive* rocks, have cooled slowly deep underground; they display larger mineral crystals than *volcanic* or *extrusive* rocks, which have cooled rapidly at or near the surface. The variation of crystal size with the rate of cooling reflects the length of time available for ions and atoms to migrate to places within crystal structures in the solidifying melt. If the temperature drops slowly enough, large crystals up to 2 cm across are possible. If the magma cools more quickly, the atoms and ions lose their energy of motion very rapidly, and many small centers of crystal growth develop instead of just a few large centers. The resulting crystal size is very small. In the extreme case, cooling is so rapid that no crystalline structures have time to develop and a glass forms. Glass has the absence of crystalline structure associated with a supercooled liquid—though it is hard and solid, its atomic structure resembles a liquid.

The size of mineral crystals is influenced by the composition of the magma, as well as by the rate of cooling. Magma rich in silica is characteristically viscous, a condition that impedes the migration of ions and atoms to sites in crystal lattices and results in smaller crystals. Highly silicic melts are therefore particularly favorable for the formation of a glass. In contrast, mafic melts are far more fluid, so ions and atoms may more easily migrate. Mafic glasses are far less common than highly silicic glasses. The presence of volatile constituents, chiefly water, within a magma also decreases the viscosity of the melt, allowing the formation of very large crystals—up to several feet across in exceptional cases. These rock bodies of exceptionally large crystals are called *pegmatites*.

Principal Igneous Textures

There are four principal igneous textures that identify the rock as either plutonic or volcanic:

1. *Crystalline granular:* The crystals are more or less uniform in size and may range up to 2 cm in diam-

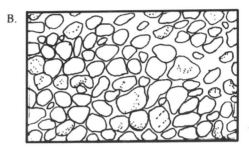

Figure 2.3
Textures magnified.
A. *Interlocking texture,* commonly seen in igneous rocks. There are no spaces between grains.
B. *Noninterlocking texture* typical of many sedimentary rocks. Note the spaces between grains.

eter (larger in exceptional cases). Color Plates 5 through 11 show crystalline granular texture. Crystalline granular texture is sometimes called *phaneritic.* It is the only primary texture of plutonic rocks. The next three textures are finer grained and indicate rapid cooling in a volcanic eruption.

2. *Aphanitic:* The individual crystals are so small that they cannot be identified in hand specimens, which typically appear massive and structureless. The characteristic interlocking texture is visible only with a microscope.

3. *Glassy:* There is no crystalline structure at all, the melt having solidified too quickly for crystals to form. Glasses typically have shiny surfaces that break along curved lines. Some glasses may also contain many small air bubbles, in which case the shiny surface may be visible only with magnification. Glassy textures are shown in Color Plates 18, 19, and 20.

4. *Pyroclastic* (*pyro,* fire or heat; *clastic,* broken): Broken angular fragments of volcanic material, rock surrounding the volcanic vent, and/or pebbles are all cemented together by very fine ash (see Figure 2.4).

Figure 2.4
Pyroclastic texture. Note the many small phenocrysts and the larger fragments of pumice (vesicular felsic glass). The horizontal streaks are structures formed by flowing of the partially solidified rock or collapse of rock fragments.

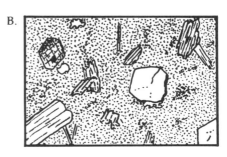

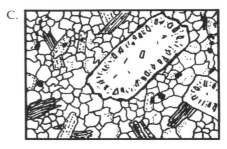

Figure 2.5
Subsidiary textures of igneous rocks.
A. *Vesicular texture* (slightly enlarged) in a felsic volcanic rock. Some of the vesicles have been stretched into elongated shapes, suggesting flow or compression while still semi-molten.
B. *Porphyritic–aphanitic texture* (enlarged) in volcanic rocks. Phenocrysts of quartz, potassium feldspar and biotite in an aphanitic matrix.
C. *Porphyritic–phaneritic texture.* Large phenocrysts of potassium feldspar in crystalline granular matrix of quartz, biotite and plagioclase (enlarged).

This texture is associated with an explosive eruption of volcanic material, which is almost invariably felsic to intermediate in composition. Pyroclastic texture (also called *fragmental*) is also shown in Color Plates 21 and 22.

Subsidiary Igneous Textures

Two subsidiary textures may occur in addition to the principal diagnostic textures listed above; these are superimposed on the principal texture. For example, a rock with an aphanitic texture, indicating that the rock cooled quickly near the surface, may also have a *porphyritic* texture as a subsidiary texture, indicating that the rock cooled in two different stages. A rock whose principal texture is glassy or aphanitic may have a subsidiary *vesicular* texture, indicating a high content of volatile material in the magma. When describing the texture of igneous rocks, you should list first the principal texture, then the subsidiary texture, as both are important in naming the rock. Subsidiary textures are illustrated in Figure 2.5 and in Color Plates 12 through 22 (except 20).

1. *Vesicular:* Vesicles—small round cavities—form when the magma is highly charged with volatile components that separate from the melt under the low pressure of the surface environment. Vesicular texture occurs only in volcanic rock, commonly in glassy or aphanitic textures. Bubbles of gas are frozen into the rock by the very rapid cooling. Since bubbles form where the confining pressure on the lava is least, their presence often indicates the top of a surface flow. Vesicular texture is shown in Color Plates 16, 18, and 19.

2. *Porphyritic:* In porphyritic texture there are two distinctly different crystal sizes. The larger mineral crystals, called *phenocrysts*, may be identified easily. The smaller crystals of the *matrix* or *groundmass* may be aphanitic or more rarely crystalline granular (in which case the rock is plutonic, not volcanic). Porphyritic texture, whether in plutonic or volcanic

rocks, indicates a complex cooling history. The magma probably began cooling deep underground in the typical plutonic environment where large crystals form. This interval of slow cooling was interrupted when the magma, with its early formed crystals, was extruded to an environment where more rapid cooling of the remaining magma produced smaller crystals. The final product is a rock with larger crystals embedded in a finer grained matrix. Porphyritic texture may be superimposed on any of the four primary diagnostic textures.

Color Plate 7 shows porphyritic texture in a coarse-grained crystalline granular rock. Color Plates 12 through 16 show the most common combination—porphyritic texture superimposed on an aphanitic primary texture. Color Plate 17 shows porphyritic-glassy texture. Color Plates 21 and 22 show fragmental texture with a porphyritic subsidiary texture.

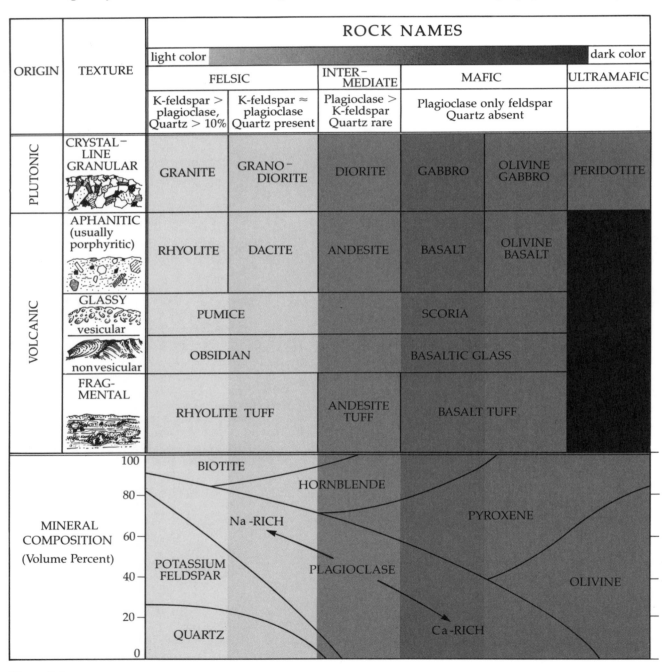

Figure 2.6
Identification chart for igneous rocks.

Problem 16

Remelt the crystalline aggregate of thymol you used in Exercise 1, making certain it is entirely melted before you allow it to begin cooling. Allow the melt to cool slowly on your laboratory table. When the single crystals of solid thymol reach about 5 mm in diameter, place the petri dish carefully in an ice bath and gently agitate the melt to initiate a period of more rapid cooling. Leave the petri dish in the ice bath until the thymol is completely solidified. Describe your results and make a sketch of the final texture of solid thymol below.

NAMING OF IGNEOUS ROCKS

Figure 2.6 relates the name of an igneous rock to its texture and its mineral composition. To use the chart, first examine the rock and determine its texture. Where more than one texture is present (a fragmental or crystalline granular rock that is also porphyritic, for example), list both textures, but choose the name of the

rock on the basis of the principal texture, which is more diagnostic than the subsidiary texture. In these cases the fragmental and crystalline granular textures are more important than the porphyritic. Locate the textural term in the vertical column labeled *texture*; the rock name will be in that same horizontal row. Be sure to include any subsidiary texture in the rock name.

Next determine the mineral composition. As a first approximation, the mineral composition may be estimated by noting the overall color of the rock. Light-colored rocks are usually felsic in composition; very dark rocks are mafic. The mineralogy of any identifiable crystals will fix the composition more precisely. Pay particular attention to the percentage of quartz and to the type of feldspar present. Observe also which dark minerals are present. Most people tend to overestimate the percentages of dark minerals in rocks. Figure 2.7 depicts various percentages of dark areas on a white background and may be helpful as a guide. Dark minerals rarely exceed 50 percent of a rock. If you have estimated accurately, the total percentage of the various minerals you have identified should not be far from 100 percent. In porphyritic aphanitic volcanic rocks, only the phenocrysts can be identified. In this case determine the relative abundance of each kind of phenocryst, and use these abundances with Figure 2.6.

Find the relevant percentages of the minerals you have identified on the graph of mineral composition at the bottom of Figure 2.6, and read up the chart to the appropriate textural line to determine the rock name. For example, a rock whose texture is crystalline

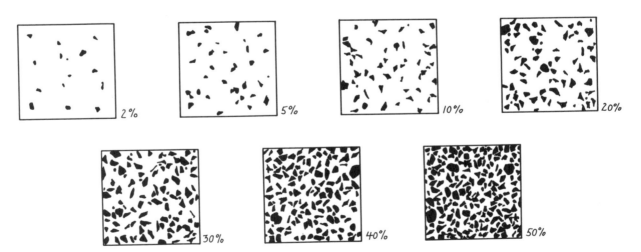

Figure 2.7
Chart for estimating percentages of dark minerals in rocks.

granular with a subsidiary porphyritic texture, which contains about 25 percent quartz, 40 percent plagioclase, 15 percent orthoclase, 10 percent hornblende, and 10 percent biotite, would be a granodiorite. It is customary when naming rocks to preface the name of the rock with the dark minerals present and with any subsidiary textures. Thus the full pedigree of the rock described above would be "porphyritic hornblende biotite granodiorite."

When assigning names to rocks, be aware that the boundaries between the different rocks are arbitrary and gradational and that few sharply defined categories exist. Granites grade imperceptibly into granodiorites, granodiorites into diorites, and so forth. As far as possible, the rocks you are given to identify will be clear-cut examples of a certain igneous rock.

The main types of igneous rocks are illustrated in Color Plates 5 through 22. The small inserts beside some of the photographs show *thin sections* of the rock.

Thin sections are slices of rock so thin (0.03 mm) that most minerals are transparent. They are examined microscopically under polarized light. The bright colors in some of these photographs are not the true colors of the mineral, but rather indicate the changes in the polarization of the light as it passes through the mineral. A second polarizer, inserted into the light path above the thin section, reveals the bright colors. These photographs are labeled *crossed polarizers*. Some textural features are best observed without the second polarizer in place. These photographs are designated *plane light*.

Problem 17

Identify the rocks provided by your instructor by filling in the chart provided below.

Sample Number	Textures	Minerals Present and Approximate Percentage	Rock Name

RELATIONSHIP OF VOLCANIC ROCKS TO PLUTONIC ROCKS

Volcanic and plutonic rocks have both solidified from the molten state and are simply different manifestations of the same phenomenon. Volcanic eruptions are only the last phase of a process that originates deep within the earth with molten rock (magma) and culminates with its solidification either at the surface (volcanism) or deep within the crust (plutonism).

The solidification of a body of magma tens or hundreds of kilometers in diameter deep within the earth produces a mass of igneous rock called a *batholith*. Smaller masses, with outcrop areas less than 100 square kilometers (40 square miles), are called *stocks*. This division is purely arbitrary since stocks are almost always upward protuberances of a batholith below. Since batholiths and stocks solidify at depth, they are exposed at the surface only after the overlying cover of rocks has been removed by erosion. Most batholiths and stocks are granitic to dioritic in composition. Their margins cut across the layers and other structural features of the rocks already in place (the *wall rock* or *country rock*); such a relationship is termed *discordant*. Other discordant igneous rock bodies are *dikes*, sheet-like intrusions that range in thickness from a few inches to hundreds or even thousands of feet.

Sheetlike bodies of igneous rock that have intruded between the layers of country rock, without cutting across them, are called *sills*. *Laccoliths* differ from sills in that they dome up the overlying layer. Sills and laccoliths are both *concordant* rock masses. Magma that reaches the surface is, of course, a volcanic feature— a cone, dome, or flow. Flows and sills may be difficult to distinguish from each other. Perhaps the best criterion of a surface flow is the presence of vesicles near the top surface.

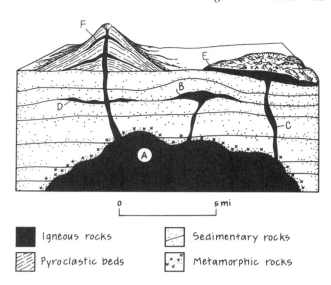

| | Igneous rocks | | Sedimentary rocks |
| | Pyroclastic beds | | Metamorphic rocks |

Figure 2.8
Cross-sectional drawing through part of the earth's crust, showing several igneous rock bodies.

General Relationships

Figure 2.8 shows a cross-sectional view of the earth's crust with some plutonic and volcanic features.

Problem 18

Identify the igneous rock masses labeled A through F in Figure 2.8, and enter names and descriptions in the chart below.

	Name of Rock Mass	Intrusive or Extrusive	Concordant or Discordant	Primary Rock Texture You Would Predict
A.				
B.				
C.				
D.				
E.				
F.				

The Palisades

Figure 2.9 includes a cross-sectional sketch of the New Jersey–Hudson River–Manhattan area, drawn in a plane approximately perpendicular to the Hudson river.

Problem 19

First Watchung Mountain might be either a buried volcanic flow or a plutonic sill from what is shown on the section. What evidence would you look for in order to distinguish between these two possibilities? Use a diagram if desired.

A.

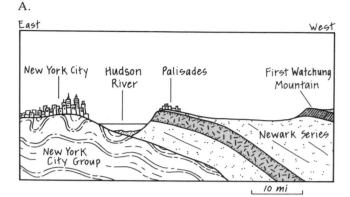

Figure 2.9 also shows an enlargement of the Palisades igneous body in cross section. It is a sill that intruded between sedimentary rock layers of the Newark Series. The size of the symbols in the diagram reflects the size of the crystals at the various levels of the Palisades sill.

B.

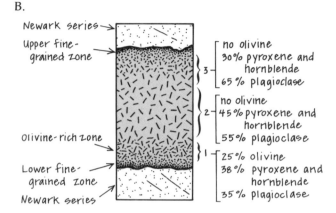

Figure 2.9
The Palisades sill, New York and New Jersey.
A. Cross-sectional sketch perpendicular to Hudson River.
B. Mineralogy of the Palisades sill.

Problem 20

How might you explain the zones of very fine crystals at both the top and the bottom of the sill? (Recall that crystal size depends largely upon cooling rate.)

Problem 22

Name and describe the process by which a large sill such as the Palisades may become nonhomogeneous in mineralogic composition.

Problem 21

Realizing that the three numbered zones shown on the section grade into each other without sharp boundaries, consider the approximate mineralogical content given in Figure 2.9. From your igneous rock chart, Figure 2.6, name the igneous rocks found in each of the three zones. (The texture of all the zones is crystalline granular.)

1.

2.

3.

TYPES OF VOLCANISM AND THEIR GLOBAL OCCURRENCE

Figure 2.10 shows the recent global occurrences of the two main types of volcanism, mafic and intermediate-to-felsic. Although there are a few exceptions to the general trend, the areas of basaltic and intermediate (andesitic) volcanism are not randomly mixed but seem to be restricted to separate regions.

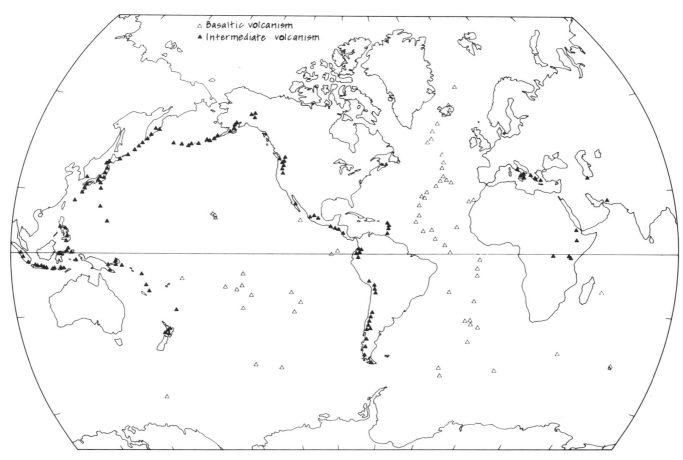

Figure 2.10
Global occurrence of some recent (<200 million years) volcanism.

Problem 23

Ignoring what seem to be anomalies, draw a smooth line separating the region of basaltic volcanism from that of andèsitic volcanism, and label the areas on Figure 2.10.

Problem 24

In what general area is basaltic volcanic activity found?

Problem 25

In what general area is intermediate volcanic activity found?

Problem 26

What general rule would you formulate for worldwide distribution of the two types of volcanism, basaltic and intermediate?

You have just set down a general *scientific law* governing the occurrence of two types of volcanism. Notice that a scientific law does not attempt to explain why volcanic activity occurs as it does; it merely describes a general observation. The process of formulating a general scientific law from many specific observations is known as inductive reasoning. It is at the heart of all scientific endeavor. The next step in the scientific process is to formulate a theory to account for the observations that are summarized in the law. We shall return to this exercise (and others, as well) in Exercise 18 and attempt to do exactly that.

Sedimentary Minerals and Sedimentary Rocks

To a person uninstructed in natural history, his country or seaside stroll is a walk through a gallery filled with wonderful works of art, nine-tenths of which have their faces turned to the wall.

THOMAS HENRY HUXLEY

Sedimentary rocks are formed by the accumulation and lithification of rock fragments, minerals, and organisms near the earth's surface or by direct precipitation of chemicals from water.

The origins of sedimentary rocks may be *detrital* or *clastic* (from broken pieces of preexisting rocks), *biogenic* (formed by the action of organisms), or *chemical* (direct precipitation from water). It is important, when attempting to reconstruct the geologic history of an area, to be able to identify these origins. The environment of formation of a biogenic limestone, for example, is very different from the environment of formation of a coarse-grained detrital rock.

The most common sedimentary rocks are the *detrital* or *clastic sedimentary rocks*. These are composed of fragments of preexisting rocks that have been broken up in the weathering process (see Exercise 9) and transported various distances by streams (Exercise 12), wind (Exercise 15), ice (Exercise 14), or the direct action of gravity (Exercise 10). This loose material is *sediment*. As sedimentary particles are transported by physical agents, they undergo a natural sorting process. The size and shape of the particles transported depends on the energy of the transporting medium and the time spent in transit. The mineralogic composition of the fragments *(clasts)* depends largely on the nature of the source rock but also on the mode and extent of transport.

The two other types of sedimentary rocks are the *chemical* and *biogenic* sedimentary rocks. Minerals enter solution primarily by chemical weathering of rocks. The dissolved chemicals are carried away by streams to lakes and oceans, where they may precipitate directly from solution as *chemical sedimentary rocks*. More commonly, they are removed from the water and concentrated as skeletal material (shells, for example) by living organisms. When the organisms die, the hard mineral remains may be incorporated into sedimentary rocks, which are said to be of *biogenic origin* (or *biogenic sedimentary rocks*).

COMPOSITION AND MINERALOGY OF SEDIMENTARY ROCKS

Sedimentary Minerals

Uniquely important to sedimentary rocks are the substances that are formed when minerals react with the atmosphere and hydrosphere. Minerals of igneous rocks crystallize from the molten state; minerals diagnostic of metamorphism form in the solid state by ionic diffusion, as we shall learn in the next exercise. Those minerals unique to sedimentary rocks precipitate from aqueous solution or form when other minerals undergo chemical reaction with water or atmospheric gases. (This process, called weathering, is the subject of Exercise 9.) They may form the bulk of the rock, as is the case with detrital clay in fine-grained detrital rocks, biogenic calcite in limestone, or the chemical sediments halite and gypsum; or they may precipitate between the grains of a clastic sedimentary rock, where they act as a cement to hold the grains together.

1. *Calcite* ($CaCO_3$) is one of the most important minerals of predominantly sedimentary origin. Calcium is derived from igneous rocks rich in calcium-

bearing minerals such as calcic plagioclase. The carbonate ion forms when carbon dioxide from the atmosphere is dissolved in water. Calcite is the most common cementing agent of clastic sedimentary rocks. More important, calcite is the major constituent of limestone, a sedimentary rock of biogenic or chemical origin. Biogenic limestone, the most common limestone, is derived from marine plants or animals and may contain identifiable fossils. Chemical limestone consists of calcite precipitated directly from solution, characteristically around springs or in caves. Calcite is distinguished from most other minerals by its effervescent reaction with dilute hydrochloric acid, HCl, which produces bubbles of carbon dioxide, CO_2.

2. *Dolomite* is a carbonate mineral containing equal amounts of calcium and magnesium ($CaMg(CO_3)_2$) and is found primarily in older rocks. Its crystal form is the same as that of calcite, and the two minerals are distinguished most reliably by the strength of reaction with HCl. Dolomite does not react readily with HCl, as calcite does, but if dolomite is powdered or if the HCl is heated, effervescence occurs. Rocks composed of the mineral dolomite are also called dolomite.

3. *Clay minerals* are perhaps the most important minerals of sedimentary origin; certainly they are the most voluminous. They are products of the chemical weathering of feldspars and ferromagnesian minerals at or near the earth's surface. The extremely small grain size of the clay particles makes it impossible to specify the type of clay with just a hand lens. Over 70 percent of clastic sedimentary rocks contain clay as the major constituent. Clay is also common as a matrix or a cement between larger fragments. Matrix clay is usually of detrital origin, carried by wind or water as solid particles. Clay cement is more often precipitated in place as sediments are buried and compacted.

4. *Halite* (NaCl) and *gypsum* ($CaSO_4 \cdot 2H_2O$) precipitate from seawater and commonly form monomineralic rocks, *rock salt* and *rock gypsum*. Extensive deposits of these minerals have formed by evaporation of saline lakes or restricted areas of the ocean and are often found at the surface in arid climates.

5. The *iron oxides* hematite (Fe_2O_3) and limonite ($FeO \cdot OH \cdot nH_2O$) form in sedimentary environments where iron and oxygen are present. Iron is derived from chemical weathering of ferromagnesian minerals. The oxides are important commercial ores of iron; they may also serve as cementing agents in clastic sedimentary rocks, where their presence is indicated by their rust color.

Sedimentary Clasts

In addition to the sedimentary minerals that form in chemical weathering or precipitate from aqueous solution, fragments that make up the clastic sedimentary rocks form in physical weathering, which is simply the mechanical breakdown of rock material into smaller pieces. They may be composed of one mineral, such as quartz, or they may be fragments of rocks, the so-called *lithic fragments*.

1. *Mineral fragments* in clastic sedimentary rocks are derived from preexisting rocks by physical or chemical weathering. The most common mineral fragments consist of the minerals that are most stable at the low pressure and temperature of the earth's surface. In the case of igneous minerals, those with the lowest crystallization temperatures (see Figure 2.1) are generally most resistant to chemical and physical breakdown. Quartz is the most abundant sedimentary mineral fragment due to its hardness and chemical stability. Orthoclase and, to a lesser extent, sodic plagioclase and biotite are also in this category. The less stable silicate minerals such as olivine, pyroxene, and calcic plagioclase are only rarely found as clasts in clastic sedimentary rocks.

2. *Lithic fragments* are fragments of rock in which the constituent minerals are not separated. Lithic fragments are the main constituents of the coarse-grained sedimentary rocks.

Problem 1

Compare the specimens of quartz, calcite, gypsum, and halite provided by your instructor. These minerals strongly resemble each other in luster and color; other physical properties provide the only means of distinguishing them. Cleavage and hardness are the most diagnostic characteristics. Examine the four mineral specimens and fill in the chart below.

Mineral	Cleavage (number of cleavage directions and angles between cleavages)	Hardness
• Quartz		
• Calcite		
• Gypsum		
• Halite		

Problem 2

Is streak a useful property to distinguish among these four minerals?

Problem 3

Hematite and limonite can assume a variety of appearances, and streak usually provides the most reliable distinction between them. Use a streak plate to determine the streak of each mineral.

• Hematite

• Limonite

Clays formed by chemical weathering of rock material, primarily feldspars and mafic minerals, are of several different types. The clay mineral formed seems to depend largely on climate. Metallic ions and finally silica (SiO_2) are progressively removed (leached) from the clay mineral structure as temperature and rainfall increase.

Problem 4

The formulas for three important clay minerals are:

• Illite: $KAl_4(Si_7AlO_{20})(OH)_4$

• Kaolinite: $Al_4(Si_4O_{20})(OH)_8$

• Gibbsite: $Al(OH)_3$

From their chemical formulas, number these minerals in order of increasing degree of leaching. Which clay would be formed in very wet tropical conditions? In more temperate climates?

TEXTURES OF SEDIMENTARY ROCKS

There are two principal textures in sedimentary rocks, *interlocking* and *clastic*.

Interlocking Texture

Interlocking texture is similar to the crystalline granular texture found in igneous rocks except that in interlocking sedimentary rocks the grains usually consist of only one mineral. An interlocking texture is formed when minerals such as calcite, silica, gypsum, or halite precipitate directly from solution. This texture may also form when clastic sediments recrystallize near the earth's surface. Clastic sediments may recrystallize when subjected to heat, pressure, or water with certain chemical impurities flowing through the sediments.

Amorphous texture, as found in opal or amber, is truly noncrystalline in nature. This texture is usually impossible to distinguish in hand sample from an extremely fine interlocking texture (sometimes called *microcrystalline* or *cryptocrystalline*). All very fine-grained nonclastic textures will be included in the interlocking category in this exercise, for simplicity's sake.

Clastic Texture

Clastic (from a Greek word meaning "broken") describes a rock made up of broken fragments of rocks, minerals, or organic remains. In describing a clastic texture, several properties of the fragments must be noted, as these are crucial in determining the environment of deposition of the rock. These properties are listed and described below.

1. *Grain size:* Usually, but not always, a dominant particle size makes up the major portion of the rock, and this size gives a clastic rock its name. The Wentworth size scale is most commonly used.

2. *Sorting* refers to the range of particle sizes in a clastic rock. In well-sorted material, 80 percent of the particles are of one size class—medium sand, for example. Poorly sorted material contains several different size ranges—pebbles surrounded by a sand and clay matrix, for example. Figure 3.1 illustrates degrees of sorting seen in sedimentary rocks. The degree of sorting depends primarily on the transporting agent, although time and the environment of deposition may also be factors. Wind and water are the best sorting agents; glaciers, mudflows, and landslides do not sort material but deposit all different sizes together.

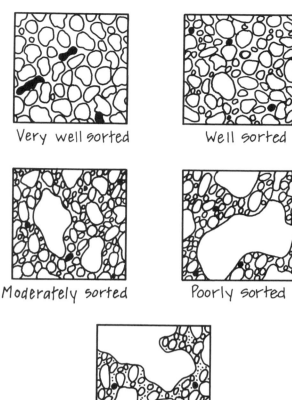

Figure 3.1
Degrees of sorting in clastic materials. The drawings represent sandstone as seen with a hand lens. Silt and clay-sized materials are shown as fine stippling.

Wentworth Size Scale

Particle Name	Diameter (mm)	Rock Name
Boulder	>256	
Cobble	64–256	
Pebble	4–64	Conglomerate
Granule	2–4	
Sand { coarse	1/2–2	
medium	1/4–1/2	Sandstone
fine	1/16–1/4	
Silt	1/256–1/16	Siltstone } Mud-
Clay	<1/256	Claystone } stone

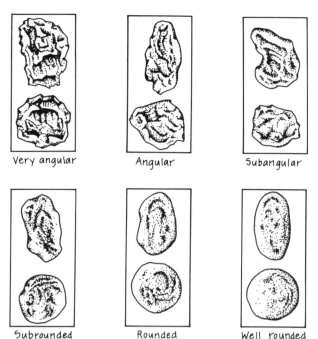

Figure 3.2
Terms to describe rounding of particles.

3. *Rounding* describes the smoothness or the angularity of a particle. It is a characteristic of the surface, not the shape of a particle. The terms used to describe rounding are illustrated in Figure 3.2. The degree of rounding is most strongly influenced by the type and duration of transport. Particles become rounded by knocking into and grinding against one another during transport, as in a stream or on a beach. Well-rounded particles have usually been in transport for a long time; angular particles have been deposited after a short transport time. The particle size and composition may also influence the degree of rounding.

4. *Cement* is material that precipitates between detrital grains during *diagenesis* (conversion of loose sediments to sedimentary rocks) and holds the rock together. Common cements are calcite, silica (SiO_2 that may be either crystalline or noncrystalline), iron oxides, and clays. Silica cement is clear and colorless. Calcite cement is also clear and colorless, but it may be distinguished from silica by its bubbling reaction to HCl. Iron oxide cements impart their characteristic yellow or orange colors to rocks. Clay cement is dark and opaque.

To examine closely some of the textures we have discussed, consider some sediments that might be lithified into detrital sedimentary rocks in the following problems.

Problem 5

Examine the sand samples provided by your instructor. What is the dominant mineral in sample 1?

Problem 6

How would you describe the degree of sorting? Rounding?

- Sorting

- Rounding

Problem 7

What minerals are present in sample 2?

Problem 8

Compare the sorting and rounding with that of sample 1.

Problem 9

Name the sedimentary rock that would be formed if this material were lithified. Use Figure 3.4.

Problem 10

Examine the sediments shown in Color Plate 35. What igneous rock contains these minerals in abundance and might therefore be the source rock of this sediment? Name the sedimentary rock that would form if this sediment were lithified. Use Figure 3.4.

Problem 11

Examine Color Plates 34, 35, and 36. Compare the size, sorting, rounding, and composition of these sediments to that of the detrital sedimentary rocks shown in Color Plates 27 through 33. Match each sediment with the rock to which it corresponds most closely, assuming the sediments were lithified.

- Color Plate 34

- Color Plate 35

- Color Plate 36

Behavior of Particles in Water. In your laboratory is a jar filled with water and sediments of various sizes. This system can serve as a crude model of how materials settle out of suspension in a natural system such as a stream or lake. Shake the jar until all the sediment is stirred off the bottom. Hold the jar upside down for a few seconds, then quickly set it upright on the table and do not disturb it again until you have completed your observations for this entire exercise.

Problem 12

During the first several seconds of settling, the water is still rather turbulent. Which size particles settle to the bottom under these conditions?

Problem 13

After several minutes, what size particles are still in suspension?

An angular particle becomes rounded during transport because protruding portions are chipped off when the particle collides with other particles (or with the stream bed or the ground). The greater the momentum (mass times velocity) of the particle, the more likely it is to break upon impact.

Problem 14

State which size clasts you would expect to be more rounded—large or small—in a poorly sorted assemblage of sediments. (Assume that all sediment sizes have traveled at the same velocity for the same amount of time.)

Problem 15

Examine the rounding of the clasts in Color Plate 36. Is your prediction accurate?

Problem 16

Water in rivers usually travels at a maximum velocity of 5 to 10 miles per hour. Wind often attains speeds of 20 or more miles per hour. Would you expect sand-sized clasts to be rounded more rapidly when transported by rivers or by wind? Why?

SEDIMENTARY STRUCTURES

One of the most diagnostic features of detrital sedimentary rocks in the field is the presence of large-scale internal and surface structures that reflect the way sediments are deposited. Geologists use these structures to characterize more fully the environment of deposition of sediments.

Internal Features

Bedding, or layering of material, is the most common large-scale structure of sedimentary rocks. The beds are marked by differences in texture, color, or composition of the sediment, and they may range in size from a few millimeters to several meters in thickness.

Graded bedding is the gradual change in particle size within a layer of sediment. Most often, coarse particles are at the bottom and finer particles at the top. In tilted and deformed layers, graded bedding may be used to identify the original top of the bed.

Crossbedding (beds that meet at an angle) reflects a change in direction of the water current or wind that deposited the sediment. Color Plate 45 shows crossbedding of wind-laid sand (inset) and water-laid silt.

Surface Features

Ripple marks can be produced by wind, by currents in a river, or by water that oscillates back and forth in a wave zone. Current ripple marks are asymmetric and are useful indicators of the direction of water flow that deposited the sediments. Oscillation ripples (Color Plate 47) are symmetric and are used to identify the original top of the bed.

Mud cracks indicate drying of sediments, then deposition of fresh material over and into the cracks. Color Plate 48 shows casts of mud cracks.

Tool marks and *scour marks* are grooves and indentations developed on the surface of a sedimentary bed when objects are dragged or bounced over the surface or when the surface is eroded by local turbulence.

Biogenic marks are tracks or other features produced and preserved on the surface of a sedimentary layer.

Problem 17

Examine Color Plate 46. Which direction was originally up in this now vertical sandstone bed?

Problem 18

Look again at your jar of sediment. Nearly all of the particles should have settled to the bottom by now. What sedimentary structure has formed? Make a sketch of a representative portion of sediment in the jar, showing the structure clearly. Show a size scale in your sketch.

Problem 19

What depositional conditions produced this structure?

ENERGY OF THE DEPOSITIONAL ENVIRONMENT

For water-deposited sediments, the sizes of particles may be used to infer the water turbulence and thus something about the environment of deposition of the sediment. Large particles are carried only by very turbulent water. When the turbulence drops even slightly, larger particles are dropped, leaving medium and small particles in suspension. Very turbulent water is most characteristic of the headwaters of streams and other high-energy environments; smoothly flowing water in main rivers is less turbulent. The water in the settling jar at the beginning of your observations constituted a relatively high-energy and high-turbulence environment as compared with the still water in the jar at the end of your observations, which was a low-turbulence and low-energy environment. Refer to the sketch you made of the settled sediments in your jar.

Problem 20

What size particles, relatively large or relatively small, are characteristically deposited in a high-energy environment (turbulent water)?

Problem 21

A low-energy, low-turbulence depositional environment (still water) is characterized by what particle size, large or small?

The presence of detrital clay in a sandstone is important for interpreting its depositional environment. Since clay matrix is easily winnowed out of sand by a small amount of turbulence, the preservation of even small amounts of detrital clay (as low as 15 percent of the rock) implies that the sediment is relatively *immature.* Such sediment has not undergone extensive sorting in turbulent water or wind. A marine basin that is being rapidly filled with sediment from nearby mountainous or volcanic land masses typically contains immature clay-rich sediments. This rapid deposition frequently takes the form of *turbidity currents,* subsea slides that mix clays in with sand-sized particles. Graded bedding is frequently found in turbidity-current deposits. A useful, if somewhat inexact, term to describe a clay-rich sandstone with abundant lithic fragments is *graywacke.* Although precise definitions

vary, graywacke is characteristically a coarse-grained sandstone consisting of poorly sorted angular grains of quartz, lithic fragments, and some feldspar embedded in a clay-rich matrix. It is typically dense, hard, and dark, and often shows graded bedding.

Problem 22

Examine the sediments shown in Color Plates 34, 35, and 36. Which of these sediments is immature? Mature?

- Color Plate 34
- Color Plate 35
- Color Plate 36

Problem 23

The degree of rounding and composition of the clasts are other indications of the maturity of a sediment. Compare the clasts in Color Plates 29 and 30. State which rock is composed of the more mature sediment, supporting your conclusion with a discussion of clast rounding and composition.

Problem 24

Your settling jar experiment was a small-scale laboratory demonstration of the correlation between particle size and the energy or turbulence of depositional environment. Now consider the different conditions under which sediment is deposited from water in nature. Where ocean waves break at the shoreline, for example, the turbulence of the depositing medium is greater than the turbulence of the water in a pond. For each of the following environments of deposition, state whether the turbulence of the water is relatively high, medium, or low and what maximum particle size you would expect to see—mud, sand, or pebble. If the maximum expected particle size is sand or larger, state whether these sediments would contain significant clay matrix; that is, whether the sediment is mature or immature.

Environment	Turbulence, Energy	Maximum Particle Size	Maturity
• Ocean beach (surf zone)			
• Lake bottom			
• Mountain stream			
• Deep ocean			
• Lowland river (near sea)			
• Swamp			

The Mississippi River carries billions of tons of sediment into the Gulf of Mexico each year. Figure 3.3 shows the distribution of this and other sediment in the Gulf.

Problem 25

As the river current laden with sediment enters the still water of the Gulf, the turbulence drops sharply. Using your settling jar experiment as a conceptual model, indicate on the map the general areas where you would expect to find pebbles, sand, and mud deposited in the Gulf of Mexico.

Problem 26

Notice the areas of carbonate deposition in the Gulf of Mexico. What rock will form if these deposits are lithified?

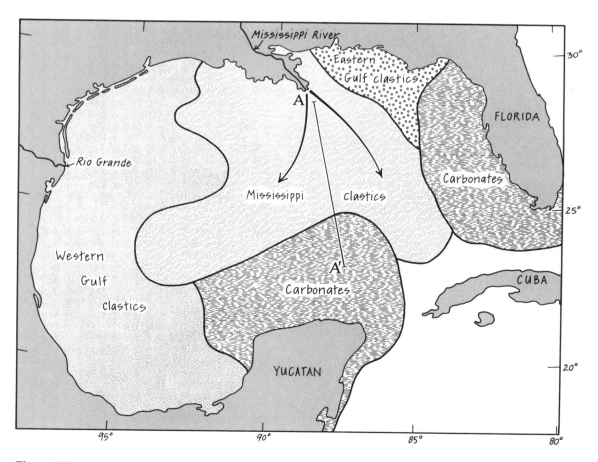

Figure 3.3
Clastic and biogenic sediments in the Gulf of Mexico.

TEXTURE	GRAIN SIZE	COMPOSITION	ROCK NAME	ORIGIN
Clastic 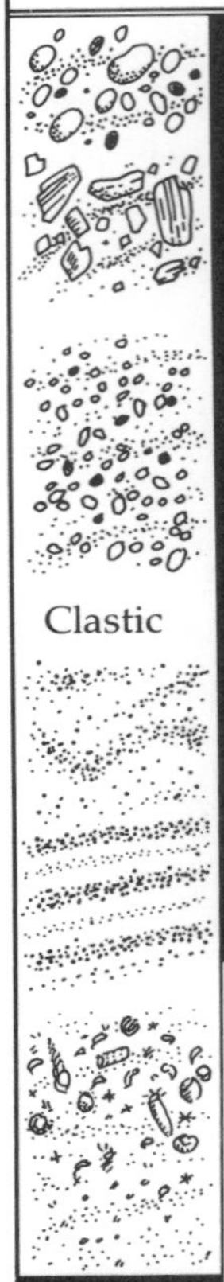	> 2 mm (grains observable)	Mixed	Conglomerate (breccia if clasts are angular)	Detrital
Clastic	2 mm – 1/16 mm (grains observable)	Mixed	Sandstone	Detrital
		90% quartz	Quartz sandstone	
		25% feldspar	Arkosic sandstone	
		15% lithic fragments	Lithic sandstone (may be called graywacke if over 20% clay is present)	
Clastic	< 1/16 mm (grains not observable)	Clay, silt very fine mica and quartz	Mudstone (shale if layered)	Detrital
Clastic; Inter-locking 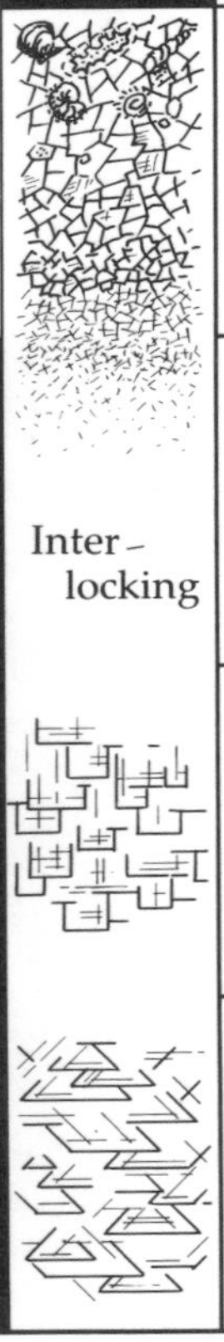	Microcrystalline to coarse	Calcite	Limestone (see Table on p.35 for varieties)	Detrital; Biogenic; Chemical
		Dolomite	Dolomite	
Inter-locking	Microcrystalline	Silica	Chert (various colors, often banded waxy luster)	Biogenic; Chemical
			Diatomite (light color, earthy luster)	
Inter-locking	Fine to coarse	Halite	Rock salt (generally colorless but may be stained)	Chemical
Inter-locking	Fine to coarse	Gypsum	Rock gypsum (white or pale colors, may be scratched by fingernail)	Chemical

Figure 3.4
Classification of sedimentary rocks.

Problem 27

The water depth increases from A to A'. From your answers to Problems 26 and 27, list the sedimentary rocks formed as water depth increases.

- close to shore (shallow water)

- moderate depth

- deep water

Problem 28

What factors might explain the absence of chemical deposition (direct precipitation from water) in the Gulf of Mexico at the present time?

Problem 29

Offshore oil-well drilling in the Gulf often encounters thick deposits of rock salt beneath the ocean floor. What does this fact suggest about the environment in this area when the salt layer was forming?

ORIGINS OF SEDIMENTARY ROCKS

The most useful way to group sedimentary rocks is by their origin. In what environment did the rocks form? The origin may be *detrital*, *chemical*, or *biogenic*. Texture, which is either *clastic* or *interlocking*, is one clue to the origin.

Sedimentary rocks of detrital origin are composed of broken rock fragments and minerals transported to the site of deposition. They have clastic texture. Ninety percent of all sedimentary rocks are detrital.

Sedimentary rocks of chemical origin have precipitated directly from water and display interlocking texture. Less than 1 percent of sedimentary rocks are of chemical origin.

Sedimentary rocks of biogenic origin are produced by biological action. Their textures may be clastic or interlocking, depending on the degree of recrystallization. Ten percent of the sedimentary rocks we see at the earth's surface are of biogenic origin.

NAMING SEDIMENTARY ROCKS

When naming a sedimentary rock, first determine its texture, whether clastic or interlocking. If the rock is clastic in texture, evaluate the grain size, sorting, rounding, and composition, and choose the appropriate name from Figure 3.4. (Limestones are classified in the table below.) Use the predominant size range of the particles when naming poorly sorted rocks. As with igneous rocks, modifying terms should be used to describe the composition of the clasts. Clay matrix

Some Varieties of Limestone

Rock Name	Color Plate	Texture	Origin	Description
Coquina	38	Clastic	Detrital and biogenic	Coarse shell fragments loosely cemented by calcite. Porous.
Fossiliferous limestone	39	Interlocking	Biogenic	Many fossils visible, but recrystallization has produced interlocking texture.
Chalk	40	Very fine grained	Biogenic	Soft, porous, very fine-grained limestone composed of microscopic organisms. Normally white.
Oolitic limestone	41	Clastic	Detrital and biogenic or chemical	Small spheres built up of concentric layers of $CaCO_3$ around a nucleus of variable composition. Spheres are cemented together by calcite.
Travertine	42	Interlocking	Chemical	Calcite deposit formed by chemical precipitation around springs or in caves. Commonly banded.
Crystalline limestone	——	Interlocking	Indeterminate unless fossils present	Interlocking crystals of calcite from fine to coarse grained. Variable appearance.

that constitutes over 15 percent of a sandstone should be noted since it indicates an immature sediment.

If a rock with interlocking texture has grains large enough to work with, use cleavage, hardness, and other appropriate properties to identify the mineral and hence the rock name. Very fine-grained sedimentary rocks present more of a challenge because it is difficult to determine whether the texture is clastic or interlocking. In these cases identify the predominant constituent, usually calcite, silica, or clay, and choose the name from Figure 3.4 and the table "Some Varieties of Limestone." Hydrochloric acid will determine whether calcite is present, and the vigor of the bubbling may suggest the abundance of calcite. The greater hardness of silica usually distinguishes it from clay.

As an example, a poorly sorted clastic rock with most particles ranging in size from 1/16 to 2 mm composed predominantly of quartz with about 30 percent feldspar would be an arkosic sandstone (or arkose). The same rock with a clay matrix constituting 20 percent of the rock would be a clay-rich arkosic sandstone.

Problem 30

Examine the rock specimens provided by your instructor and complete the chart below. (Notice that sorting and rounding do not apply to a rock with an interlocking texture.)

Specimen Number	Textures	Grain Size	Composition (percent)	Clast Properties		Fossils (percent)	Rock Name	Origin
				Sorting	Rounding			

Metamorphic Minerals and Metamorphic Rocks

Some days are diamonds; some days are stones.

JOHN DENVER

Metamorphic rocks are formed by transformation of other rocks of the earth's crust by heat, pressure, and chemically active fluids *while in the solid state*. Most metamorphic rocks may be recognized by their unusual textures and their distinctive mineral assemblages. The degree of change in the rock depends on the pressures and temperatures to which the rock was subjected.

There are three main types of metamorphism:

1. *Contact metamorphism* occurs in rocks subjected to high temperatures without high directed pressure, such as rock that is intruded by a dike or sill. Minerals characteristic of high temperatures, such as pyroxenes and aluminosilicates, are typically found in these rocks. The rock is most commonly fine grained and lacks any directional textures; it is called a *hornfels*.

2. *Dynamic metamorphism* is produced mainly by high pressure at low temperature. Fault zones, for example, display rock that has been sheared, crushed, and sometimes recrystallized, although the mineralogy remains unaffected. Rocks such as *mylonites* and *cataclasites* are produced under such conditions. This is a special class of rocks that can form in a wide variety of environments. They will not be considered in any further detail in this laboratory.

3. *Regional metamorphism* accounts for the vast bulk of metamorphic rocks. In this process rock is subjected to high temperature and high pressure, resulting in changes in both mineralogy and texture.

To reach the conditions typical of regional metamorphism, a rock must be deeply buried; to reach the surface again where we see it, the rock must be raised by the folding and uplifting process of mountain-building, or *orogenesis* (from the Greek *oro-*, mountain, and Latin *genesis*, birth).

In active areas of the earth's crust where rocks are folded into mountain belts, rocks are subjected to directional pressures. In the roots of these *orogenic belts*, temperatures are high due to burial beneath great thicknesses of overlying rocks. The rise in temperature with increasing depth in the earth's crust is called the *geothermal gradient*. The normal geothermal gradient, shown in Figure 4.1, is 25–30°C/km.

The deep burials and vast uplifts that are recorded by metamorphic rocks must obviously take enormous lengths of time and equally great amounts of energy. James Hutton in 1795 was the first to show that the earth has had that much time: "We find," he wrote, "no vestige of a beginning—no prospect of an end." But the energy source and the mechanism of mountain-building were not well explained for nearly 200 years more. Today, most geologists are convinced that plate tectonics (Exercise 18) is the main engine of orogenesis.

Figure 4.1 shows the two main types of metamorphic rocks, contact and regional, in relation to pressure and temperature. As is the case in all rock classifications, the boundaries here are gradational, and the solid lines in the chart are general guidelines only.

METAMORPHIC MINERALS

Historically, one of the first criteria used to identify metamorphic rocks was their unusual mineral content. In addition to quartz, micas, amphiboles, pyroxenes, and feldspars (which, of course, are found also in igneous rocks), minerals such as garnet, glaucophane, andalusite, and kyanite often appear in meta-

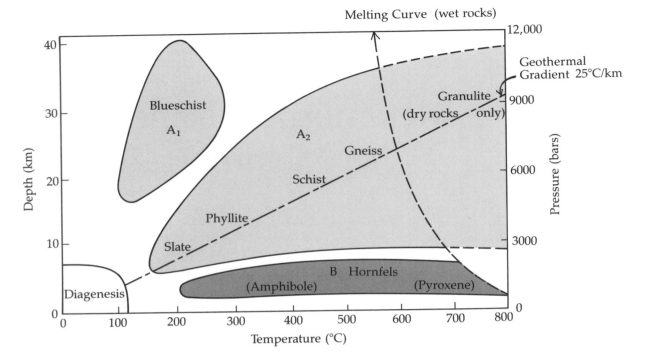

Figure 4.1
Approximate temperature-pressure fields of various metamorphic grades.
Field A₁. Pressure is relatively high, temperature relatively low. Regional metamorphism of mafic rocks under these conditions produces the characteristic mineral glaucophane, an amphibole with a distinctive blue color. Glaucophane is diagnostic of regional metamorphism where the geothermal gradient is relatively low.
Field A₂. These are the high-pressure, high-temperature conditions of normal regional metamorphism. The geothermal gradient is about 25–30°C/km.
Field B. Low pressures and high temperatures are typical of contact metamorphism.

morphic rocks, and these minerals are usually diagnostic of metamorphic rocks.

Some common igneous minerals may look slightly different in metamorphic rocks. Metamorphic plagioclase, for example, lacks the twinning so common in igneous plagioclase.

Problem 1

Your instructor has provided two specimens of garnet. One is grossular garnet ($Ca_3Al_2Si_3O_{12}$); the other is almandine garnet ($Fe_3Al_2Si_3O_{12}$). From what you have learned about color in minerals, which is the pale garnet and which is the dark garnet?

Problem 2

Examine a rock that contains garnet. (See also Color Plates 51 and 52.) What is the typical crystal habit of garnet? (prismatic? equant? tabular? platy?)

Problem 3

Which kind of garnet, grossular or almandine, is in the rock?

Problem 4

Examine the specimen of glaucophane, a metamorphic amphibole. The thin section of Color Plate 57 shows glaucophane in plane light, which gives an approximate color. What is the most obvious way to distinguish glaucophane from the common igneous amphibole, hornblende?

Problem 5

Serpentine is a metamorphic product of olivine- and pyroxene-bearing rocks such as peridotite. It may have wildly different appearances, ranging from a very dark finely crystalline rock to a pale green or blue rock with a very slick surface, depending on how it was emplaced. Examine your sample of serpentine and note its distinguishing characteristics. Color Plate 60 shows three varieties of serpentine.

Problem 6

Quartzite and marble are two monomineralic metamorphic rocks composed of quartz and calcite, respectively, which may be similar in appearance. See Color Plates 58 and 59. What physical properties of quartz and calcite would be most valuable in distinguishing between quartzite and marble?

Problem 7

Chlorite is a sheet silicate typical of low-grade metamorphism. See Color Plate 55. What simple physical property is useful to distinguish chlorite from biotite and muscovite?

Metamorphic minerals have been studied very successfully in the laboratory, where large instruments that simultaneously apply heat and pressure to specimens are used to simulate the natural conditions of metamorphism. Many metamorphic minerals have been synthesized in this way, demonstrating that metamorphic rocks are indeed products of elevated temperature and pressure.

Andalusite, kyanite, and sillimanite are three different minerals with the same chemical formula, Al_2SiO_5. This phenomenon is called *polymorphism* ("many forms"). The three different crystal structures of these minerals form under different pressure–temperature conditions, making these minerals useful in defining conditions of metamorphism. The Al_2SiO_5 system has been studied carefully in the laboratory, and the conditions under which each of the three polymorphs forms are well known. This information is presented in Figure 4.2, at the same scale as that of Figure 4.1. In *phase diagrams* such as this one, the only conditions under which two minerals in adjacent fields can coexist are defined by the line separating the fields; otherwise only one of these three minerals can exist in any given rock. Pressures are given by the vertical scale on the right; units are kilobars (1 kilobar is roughly 1000 times atmospheric pressure).

Problem 8

The three minerals can occur together at only one pressure and temperature: at what pressure and what temperature?

Problem 9

The *specific gravity* (the density relative to that of water, given by the Greek letter rho, ρ) of andalusite is 3.14; that of sillimanite is 3.25; that of kyanite is 3.63. What do these values tell you about the interatomic distances in the three minerals?

Problem 10

How might the difference in specific gravity be related to the pressure conditions under which each one forms?

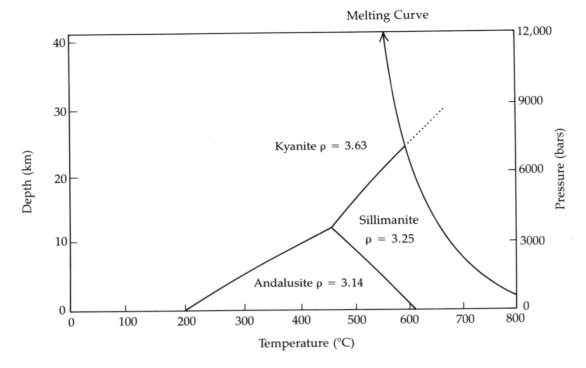

Figure 4.2
Pressure and temperature fields of stability of Al_2SiO_5.

Problem 11

Observe hand specimens of the three polymorphs. How can you differentiate between andalusite, kyanite, and sillimanite?

TEXTURES OF METAMORPHIC ROCKS

Rocks of the regional metamorphic type have been subjected to directed pressure deep underground. They develop characteristic textures as a result of (1) recrystallization of their minerals or (2) plastic flow, while solid, under this consistently oriented pressure. These new textures take the form of *foliation* (Latin *folia*, leaves), an orientation of planar minerals such as clay minerals and micas such that the rock tends to cleave into leaf-like plates or the minerals segregate into dark and light bands. Color Plates 49 through 57 show foliated metamorphic rocks.

Some regionally metamorphosed rocks do not always develop the characteristic foliated textures. These are the monomineralic rocks consisting only of quartz (quartzite) or calcite (marble). These rocks usually are simply recrystallized into coarser crystal sizes. See Color Plates 58 and 59.

PARENT ROCKS OF METAMORPHIC ROCKS

Because metamorphic rocks form from preexisting rocks, it is sometimes possible (and always desirable) to determine the parent rocks of metamorphic rocks. This information enables us to reconstruct the geologic history of the ancient mountain belts in which these rocks occur. Igneous or sedimentary parent rocks can sometimes be identified by the bulk chemical composition of the metamorphic rock because often very little material is gained or lost in metamorphism (except for the volatile constituents, H_2O and CO_2). The correlation between parent and metamorphic rock is most apparent in the case of quartzite or marble. Quartzite is coarsely crystalline silica, derived by recrystallization of a pure quartz sandstone. Marble, although its composition may vary, is recrystallized from a more or less pure limestone.

Parent rocks of metamorphic rocks with more complex combinations of minerals are, of course, somewhat harder to diagnose, but certain generalizations are still possible, based on the bulk chemical composition of the rock. Consider, for example, the fate of mudstone, a common sedimentary rock composed of very fine-grained clay and quartz. As mudstone is subjected to increasing temperature and pressure during regional metamorphism, the clay minerals recrystallize into small grains of micas, commonly biotite or muscovite, producing the metamorphic rocks *slate* and *phyllite*. These rocks demonstrate the parallel alignment of mica flakes within them by cleaving into thin flat sheets, which makes slate so useful as roofing or paving material. They are strongly *foliated* rocks.

As temperature and pressure increase further, the micas themselves recrystallize into coarser grains, producing *schist*, a metamorphic rock that may contain visible grains of quartz, feldspar, and amphibole. Under conditions of very high temperature and pressure, mineral recrystallization in the solid state proceeds to the point where the micas have largely disappeared, giving rise to quartz, feldspar, and amphibole as the major constituents of the foliated rock. Lacking the aligned sheet silicates, the rock—*gneiss*—has little or no tendency to split into sheets, and foliation is expressed instead by separation of minerals into light and dark bands. Gneiss is a *banded* rock (still in the foliated category). Dry rocks may proceed to the *granulite* grade, but this happens only rarely.

Fine-grained felsic to intermediate volcanic rocks undergo a similar series of metamorphic steps, but their coarse-grained equivalents (granite and diorite) proceed directly to a gneissic rock when temperature and pressure become high enough. The progression to gneiss is shown in Figure 4.1.

Mafic rocks subjected to increasing temperature and pressure in regional metamorphism undergo similar textural changes, but because of their higher content of mafic minerals, the rocks formed do not usually contain quartz. Instead there is generally a higher content of amphibole and pyroxene minerals in the schist and gneiss stages. (See Color Plate 56.) The mica of the mafic schists is often chlorite, a sheet silicate that contains magnesium and iron. (See Color Plate 55.)

An important kind of regional metamorphism involves exceptionally high pressures, producing rocks of *blueschist* grade. See Figure 4.1. When mafic rocks are subjected to blueschist grade metamorphism, a distinctive blue amphibole, glaucophane, is produced, and the rock is called glaucophane schist or blueschist. The occurrence of these rocks is important for the geologic history of an area since it indicates a period of anomalously high directed pressure.

Ultimately, if the temperature becomes high enough, the rock melts. In a sense, the ultimate product of increasing metamorphic grade is a magma, the molten parent material of igneous rocks. *Migmatites* are rocks that seem to be gradational between gneiss and granite. Parts of the rock are clearly foliated, having been converted to that state while solid, but mixed intimately with the gneiss are swirls and veins of rock that display the crystalline textural pattern of granite. Migmatites are not often recognizable in small hand specimens, but the mixture of rock types may be seen clearly in outcrops in the field.

Problem 12

Arrange the foliated specimens provided by your instructor in order of increasing pressure and temperature of formation (metamorphic grade). Describe the differences in texture and mineralogy upon which you base your order.

Problem 13

Identify the rocks furnished by your instructor by filling in the chart. For those rocks designated by your instructor, suggest the most likely parent rock and the metamorphic process involved. The metamorphic rock classification chart, Figure 4.3, is used in much the same way as the other rock charts you have used. First determine the texture of the rock (foliated or nonfoliated), then its grain size. Next determine the mineral composition, and then the rock name. Preface the rock name with the names of any mafic minerals or unusual minerals. For example, a rock with medium-grained foliated (but not banded) texture, composed of muscovite, quartz, feldspar, biotite, and garnet, would be a biotite garnet schist.

Rock Number	Texture	Minerals Present	Rock Name	Parent Rock and Metamorphic Process

TEXTURE	GRAIN SIZE	COMPOSITION	ROCK NAME	METAMORPHIC PROCESS		PARENT ROCK
Foliated	Very fine	Clay	Slate	Regional Metamorphism		Mudstone
	Fine	Micas, clay	Phyllite			Mudstone
	Medium to coarse	Micas, chlorite, talc, hornblende	Schist			Mudstone, intermediate to mafic igneous rock
Banded	Medium to coarse	Feldspar, quartz, micas, pyroxenes, hornblende	Gneiss			Graywacke, arkose, felsic–intermediate igneous rock
Nonfoliated	Fine to coarse	Quartz	Quartzite		Contact Metamorphism	Quartz sandstone
	Medium to coarse	Calcite, dolomite	Marble			Limestone
	Fine	Dark silicate minerals	Hornfels			Mudstone, basalt, andesite

Figure 4.3
Classification of metamorphic rocks.

FIELD RELATIONSHIPS IN REGIONAL METAMORPHISM

The distribution at the earth's surface of zones containing certain diagnostic metamorphic minerals is frequently mapped by geologists as an aid to understanding the metamorphic process. The extent of these zones is indicated by different colors or patterns on a *geologic map* in the same way that political areas such as states are represented on conventional maps.

Figure 4.4 shows geologic maps of two metamorphic zones surrounding granitic plutons. From the information given in Figure 4.2, you may make some inferences about the depth of burial in the two cases shown.

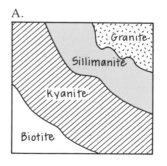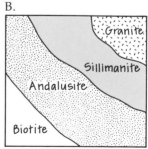

Figure 4.4
Maps of metamorphic mineral zones.

Problem 14

What is the minimum depth of burial of the area shown in A during metamorphism? (Assume that the entire region was buried to the same depth throughout.)

Problem 15

Was the region shown in B buried more deeply or less deeply during metamorphism than area A? Explain your answer.

Figure 4.5 is a map of an area in northern Vermont showing the occurrence of metamorphic mineral zones. Note the concentric nature of the mineral zones around several granitic plutons in the eastern half of the map.

Problem 16

In what zones might you expect to find migmatites?

Problem 17

Large-scale intrusion of granitic magma is commonly associated both spatially and temporally with regional metamorphism. Where in the map area might other granitic plutons lie at depth, as yet unexposed by erosion?

Problem 18

Using Figure 4.2, estimate the minimum depth of burial needed to produce the mineral zones in the northeast corner of the map.

Coal is widely known as a valuable energy resource. It is perhaps less widely known as a product of metamorphism. Coal begins as lush plant growth in still water environments such as swamps. When plants die and fall into the water, some of the plant tissues may be preserved from decay due to the lack of free oxygen in the stagnant water. In such an environment, plant remains accumulate beneath the water to form *peat*, a brown fibrous matted deposit of easily recognizable plant remains. If burial continues, increasing temperature and pressure at depth convert

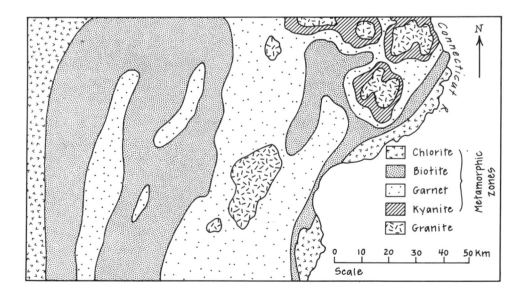

Figure 4.5
Metamorphic zones in northern Vermont.

peat to *lignite*. If the process continues, *bituminous coal*, then *anthracite* form. *Graphite*, the material of pencil leads, is the ultimate product of the progression to higher pressure-temperature conditions.

Problem 19

Examine Figure 4.6, which shows the content of moisture and volatile material in the different grades of coal. Explain these trends in terms of the metamorphic process.

Problem 20

Methane gas (CH_4) is often found with coal and is plentiful enough in some coal-producing areas to be an economically valuable resource. Explain in terms of metamorphism why methane, carbon dioxide, and water might be generated from the complex hydrocarbons of plant tissues when coal is formed.

Problem 21

Examine the coal deposit map of the eastern United States, Figure 4.7. What is the regional trend in coal grades along line A–A'? Explain this trend in terms of past metamorphic grades in the region.

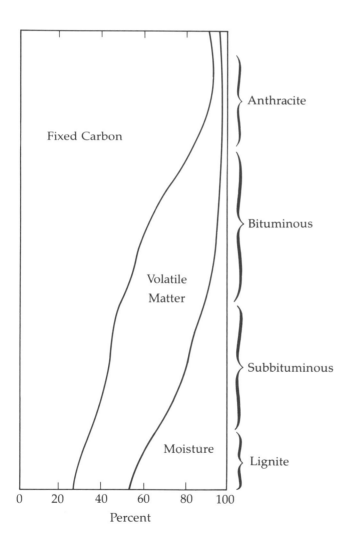

Figure 4.6.
Fixed carbon, volatile matter, and moisture content of coal grades.

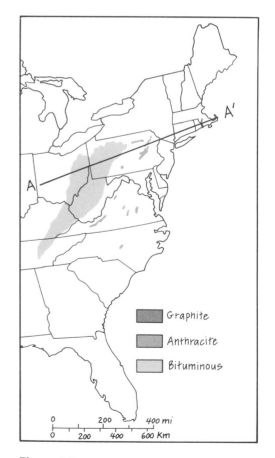

Figure 4.7
Distribution of coal in the eastern United States.

Geologists have determined that heat is the most important metamorphic agent in coal formation.

Problem 22

Physical and chemical descriptions of three coal samples are given below. They were found at the same depth in an area of variable geothermal gradient. Arrange the samples according to their proximity to a granitic intrusion in the area, and name the type of coal.

1. Definite and smooth surface when broken and a pronounced shine to the surface. Fixed carbon content, 65 percent.

2. Powdery detachable fibrous strands. Dull earthy luster. Fixed carbon content, 30 percent.

3. Coherent and uniform solid. Breaks with a clean conchoidal fracture, has a vitreous luster.

• Closest to intrusion

• Farthest from intrusion

REGIONAL AND GLOBAL PATTERNS IN RECENT BLUESCHIST METAMORPHISM

Figure 4.8 shows a simplified map of part of northern California. The hatched areas designate rocks under 200 million years old (relatively young, as geologic things go) that contain glaucophane, a product of blueschist metamorphism. See Color Plate 57.

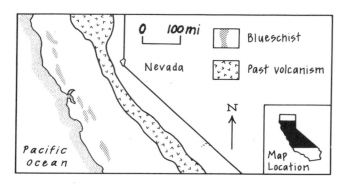

Figure 4.8
Zones of recent blueschist metamorphism and volcanism in northern California.

Problem 23

What is the average geothermal gradient for regional metamorphism given in Figure 4.1?

Problem 24

From Figure 4.1, calculate a typical geothermal gradient for regions of blueschist metamorphism.

Problem 25

Describe in words the temperature-pressure conditions in which blueschist metamorphism occurs relative to the average geothermal gradient.

Problem 26

What is the spatial relation of the zones of past volcanism and blueschist metamorphism relative to the coastline in Figure 4.8? This relationship is also true of the other areas of blueschist metamorphism.

Figure 4.9 shows the global distribution of blueschist metamorphism.

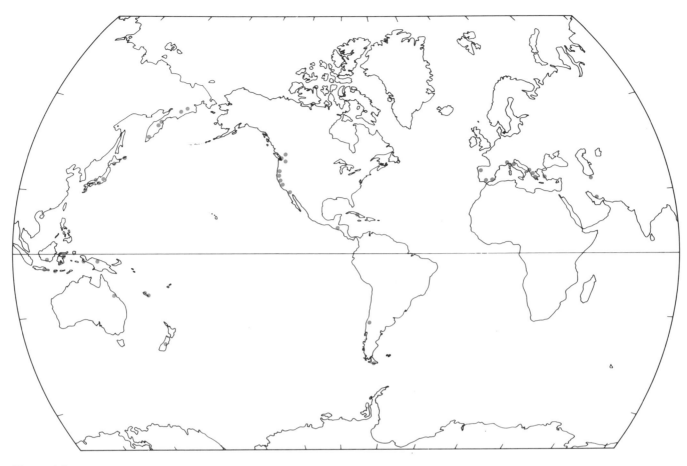

Figure 4.9
Global occurrence of glaucophane (blueschist metamorphism) less than 200 million years old.

Problem 27

What pattern, if any, do you observe in the geographical distribution of glaucophane with respect to continents and ocean basins?

Problem 28

Refer to Exercise 2 (Igneous Rocks), Figure 2.10, which shows the global occurrence of volcanism of roughly the same age as the blueschist metamorphism shown in Figure 4.9. Does there seem to be a similarity in the two patterns? Elaborate.

48

REFERENCES FOR PART A

Books

Best, M.G., *Igneous and Metamorphic Petrology*. W.H. Freeman, New York, 1982.

Blatt, Harvey, Gerald Middleton, and Raymond Murray, *Origin of Sedimentary Rocks*. Prentice-Hall, Englewood Cliffs, N.J., 1972.

Carmichael, I.S.E., *Igneous Petrology*. McGraw-Hill, New York, 1979.

Chesterman, C.W., *The Audubon Society Field Guide to North American Rocks and Minerals*. Alfred Knopf, New York, 1978.

Decker, R.W., and B. Decker, *Volcanoes*. W. H. Freeman, New York, 1981.

Dietrich, R.V., and B.J. Skinner, *Rocks and Rock Minerals*, John Wiley & Sons, New York, 1979.

Ehlers, E.G., and H. Blatt, *Petrology: Igneous, Sedimentary and Metamorphic*. W. H. Freeman, New York, 1982.

Harris, S.L., *Fire and Ice: The Cascade Volcanoes*. The Mountaineers, Pacific Search Books, Seattle, 1980.

Klein, Cornelis, and Cornelius Hurlbut, *Manual of Mineralogy*, 20th ed., John Wiley & Sons, New York, 1985.

Leveson, David, *A Sense of the Earth*. Doubleday, Garden City, N.Y., 1972.

Lipman, P.W., and D.R. Mullineaux, eds., *1980 Eruptions of Mount St. Helens, Washington*. U.S. Geological Survey Professional Paper 1250, 1981.

MacDonald, G.A., *Volcanoes*. Prentice-Hall, Englewood Cliffs, N.J., 1972.

McPhee, John, *Basin and Range*. Farrar, Strauss, and Giroux, New York, 1981.

Pettijohn, F.J., *Sedimentary Rocks*, 3rd ed. Harper & Row, New York, 1975.

Press, F., and R. Siever, eds., *Planet Earth: Readings from Scientific American*. W.H. Freeman, New York, 1975.

Pring, M., G. Harlow, and J. Peters, *Simon and Schuster's Guide to Rocks and Minerals*. Simon and Schuster, New York, 1978.

Reading, H.G., ed., *Sedimentary Environments and Facies*. Blackwell Scientific Publications, Oxford, 1978.

Reineck, H.E., and I.B. Singh, *Depositional Sedimentary Environments*. Springer-Verlag, New York, 1973.

Rhodes, G.H.T., and R.O. Stone, *Language of the Earth*. Pergamon, New York, 1981.

Tennissen, Anthony C., *Colorful Mineral Identifier*. Sterling, New York, 1985.

Time–Life Books, Planet Earth Series: *Gemstones, Noble Metals*, and *Volcano*. Time–Life Books, Alexandria, Va., 1982.

Turner, Francis J., *Metamorphic Petrology: Mineralogical, Field, and Tectonic Aspects*, 2nd ed. McGraw-Hill, New York, 1981.

United States Geological Survey, *Atlas of Volcanic Phenomena*. U.S. Government Printing Office, Washington, D.C., 1982.

Williams, H., F.J. Turner, and C.M. Gilbert, *Petrography*, W.H. Freeman, New York, 1974.

Government Publications

U.S. Geological Survey Information Pamphlets: Free upon request from

Distribution Branch
U.S. Geological Survey
Box 25286
Denver, CO 80225

Building Stones of Our Nation's Capitol, 1981.
Collecting Rocks, 1983.
Eruptions of Mount St. Helens: Past, Present and Future, 1981.
Geysers, 1983.
Man Against Volcano: The Eruption on Heimaey, Vestmannaeyjar, Iceland, 1983.
Monitoring Active Volcanoes, 1983.
Volcanoes, 1985.
Volcanoes of the United States, 1982.

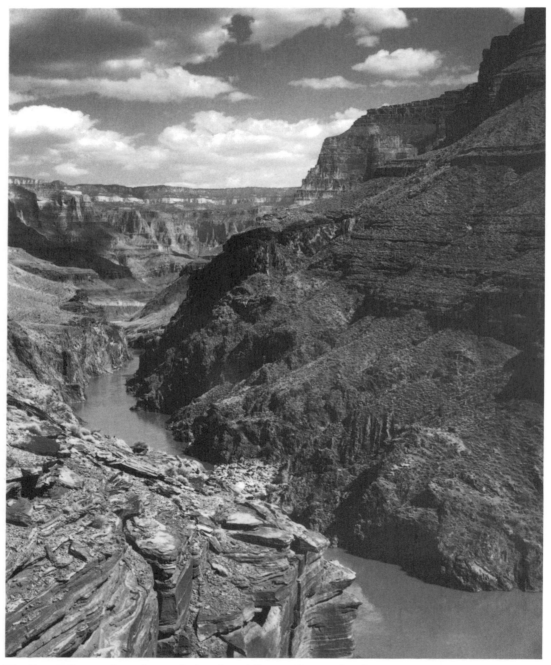

The Grand Canyon from the south rim, Grand Canyon National Park, Arizona. The Grand Canyon is carved over 5000 feet deep into rocks that record half of our planet's history.

Geologic Time, Maps, and Structures

The rock formations present at the earth's surface are vital clues to the geologic history of an area. You have learned to infer from different kinds of rocks the environments and processes that produced them. For example, the presence of limestone suggests a warm shallow sea; basalt indicates volcanic activity; schist is produced deep within the crust and is exposed at the surface only after extensive erosion. To be useful, however, information from the rocks must be organized and presented in a clear and simple manner. The most straightforward approach to organizing this kind of data is through the use of maps.

Most maps of the extent of rock formations at the earth's surface are superimposed on a topographic map. This kind of map represents the three-dimensional forms of the earth's surface—hills, valleys, plains, lakes, and mountains—in two dimensions by the use of topographic contours—lines connecting points of equal elevation of the land surface. When topographic information is combined with the areal extent of rock units at the surface, a three-dimensional picture of the earth begins to emerge because we can see how the rocks change at different elevations.

Structural geology adds more data to the picture; by measuring the orientation of rock layers and other features, we can project their position and extent at depth in the crust. These three kinds of information—the topography, the areal extent of rock units, and the geologic structures—are the basic data presented on geologic maps.

No picture of the rocks of the earth's crust is complete without some notion of geologic time. Perhaps the most important contribution of modern geology is the knowledge that the earth is several billion years old. The formation of rocks, including economically vital metal ores and hydrocarbons, requires many millions of years, as do the surface processes of weathering and erosion. Part B begins in Exercise 5 with a study of geologic time, the framework within which all geologic processes occur.

Exercise 6 will acquaint you with topographic maps and their uses. Geologic structures and their representation on geologic maps are the subject of Exercises 7 and 8. In a sense, this part introduces you to the basic tools of the geologist, and you must understand their use before proceeding to other exercises.

Time and Correlation

The result, therefore, of this physical inquiry [into the age of the earth] is that we find
no vestige of a beginning—no prospect of an end.

JAMES HUTTON, 1795

One aspect of geology that sets it apart from many of the other natural sciences is the importance of time. Geology is a historical science. It is not enough to look at the length, width, and depth of a body of rock; for a fuller understanding of that rock body we need some knowledge of the fourth dimension, time. "How old is that rock?" is a fundamental question geologists continually ask. Geologists wish to learn the sequence of events in the past, and this can be achieved only by determining which rocks are older and which are younger. Our usual notion of time implies a periodically repeating process; by counting the repetitions we measure the passage of time. The ticking of a clock mechanism is one such process; the planetary motions that mark the day, the month, and the year are others.

We shall consider three different time scales. The *human time scale*, the most familiar in everyday experience, measures short intervals of time—minutes, hours, days. The *historical time scale* is also familiar to us because it deals with historical events in terms of years or thousands of years. The *geologic time scale* is unfamiliar to most people because of the vast spans of time involved, typically millions or billions of years. The magnitude of geologic time is difficult to comprehend, yet most of the processes of geology occur so slowly that millions of years are required to form the landscape—to carve the Grand Canyon, for example, or to raise the Himalayas to 29,000 feet above sea level, or to deposit wedges of ocean sediments thousands of feet thick. When one considers that very little change in the geologic features around us—mountains, plains, lakes—is discernible over the course of a human lifetime or even over the several thousand years of recorded history, one begins to realize that the earth is indeed an ancient body whose history cannot be confined to mere thousands of years.

HUMAN TIME SCALE

Since the human time scale of hours and days is the most familiar, we shall begin with it. Many periodically repeating processes are used to keep time day to day: the ticking of a clock, the vibrations of a quartz crystal in a watch, and the reversing of an electrical current 60 times in a second are all used to keep track of minutes and hours. The periodically repeating process that in nature defines a day is the earth's rotation on its axis.

Problem 1

The earth's equatorial radius is 6378 km (3960 mi). Calculate its circumference at the equator (circumference = πd). What is the average width in kilometers of a one-hour time zone at the equator? Calculate the speed of a point on the equator in kilometers per hour.

The length of the day—the earth's rotation period—is known today to within microseconds. That makes it possible to calculate the exact local time of eclipses far into the future and equally far back into the past. When the calculations are checked against the historical record, however, a small but definite change is found in the length of the day.

A clay tablet from ancient Babylon, now in the British Museum, contains a record of an eclipse of the moon about $1\frac{1}{2}$ hours after local sunset on October 3, 685 B.C. According to modern calculations, that eclipse should have taken place over 5 hours earlier, before the sun had set at Babylon, and the Babylonian astronomers should not have seen it at all.

Problem 2

Given that the moon's orbit and the earth's orbit have not changed, what does this imply about the length of the day? About changes in the earth's rotational speed?

The periodically repeating process that defines our year is the completion of one circuit of the earth around the sun. The division of each year into distinct seasons, caused by the tilt of the earth's rotational axis, leads to another way to tell time. The time scale important to organized society—the historical time scale—counts not in days but in years or thousands of years.

HISTORICAL TIME SCALE

One natural clock that records the passage of years is familiar to all of us—tree rings. In a growing season, a tree forms a wide layer of light wood and a narrow dark layer. By counting these rings back from the outer layer of a tree that has recently been cut or cored, one may determine its age. In addition to the age of the tree, rings record environmental conditions during the lifetime of the tree. A period of hardship—several years of drought or excess cold, when growth is hindered—may be recorded as a group of closely spaced rings that stands out from the normal pattern. Thicker rings suggest a period of favorable conditions, possibly a warmer, wetter climate. Trees growing under similar conditions within an area form similar patterns of wide and narrow rings, which may be matched, or *correlated*, in trees of slightly different ages in any given area. Scientists have been able to piece together a record of time and environments extending back 10,000 years in some regions.

Problem 3

Study the tree ring records of a bristlecone pine from Hermit Lake, Colorado, shown in Figure 5.1. What is the average yearly growth thickness?

Problem 4

Mark the time periods of unusually thin and thick rings in the Hermit Lake record—thin rings blue, thick rings red. Do the same for graphs 2 and 3. Using the Hermit Lake ring widths as an index, correlate bristlecone pine trees from northern New Mexico to the known time scale of Hermit Lake. Do this by drawing lines between the Jicarita Peak and Red Dome data and the Hermit Lake data for the years 1300, 1500, and 1700.

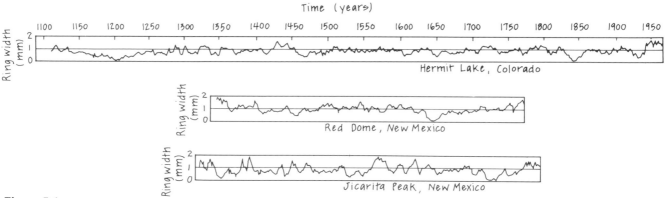

Figure 5.1
Tree ring data for age correlation.

Problem 5

When did the New Mexico trees begin growing?

• Jicarita Peak

• Red Dome

GEOLOGIC TIME SCALE

The human and historical time scales are familiar to all of us; we deal with these time spans routinely in our daily lives and in the study of history. The geologic time scale, however, is another matter. Only within the last several decades has it been possible to determine reliably the *absolute age* of a rock in terms of years. The reason is quite simple. Until the discovery and use of radioactivity, there was thought to be no periodically repeating process that accurately recorded the great lengths of time needed to measure the ages of the rocks. For many years geologists could determine only the *relative ages* of rocks, a simple statement that one rock was older than another. Relative age is still the geologist's most useful tool.

Relative Geologic Time

Problem 6

Recall the experiment you performed with the jar of sediments and water in Exercise 3 on sedimentary minerals and rocks. Which of the layers in the jar was deposited earliest, the bottom or the top layer?

Problem 7

Extrapolate this result to the general case of layers of sediments deposited from water and write a general statement that governs the relative ages of such layers lying one on top of the other.

Problem 8

Recall also the orientation of the crude layering in the sediment jar. Were the layers nearly horizontal? Vertical? Generalize your observation into a law that governs the orientation of sedimentary beds as they are formed.

These two laws you have devised are called the *law of superposition* and the *law of original horizontality*. Although they seem obvious to you today, they constituted a major breakthrough in geologic thought when they were first formulated and applied to layers of rock by Nicolaus Steno in 1669. These two laws were the first step in the understanding of geologic time, and they govern the study of layered rocks, the branch of geology called *stratigraphy*.

If the rocks one wishes to compare are hundreds of miles apart, or even on different continents, it is still possible to determine which are older and which are younger. This important step in comparing geologic ages was developed in the 1790s by an English engineer, William Smith, and slightly later by the French biologist Baron Georges Cuvier. Both men observed while mapping layered rocks that the fossils found in lower layers were not the same as those in upper layers. The fossils changed slowly through a vertical sequence of sedimentary rocks. From the work of Smith and Cuvier, two important principles emerged: the laws of faunal succession and faunal assemblage. The *law of faunal succession* is similar to Steno's law of superposition and states that in a sequence of fossil organisms in rock layers the oldest are the lowest. Fossil populations change form with time, and once changed, they never reappear in younger rocks in their previous form. The *law of faunal assemblage* states that similar assemblages of fossil organisms are of similar age and that the rock layers containing them are thus of similar age, no matter where they occur. Using these principles, rock layers from different areas may be matched or correlated with respect to age if their fossil assemblages are similar. A limestone in New Mexico, for example, is said to correlate with a limestone hundreds of miles away in Utah if the fossil organisms in each limestone are the same.

There is another way to unravel the relative ages of some bodies of rock, particularly in a sequence of igneous and sedimentary rocks. The *law of crosscutting relationships* states that, if a rock layer B cuts across another rock body A, then the rock that does the crosscutting, B, is the younger rock (see Figure 5.2).

During the nineteenth century, geologists developed the framework for stratigraphic correlation and pieced together the history recorded in rocks from different areas of the earth. In any given area, the bottom-to-top succession of rocks is called the *stratigraphic sequence.* Geologists illustrate the sequence of rocks present by a *stratigraphic column,* a vertical column with the oldest rocks at the bottom, the youngest at the top. In a stratigraphic sequence, rocks with similar lithologic characteristics are classified as *formations,* the fundamental unit in stratigraphy. Individual distinguishable layers within a formation are *beds* or *strata* (singular, *stratum*). A *group,* a larger rock subdivision, includes several formations that have similar characteristics. The boundary surface separating two rock units is the *contact.* Strata, formations, and groups are named for the geographic setting where the rocks were first studied or where they are exceptionally well exposed.

Using all these stratigraphic tools, geologists of the nineteenth century pieced together the *geologic time scale.* The present scale has three main divisions: eras, periods, and epochs. The *era* is subdivided into *periods,* which in turn are divided into *epochs.* The three most recent eras, Paleozoic ("old life"), Mesozoic ("middle life"), and Cenozoic ("recent life"), are separated by major extinctions of life forms; the division between the Mesozoic and Cenozoic Eras, for example, is marked by the extinction of the dinosaurs. The distinction between periods is based on less widespread extinction events.

The geologic time scale is illustrated in Figure 5.3. This diagram lists only those divisions of time that are commonly used by most geologists. Notice that the early history of the earth is divided into very broad outlines because so little evidence remains of events that shaped the earth before 570 million years ago. As we approach the present, the divisions of the time scale become finer and finer. The epochs of the Tertiary and Quaternary Periods last only millions or tens of millions of years, whereas the first 4000 million years of earth history are lumped together as Precambrian time (the term most often used by geologists to refer to the Archean and Proterozoic Eons). In this propensity to discuss the recent past in minute detail, geologists are like historians. A semester-long course in world history may dispense with several thousand years of ancient Egyptian history in one lecture, whereas the years since World War I might occupy two or three weeks. We discuss in much more detail those events that we know most about.

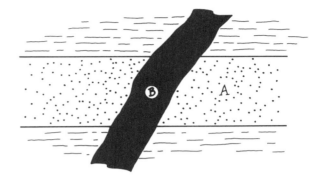

Figure 5.2
The law of crosscutting relationships. Here B is younger than A.

It often happens that a particular stratigraphic sequence may lack rocks of a certain age. Rocks of Cenozoic age may lie directly over those of Paleozoic age. Perhaps the missing Mesozoic layers were deposited over the Paleozoic layers but were eroded away before the Cenozoic layers were deposited. Or perhaps they were never deposited at all. In either case, the boundary between the Cenozoic rocks and the underlying Paleozoic rocks is called an *unconformable contact* or an *unconformity.* The stratigraphic symbol for an unconformity is a wavy line.

When describing a stratigraphic sequence, geologists look not only at the individual layers but also at the relationships between layers. Four general properties of each rock unit must be described:

1. *The rock type.* Is it a volcanic flow, a sandstone, a shale? What is the texture, the mineral composition, the color? Are there any large-scale features such as graded bedding?

2. *The fossil content.* Are there land *(terrestrial)* or *marine* fossils present? This might suggest the relative age of the rock and the environment of deposition.

3. *The thickness.* How thick is the rock layer? Does the thickness vary from place to place, or is it constant over a wide area? The thickness of a sedimentary rock bed or unit is a function of several factors: the rate at which sediment accumulated, the length of time during which sedimentation took place, and the amount of compaction that occurred during the conversion of sediments to rock.

4. *The nature of the contacts.* Each rock layer is bounded by an upper and a lower *contact.* These surfaces are very important because they record the transition between different rock-forming processes. A tran-

56

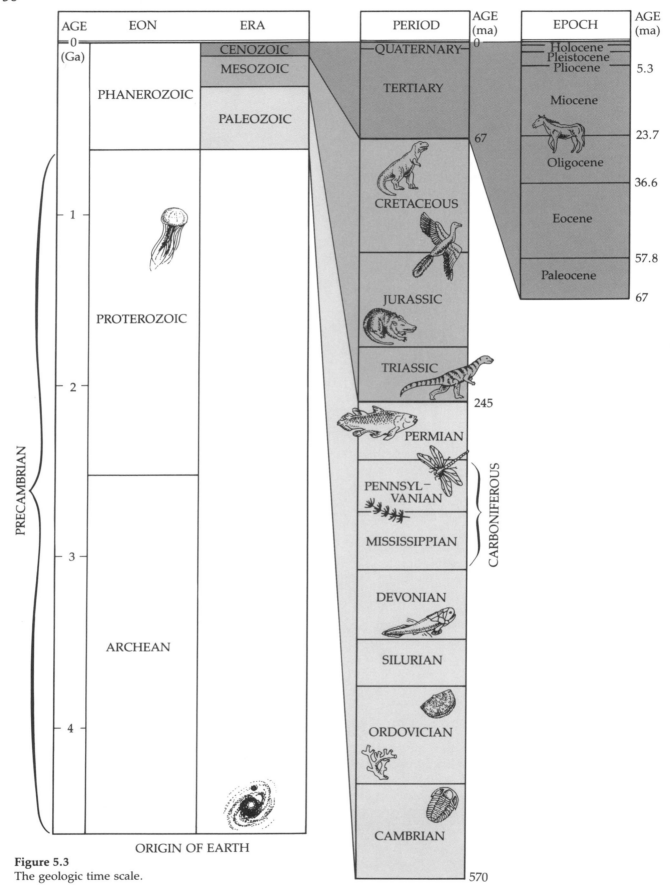

Figure 5.3
The geologic time scale.

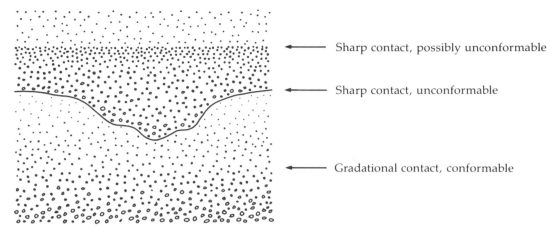

Figure 5.4
Types of contacts in layered rocks.

sition between rock types that is not sharp but seems to occur gradually over some small vertical distance suggests a gradual change in the environment of deposition. A gradual contact of this sort is said to be *conformable*. An abrupt transition between beds suggests a rapid change in depositional conditions. Such a contact may also be conformable; a lava flow may cover a lake bed without a gap in time, for example. If the surface of the sharp contact is irregular or undulating, however, then there is strong evidence for a break in time between the deposition of the two beds during which the lower bed was eroded. In this case the contact is *unconformable*. Examples of different contacts are shown in Figure 5.4.

Geologic correlation between stratigraphic columns of different locations is done in the same way as the correlation of tree rings—by matching distinctive patterns. These patterns may exist in rock or fossil sequences or both. Even more helpful is the occurrence of a unique rock layer in a sequence—a volcanic ash layer in a sandstone–limestone sequence, for example. These unique layers serve as *marker beds* that help establish a correlation from one area to another.

Problem 9

The stratigraphic columns in Figure 5.5 illustrate hypothetical rock sequences in four different areas. Correlate these columns by drawing lines that connect contacts between similar rock strata. By what evidence in the rocks did you make the correlation?

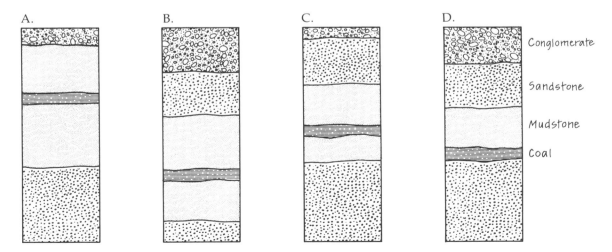

Figure 5.5
Hypothetical stratigraphic columns.

Problem 10

Figure 5.6 shows two stratigraphic columns. Consider the laws of superposition and crosscutting relationships and the nature of the contacts carefully, and list on the figure the geologic events (deposition of sediments, intrusion of dikes, erosion of the surface, metamorphism, etc.) that occurred in each area with the oldest event at the bottom and the youngest event at the top.

Radiometric Time

The absolute ages of rocks and the age of the earth itself have been the subject of spirited debate for centuries.

For many years biblical scholars argued from scriptural evidence that the earth was no more than 6000 years old. Starting in the eighteenth century, however, scientists began looking for a more objective way to determine the age of the earth. Using various physical phenomena, such as the earth's cooling rate or the deposition of sediment, which were thought to be occurring at known and constant rates, scientists came up with age estimates as low as 75,000 years and as high as 30 million years. This was too old to satisfy the traditionalists and too young to satisfy most geologists, who realized that vast amounts of time were needed first to form the rocks and then to carve them into the landforms we see today.

The discovery of radioactivity in rocks by Henri Becquerel in 1896 revealed to scientists a new clock, slow enough and constant enough to record the most ancient rock-forming events in our planet's history. The first radiometric ages determined for rocks were around 500 million years, measured in the early 1900s by Ernest Rutherford and Bertram Boltwood. Today, the first rocks on Earth are thought to have solidified about 4.6 billion years ago.

Several radiometric "clocks" are used by geologists to measure time, but they all have in common the fact that the radioactive decay of any unstable isotope, such as uranium-235, uranium-238, potassium-40, or rubidium-87, takes place at a uniform rate. (The number after the element name is its atomic weight, the number of protons and neutrons in the nucleus.) Furthermore, these rates are for all practical purposes unaffected by temperature, pressure, chemical state, or any other known factor.

The rate of decay of a radioactive substance is expressed as its *half-life*, the length of time required for half the so-called *parent* radioactive atoms to disintegrate into *daughter* atoms. If a radioactive species

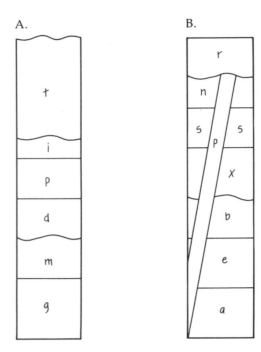

Figure 5.6
Two stratigraphic columns for sequence determination.

has a half-life of 40 years, and one starts with 12 grams of the pure parent substance, then at the end of 40 years 6 grams of parent remain and 6 grams of daughter have been formed. At the end of another 40 years, 3 grams of the parent (one-fourth of the original amount) remain and 9 grams of daughter are present.

Problem 11

The techniques and calculations by which radiometric "clocks" are used to tell geologic time can be rather complex, but a simplified example will illustrate the basic theory. Using the same reasoning as in the preceding paragraph, complete the chart in Figure 5.7A.

Problem 12

Present these results in a graph in the coordinate system in Figure 5.7B, and draw a smooth curve connecting the points.

The minerals used in dating rocks must have a high content of the radioactive parent atom. Zircon ($ZrSiO_4$), a mineral in which uranium may substitute for zir-

A.

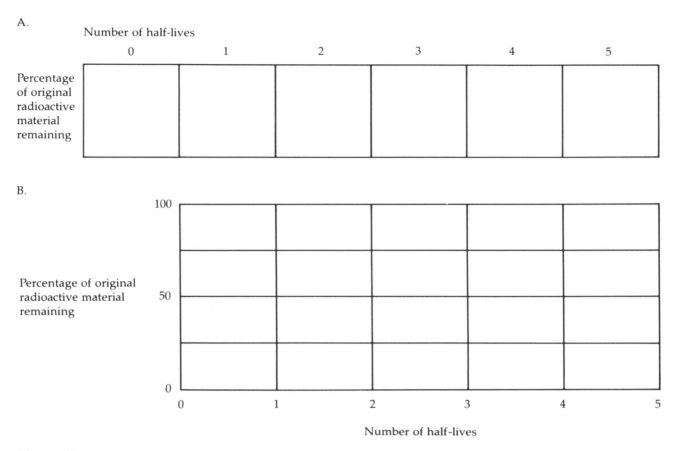

Figure 5.7
A. Amount of radioactive parent material remaining as a function of the number of half-lives elapsed. B. Decline in quantity of radioactive parent material with time.

conium in the crystal structure, is the preferred mineral for uranium–lead dating systems (parent uranium-235 decays to daughter lead-207; uranium-238 decays to lead-206). Two other widely used systems are the decay of potassium-40 to argon-40, and rubidium-87 to strontium-87. Chemically, rubidium is very similar to potassium and is usually present in potassium-bearing minerals.

Problem 13
What minerals have a high potassium content and would be well suited to use in these dating methods? (Refer to Exercise 2.)

When measuring the absolute age of a rock by one of the radioactive substances in it, the geologist makes the assumption that none of the parent or daughter atoms has escaped from the rock or mineral being used. If the rock is heated (as in metamorphism), melted, or badly weathered, this assumption is no longer valid, as heating or leaching by water are both known to change the chemical balance of the system. Material can enter or leave the rock, and the amounts of parent and daughter change, which results in a resetting of the "geologic clock."

Problem 14
If you were to date the feldspar grains in an arkose using the decay of radioactive ^{40}K to ^{40}Ar, would the age be that of the time of deposition? Why or why not?

FIELD STUDY: THE COLORADO PLATEAU

The Colorado Plateau is a remarkable geologic feature. Formerly a low-lying area of the North American continent, it has slowly risen over a mile during the past 60 million years, and it is presently rising at the geologically rapid rate of 2–3 mm/yr. As the land rises, rivers cut down through the layers of sediments and rocks, exposing the depths of a large area of the earth's crust and providing geologists with an unparalleled view of earth history extending back through almost half the lifetime of our planet. The Colorado Plateau area in Arizona, New Mexico, Utah, and Colorado provides a natural laboratory of rocks, depositional environments, and geologic structures that you will use in this and other laboratories. Many of the most spectacular river gorges have been preserved as national parks, and stratigraphic columns are readily available.

For the field study you will correlate the stratigraphic columns of five national parks and one national monument: Bryce Canyon, Zion, Grand Canyon, Capitol Reef, and Canyonlands National Parks and Colorado National Monument. See Figure 5.9 for the locations of these areas. The tools at your disposal include rock type, fossil content, thickness, and the nature of the contacts.

Problem 15

In each national park a river has cut canyons, exposing the underlying rocks. From the general elevation at the base of the canyons listed below and your knowledge of sedimentary deposition, which park probably exposes the oldest rocks? The youngest rocks? Explain your answer.

National Park	Elevation of Canyon Base
Bryce Canyon	6900 feet
Grand Canyon	2500 feet
Zion Canyon	3900 feet

Correlation by Rock Types

Problem 16

Examine the stratigraphic columns for Bryce Canyon, Zion, and Grand Canyon National Parks shown in Figure 5.8. From a consideration of rock type *only*, which of these three stratigraphic sequences is probably the lowest (oldest) in the section? (*Hint:* Would you expect to find a limestone stratigraphically below a schist?)

Recall the use of marker beds to correlate stratigraphic columns.

Problem 17

There is very little overlap in the stratigraphic columns of Zion and the Grand Canyon, but there is a distinctive marker bed, a chert-bearing limestone called the Kaibab Formation, that allows a correlation. What formation at Zion correlates with the Kaibab limestone at the Grand Canyon?

Problem 18

What further evidence in the rocks (in addition to rock type) would you look for to substantiate your correlation?

Problem 19

Leaving wide margins for notes, cut the Grand Canyon, Zion, and Bryce Canyon stratigraphic columns from the page. Trim the top of each column even with the upper contact. Correlate the three stratigraphic sequences by taping the columns together so that one stratigraphic column is formed. Overlap the columns vertically where necessary. Check with your instructor before proceeding.

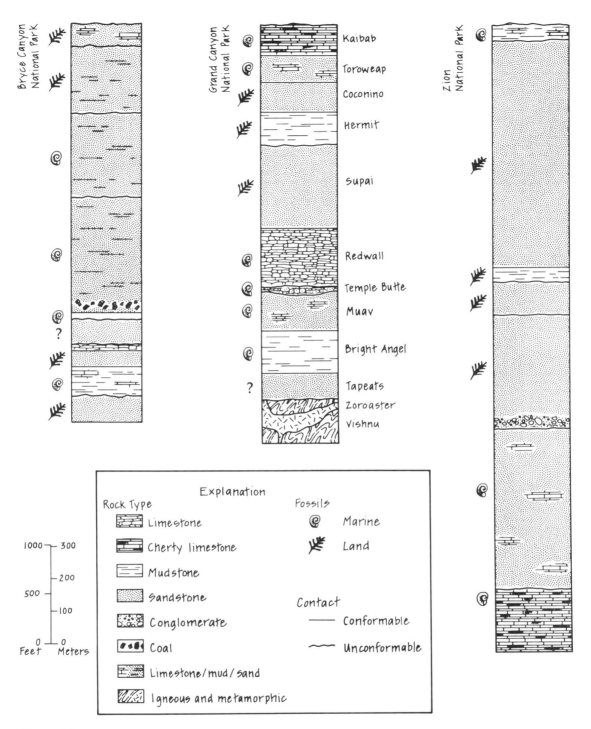

Figure 5.8
Stratigraphic columns for Bryce Canyon, Zion, and Grand Canyon National Parks.

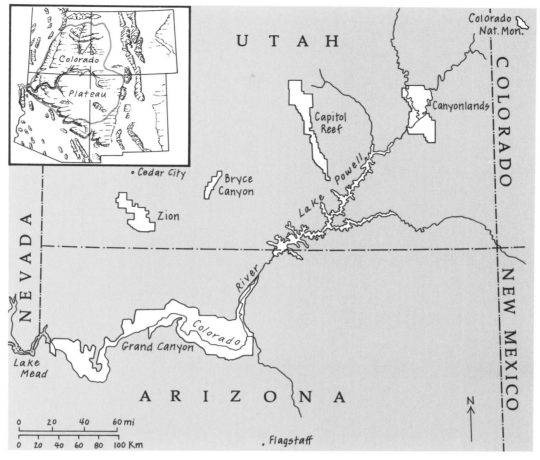

Figure 5.9
Locations of National Parks on the Colorado Plateau.

Correlation by Fossils

Using fossils as indicators of environment, one may distinguish rocks deposited on land *(terrestrial)* from those deposited in the oceanic *(marine)* environment. A ⓠ or a 🌿 beside the stratigraphic column indicates the type of fossils found in the rock formation.

Problem 20

What rock types are commonly deposited in the marine environment? In the continental environment?

Problem 21

How many times have marine waters inundated this area?

Bed Thickness

Problem 22

Based on your correlated stratigraphic sequence, what is the total thickness of sedimentary rock in the area of the Colorado Plateau near the Grand Canyon?

Problem 23

Assuming that the rocks were compacted during burial to three-fifths of their original thickness, estimate the original thickness of sediment represented in the combined stratigraphic sequence.

Problem 24

A reasonable average sedimentation rate, determined from modern sedimentation rates, is about 0.1 foot per 1000 years. Determine the time required to deposit the original thickness of the sediment before burial.

Problem 25

What features in the stratigraphic column suggest that the total thickness of sediments originally deposited was even greater than the number you calculated?

Nature of the Contacts

Refer to Figure 5.10, the stratigraphic columns of Capitol Reef, Canyonlands, and Colorado National Monument for the next set of questions.

Problem 26

Of the four common sedimentary rock types—limestone, shale, sandstone, and conglomerate—what rock types are most often found immediately above the unconformities?

Problem 27

In what energy environments—high, medium, or low—are these particle sizes deposited?

In general, both erosion and high-turbulence depositional environments occur above sea level (headwaters of streams or ocean beaches, for example). The offshore marine environment, in contrast, is characterized by low-turbulence deposition.

Problem 28

Is the Colorado Plateau undergoing erosion or deposition at the present time? Explain.

Problem 29

Refer to Figure 5.8. What type of contact separates the sedimentary Tapeats Formation from the Vishnu Formation? Would you expect that this type of contact always separates metamorphic rocks from overlying sedimentary rocks? Explain.

Radiometric (Absolute) Time

Geologists have dated rocks at the base of the Grand Canyon radiometrically using the uranium–lead method. Zircons from the Zoroaster granite were chemically analyzed, with the following results:

$$^{238}U = 0.170 \text{ mg/g} \qquad ^{206}Pb = 0.032 \text{ mg/g}$$

Problem 30

Assume that all the ^{206}Pb was produced by the decay of ^{238}U. What percentage of the original parent substance (^{238}U) remains? (If you wish, assume that the weights of the two substances may be used instead of the number of atoms or moles. The atomic weights of uranium and lead are close enough that only a small error is introduced by this simplification.) Show your calculations.

65

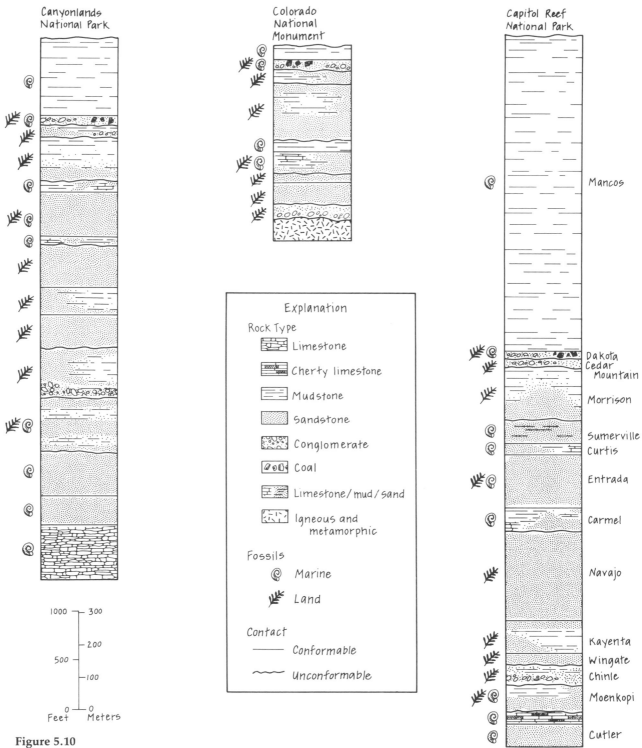

Figure 5.10
Stratigraphic columns for Capitol Reef and Canyonlands National Parks and
Colorado National Monument.

Problem 31

Now return to the graph you prepared in Figure 5.7B. The percentage of original parent substance remaining, which you have just calculated, corresponds to the passage of how many half-lives?

Problem 32

The half-life of ^{238}U is 4.56×10^9 yr (4.56 billion yr). How old is the Zoroaster granite? Show your calculations.

Problem 33

What specific geological event are you dating when you make this calculation?

Problem 34

What geological event would you be dating by making U–Pb analyses of the Vishnu schist?

Problem 35

Consider the conglomerate that rests directly over the schist and granite basement rocks. It contains cobbles of the Zoroaster granite. What age would you determine from a U–Pb analysis of these cobbles?

Problem 36

Is this the true age of the conglomerate?

Problem 37

Summarize the depositional and erosional history of the Grand Canyon, Zion, and Bryce Canyon National Parks.

Problem 38

Using the stratigraphic concepts and procedures you have learned, correlate the stratigraphic sequences of Capitol Reef National Park, Canyonlands National Park, and Colorado National Monument. Draw lines between the contacts to show your correlation.

Problem 39

Try to correlate the stratigraphic sequences of Capitol Reef, Canyonlands, and Colorado National Monument (Figure 5.10) with the stratigraphic sequences of the Grand Canyon, Zion, and Bryce Canyon. Use Roman numerals to identify similar rock units or contacts. In the space below summarize your correlation (or lack of correlation) of the sedimentary strata across the Colorado Plateau. Pay particular attention to the matching of marine and land fossils from place to place. If the match is not exact, what is a possible explanation for the difference in terms of the ancient shorelines of the region?

Dimensions on the Earth

Down, down, down. Would the fall never come to an end? "I wonder how many miles I've fallen by this time?" she said aloud. "I must be getting somewhere near the center of the earth. Let me see: that would be four thousand miles down, I think—" (for, you see, Alice had learnt several things of this sort in her lessons in the schoolroom, and though this was not a very good opportunity for showing off her knowledge, as there was no one to listen to her, still it was good practice to say it over) "—yes, that's about the right distance—but then I wonder what Latitude or Longitude I've got to?" (Alice had not the slightest idea what Latitude was, or Longitude either, but she thought they were nice grand words to say.)

LEWIS CARROLL

Dimensions of the earth's features and properties are usually represented on maps. A map is a representation of natural and manmade features present on or near the earth's surface. The development and use of maps appear to be a fundamental component in the growth of human civilization. Maps are common to all but the most primitive cultures, and anthropologists have discovered map-making even among tribes lacking a written language. The oldest record of map-making is from the Mesopotamian civilization, about 6000 years ago. These ancient maps were made of clay and illustrated small areas along the Euphrates River. The art and science of map-making, *cartography*, has progressed over the centuries, and today paper maps illustrate all areas of the planet's surface. Maps representing weather conditions, political boundaries, and roads are common everyday features in the modern world. In geology, maps are a vital tool used to illustrate and analyze the earth's surface features.

REFERENCE POINTS AND ORIENTATION

The position of the sun and all the fixed stars, particularly Polaris (the North Star), makes it possible to determine the four basic geographic directions, north, south, east, and west. The sun is located to the east in the early morning and to the west in late afternoon. Polaris is nearly aligned with the earth's rotational axis and is therefore always seen in the northern sky at night in the Northern Hemisphere.

Problem 1

Imagine yourself facing a beautiful June sunrise on the beach in Key West, Florida.

- Which way is north, to the right or left?
- Which way is south?
- Which way is east?

Problem 2

Imagine yourself looking at the earth from Polaris. Does the earth rotate clockwise or counterclockwise?

An interesting and useful method for determining compass direction is available to those who wear watches with hands and numbered faces. Holding the watch flat (horizontal), point the hour hand at the sun. Geographic south is halfway between the hour hand and the 12.

Problem 3

Try this method for determing geographic south. Draw a clock face with the hour hand at 8. Imagine where the sun would be at 8:00 A.M. and at 8:00 P.M. on a summer day, and see whether your determination of geographic south is approximately correct according to the above method. Check your result with a compass. What futher refinement would you make in the instructions for locating south? [*Hint:* Which angle between 8 and 12, the acute (<90°) or the obtuse (>90°), must you use after 6:00 P.M. or before 6:00 A.M.?]

true (geographic) north direction. The north arrow is the fourth component of a map. Technical maps, in addition, show the direction of magnetic north, which differs in most areas from the true geographic north. If there is no north arrow, north is assumed to be at the top of the page.

Problem 4

Sketch and label seven different symbols shown on the map provided by your instructor.

MAPS

Maps are perhaps the geologist's most important and useful tool because of all the information that can be represented on them. All maps have four components: a title, symbols, a scale, and an orientation. These components are shown on all the maps in this chapter and *should be shown on any map you make in this or any other exercise.* The map title states the location and subject of the map. The map symbols represent different natural and manmade features, such as water bodies, mountains and hilltops, roads, and buildings. The map legend or key identifies the symbols on the map. Keys for maps used in geology are shown in Appendix 5.

The third component, the map scale, expresses the relation between the linear distances shown on the map and the corresponding distances on the earth. The scale is expressed in three ways: as an equation, a bar, or a fraction (or ratio). Popular maps, such as road and location maps, record the scale as a bar or an equation. In geology, map scale is generally given as a dimensionless fraction, for example, 1/24,000, or as a ratio, 1:24,000. The numerator, usually 1, represents map distance, and the denominator, a large number, represents horizontal ground distance. In this case 1 inch on the map represents 24,000 inches (2000 feet) at the surface. Large-scale maps, which show a small area in great detail, are useful for hiking and engineering planning; small-scale maps, which cover large areas in less detail, are useful for regional planning or overviews of large areas.

The map perspective in most cases is from above, as if seen looking straight down from an airplane, a perspective called *plan view.* The orientation of maps may vary, and an arrow on the map designates the

Problem 5

Determine the scales of the maps provided by your instructor, and measure the distances indicated by your instructor.

REFERENCE GRIDS

Reference grids are networks of fixed imaginary lines placed on a map to designate positions or areas on the earth's surface. The earliest known use of reference grids is from the ancient Greek civilization, over 2000 years ago. Since then, both national and global grid systems have been devised. Reference grids appear on a map as thin black or red lines that are labeled along the margins.

Since the earth is approximately spherical in shape, positions on the earth's surface are given by angular measurements from a fixed reference. The earth's surface is divided into areas by imaginary lines of longitude and latitude.

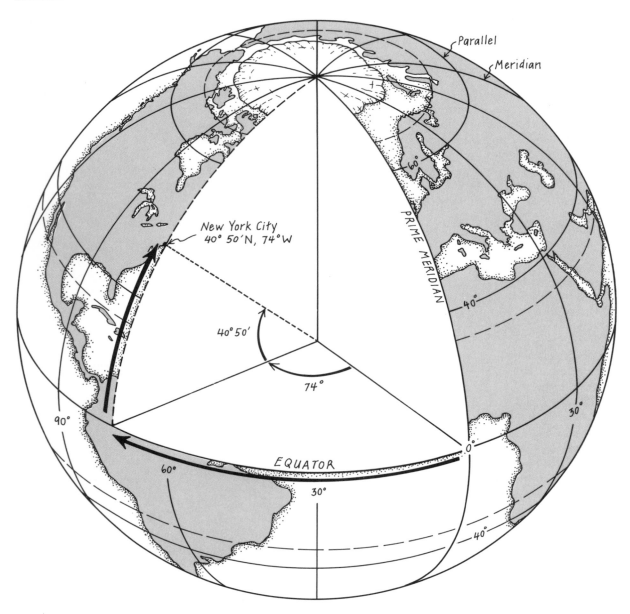

Figure 6.1
Longitude and latitude system showing New York at 40°50′N, 74°W.

Lines of *longitude,* called *meridians,* radiate north and south from the poles. The spacing between meridians is 1/360 of a complete circle (1° of longitude). The line of 0° longitude (the prime meridian) by international agreement runs through Greenwich, England; east–west distances are given in degrees east or west of the prime meridian. New York, for instance, is 74° west of Greenwich.

Lines of *latitude,* called *parallels,* circle the earth in an east–west direction and divide it into 180 degrees from north pole to south pole. The 0° parallel is the equator, and positions north or south of the equator are given in degrees north or south. The north pole is at 90°N; the south pole is at 90°S. New York is at latitude 40 degrees, 50 minutes north (written 40°50′N).

Figure 6.1 shows how the system of longitude and latitude is used to specify locations on the earth's surface.

The United States has its own national grid system to describe the position and size of areas on the surface. The land ordinance of 1785 established a coordinate system called township and range to designate both size and location of land parcels in areas surveyed after 1785, principally in the midwestern and

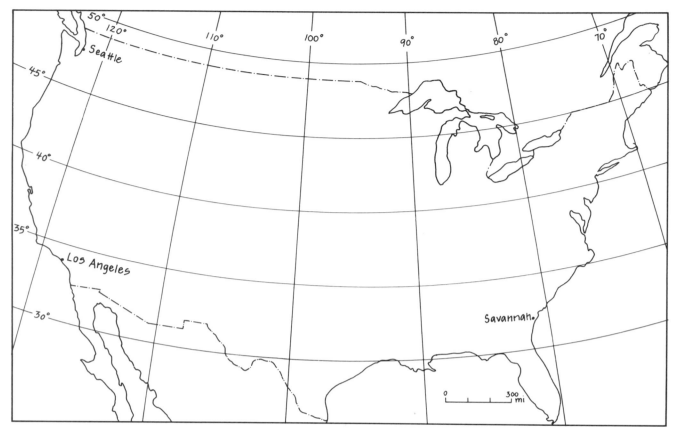

Figure 6.2
Longitudes and latitudes in central North America.

Problem 6

From the latitude and longitude coordinates given below, locate and label the cities on the map (Figure 6.2).

City	Latitude	Longitude
San Francisco	37°47′N	122°25′W
Washington, D.C.	38°54′N	77°01′W
New York	40°50′N	74°00′W

Problem 7

Determine the approximate latitude and longitude coordinates of the cities shown in Figure 6.2. Record the coordinates below.

City	Latitude	Longitude
Los Angeles		
Seattle		
Savannah		

western states. Land is divided into squares 6 miles on a side called *townships*. Townships are identified by their distance from surveyed lines called *base lines* and *meridians*, shown in Figure 6.3. East–west distance is given by the number of *ranges* east or west of the nearest meridian; north–south distance is given by the number of *townships* north or south of the nearest base line. Note that the term *township* has two

distinct meanings: it is an area of land 6 miles on a side and a unit of distance north or south of a base line.

The area of a township is divided into 36 squares, each 1 mile on a side, called *sections*. Sections (1 square mile, or 640 acres) are divided into four smaller squares called *quarters*. The section quarters are further divided into four 40-acre tracts. These small areas are identi-

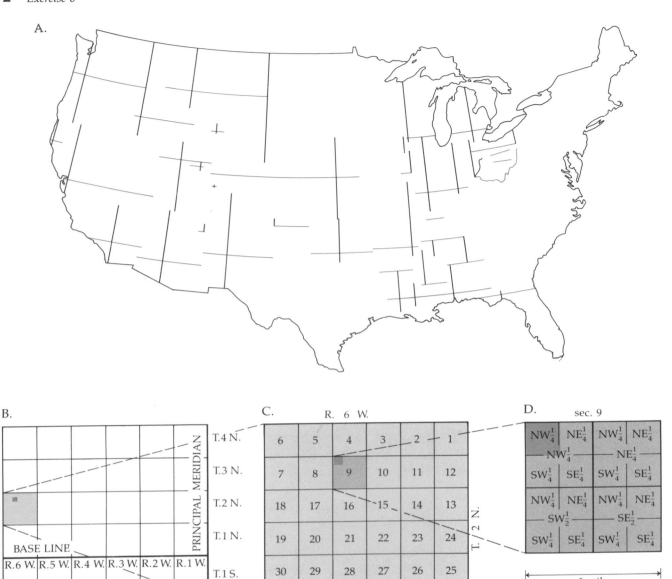

Figure 6.3
A. The location of some principal meridians and base lines. B. Township and range lines. C. A township area. D. A section area.

Problem 8

Determine the township and range coordinates of locations 1, 2, and 3 on the map provided by your instructor. Include section and quarter section.

fied by their location within the quarter and section. For example, the 40-acre cross-hatched tract in Figure 6.3 is described as the northwest quarter of the northwest quarter of section 9, township 2 north, range 6 west (abbreviated NW$\frac{1}{4}$ NW$\frac{1}{4}$ sec. 9, T. 2 N., R. 6 W.). Thus it is possible to define the size and location of an area by the township and range coordinate system.

CONTOUR LINES

Another type of reference line used on maps is the *contour line,* a line determined by points representing equal values of some quantity. The contour line was invented in the early 1700s by the astronomer Edmund Halley. It was a major step in the development of cartography as it permitted the use of thematic maps. Thematic maps describe the distribution of a particular property over the earth's surface and will be used in several subsequent exercises. A commonly used contour line is an isotherm, a line of equal temperature shown on weather maps. Every point on an isotherm is at the same temperature. Other values commonly described by contour lines are atmospheric pressure (isobars) and precipitation (isohyets).

One of the most useful types of contour lines in geology is the *topographic contour,* a line of equal elevation. Topographic contours may be visualized as the intersection of the earth's surface with a set of imaginary horizontal planes that are separated by a particular vertical distance, the *contour interval* (see Figure 6.4). The contour interval depends on the difference between the maximum and minimum elevations in the map area, sometimes called the *relief* of an area. Maps of areas with a large relief require a large contour interval, so as not to have too many lines on the map. Conversely, maps of areas with low relief have small contour intervals so as to show the subtler details of flat country. On a topographic map, every fourth or fifth contour is darkened and labeled with the elevation. Such lines are *index contours.* Topographic contours having hachures indicate closed depressions on the surface. Notice the left side of Figure 6.4; the hachures point in the direction of decreasing elevation.

For the following problems, use the topographic map of the Bright Angel quadrangle in Exercise 8, Figure 8.6.

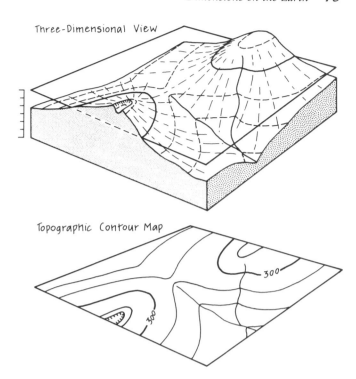

Figure 6.4
Topographic contour lines.

Problem 9

What is the average elevation of the Kaibab Plateau near the Grand Canyon Lodge?

Problem 10

What is the elevation of the Colorado River near Phantom Ranch?

Problem 11

In what direction is Bright Angel Canyon oriented? The Phantom fault?

Problem 12

How wide is the Grand Canyon, rim to rim, from the Grand Canyon Lodge to the El Tovar Lodge?

Problem 13

What is the relief of this area?

Figure 6.5 is a perspective view of the area north and south of the Golden Gate Bridge, California. San Francisco is at the right side of the figure; San Francisco Bay is at the top, and the Pacific Ocean is at the bottom, toward the viewer. Figure 6.6 is a topographic map of part of the same area. To understand contours, think of them as imaginary lines on the ground that take whatever shape is required to maintain the same elevation. Because the water surface is horizontal, the shoreline is the contour representing an elevation of zero feet above sea level. If the sea rose 50 feet, that

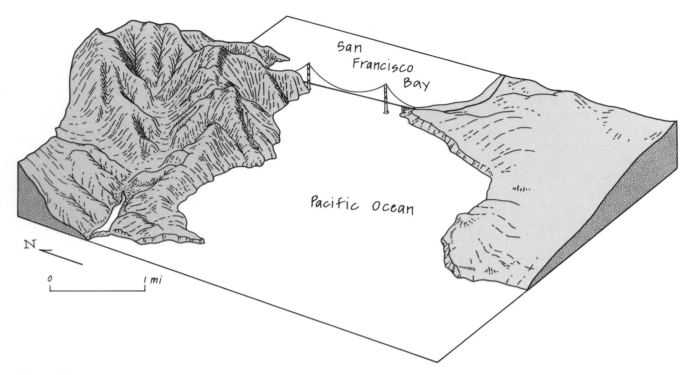

Figure 6.5
Perspective view of the Golden Gate region, California.

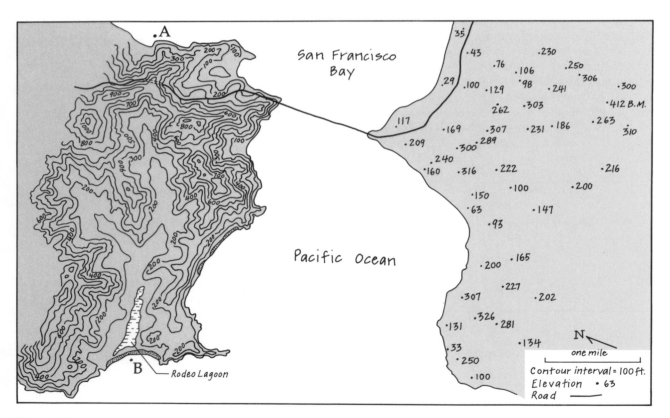

Figure 6.6
Topographic map of part of the Golden Gate region.

new shoreline would trace out the 50-foot contour of the old landscape. If the sea rose 100 feet, the shoreline would match the 100-foot contour, and so forth. Note that index contours are marked with their elevation and indicated by a heavier line.

Problem 14

What is the approximate surface area of Rodeo Lagoon? (Appendix 8 gives formulas for the areas of several geometric shapes. Figure 6.8 shows measurements of size.)

Problem 15

The Golden Gate Bridge's towers are 750 feet high, and its roadway is at 200 feet elevation. If you were standing on the eastern edge of Rodeo Lagoon, could you see the Golden Gate Bridge? Why or why not?

Problem 16

Complete the right side of the topographic map by drawing in contour lines at a 100-foot contour interval. Start at elevation zero feet (sea level), and draw contour lines at 100, 200, 300, and 400 feet. Label the elevations of your contour lines. Keep in mind that the hills have rounded forms cut by V-shaped stream-eroded gullies. Some elevations are shown as an aid. The designation BM followed by a number designates a *bench mark*, a point whose elevation has been surveyed accurately. Bench marks are often placed at conspicuous summits. As a guide in drawing topographic contours that express the topography accurately, use the stereoscopic pairs of photographs of the San Francisco area, Figure 6.7.

Stereographs

The use of pairs of photographs taken from slightly different points makes it possible to view the topography in three dimensions, in the same way that the use of our two eyes in their slightly different positions enables us to see in three dimensions. In the case of the aerial photographs, the two "eyes" are represented by the lens of a camera in two different positions as a plane flies over the area. Photographs are taken at set intervals along the flight path, and any two consecutive frames in the flight line can be used to give a three-dimensional view of the land surface below. Because the photographs are taken at relatively long intervals, the "eyes" are very far apart, resulting in a view with substantial vertical exaggeration. Trees and buildings appear abnormally tall, and hills, abnormally steep.

The technique for viewing pairs of stereo photographs varies slightly, depending on the type of viewer used, but in general you will need to separate the two photographs by several inches. Place them under the viewer in such a way that the two views of a chosen feature in both photographs line up parallel to the long edge of the viewer, which in turn should line up with your eyes. Adjust the photographs as you look through the viewer, being careful to maintain the parallel orientations, until you reach the correct separation distance and the topography appears in three dimensions. If you have trouble doing this, try placing a finger on the same feature in each photograph, then move the photographs until your two fingers appear one on top of the other. Do not mark the photographs or touch the silvered mirror of the stereoscope (if your viewer is the mirror type).

TOPOGRAPHY

Topography is the general term that denotes the shape of the earth's surface, the lay of the land. Mountains, hills, plateaus, and plains are examples of the various types of topography present on our planet.

The size, slope, patterns, and material of the surface are important characteristics of an area's topography. One may measure the *size* of various features from the map scale. Surface area and relief are horizontal and vertical measurements of the size of a feature.

The *slope* of the topography is measured as the change in the vertical elevation of the surface over a given horizontal distance. On a topographic map, the amount of space between adjacent contour lines reflects the slope. Widely spaced contour lines indicate gentle slopes, and closely spaced contour lines indicate steep slopes.

The *pattern* of the topography is the geometrical arrangement of various elements of the surface, such as stream valleys or ridges. In plan view, for example, streams commonly have a randomly branching (dendritic) pattern.

Figure 6.7
Stereoscopic photographs of the San Francisco Bay area.

N↑   A.

The natural materials present on the surface are a very important characteristic of an area's topography. On earth, the primary *surface materials* are rock, water, ice, and vegetation. (Exercise 19 describes other planets.) The presence or absence of these materials affects the topographic size, slope, and patterns of the surface. Maps generally specify the different surface materials by color—brown contours denote rock and soil, blue areas are water, white areas with blue contours are ice, and green patterns indicate vegetation.

The shape of the surface is often better visualized from a side view called a *topographic profile* or *cross section*. To be precise, the profile is the line of inter-section between a vertical plane and the land surface. In a general sense, it is the silhouette of the earth's surface. The profile may be oriented in any direction and is drawn simply as the graph of surface elevation plotted against horizontal distance. To avoid distortion, profiles should be drawn with equal horizontal and vertical scales. Cross sections of flat areas or of great length, however, may require the vertical scale to be expanded to emphasize topographic details. This distortion is called *vertical exaggeration*.

Figure 6.9 illustrates the method for constructing an accurate profile. Place a strip of paper along the line of the profile on the map. Where a contour line crosses the line, place a small mark on the strip. Note

Figure 6.7
Stereoscopic photographs of the San Francisco Bay area.

B.   C.

elevations of any index contours; also note positions of ridges, rivers, or roads. Next, lay the strip at the bottom of the profile line, and transfer the positions of the contour lines and other features to the appropriate elevations. Remove the strip, and connect the points on the profile with a smooth curve. Profiles reflect the jagged, rolling, or flat character of the topography. A profile across a mountain stream, for example, generally has a sharp V-shape. The arrangement of the topography, both in plan and in profile, can be important in interpreting the surface processes active in an area.

Surface characteristics define five major types of topography, mountains, hills, plains, tablelands, and plains with mountains or hills. Mountains have steep slopes and high relief, whereas hills have moderate slopes and moderate to low relief. Plains have dominantly flat or gently sloping land. Tablelands and plains with mountains or hills are topographies with both gentle and steep slopes; the location of the flat surface differentiates the two. Tablelands are areas where the flat surface lies above the steep slopes, which are cut into the flat surface. In plains with mountains or hills, the flat surfaces lie below the steep slopes. Figure 6.10 illustrates the five major types of topography. Color Map 6 shows the distribution of these five land surface forms in the conterminous United States.

Plan View   Three-Dimensional View

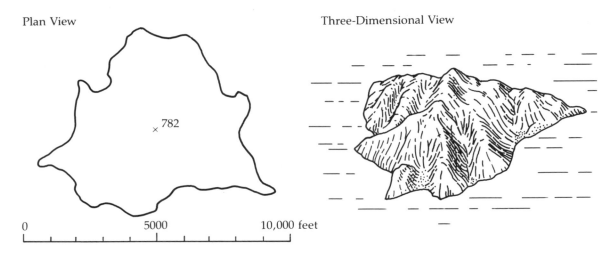

0   5000   10,000 feet

Estimation of Area

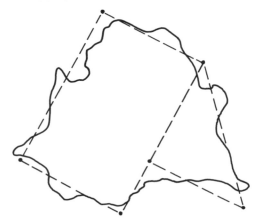

Area of rectangle =
    base × height
Area of triangle =
    $\frac{1}{2}$(base × height)

Estimated area of rectangle
    = 28,050,000 ft^2
Estimated area of triangle
    = 8,600,000 ft^2

Estimated area of island
    = 36,650,000 ft^2
    = 1.3 mi^2

Measurement of Relief

782

0

$$\frac{\text{Maximum}}{\text{Elevation}} - \frac{\text{Minimum}}{\text{Elevation}}$$

782 − 0 = 782 feet

Figure 6.8
Measurement of topographic dimensions, surface area, and relief.

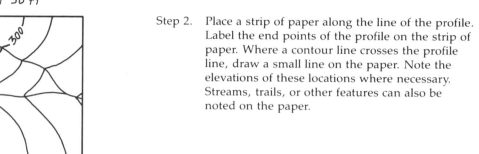

1500 feet

Contour interval 50 ft

Step 1. Determine the line of the profile. Label the end points, and trace the profile line on the map. The profile line can curve and have any length.

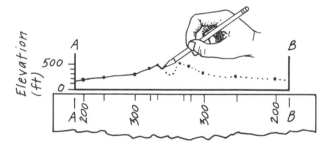

Step 2. Place a strip of paper along the line of the profile. Label the end points of the profile on the strip of paper. Where a contour line crosses the profile line, draw a small line on the paper. Note the elevations of these locations where necessary. Streams, trails, or other features can also be noted on the paper.

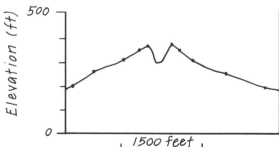

Step 3. Now lay the paper strip across the base of a graph of elevation versus horizontal profile distance (map distance). Transfer the noted positions to the graph. Connect the points with a smooth curve.

Step 4. If more profile detail is desired, vertically exaggerate the profile by increasing the vertical axis. The amount of vertical exaggeration is determined by dividing the value of the horizontal scale into the vertical scale.

Vertical scale: 1/4000 = 0.00025
Horizontal scale: 1/18,000 = 0.000055
0.00025 ÷ 0.000055 = 4.5
Vertical Exaggeration = 4.5×

Figure 6.9
Steps in the construction of a topographic profile.

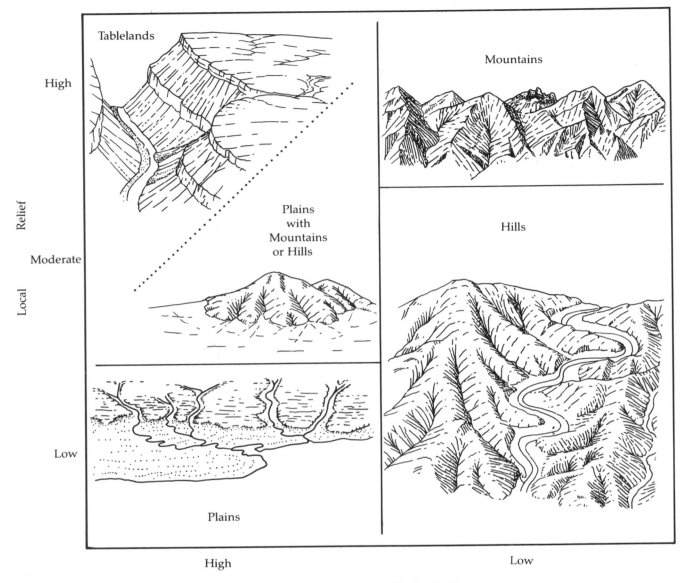

Figure 6.10
A classification of topography based on the characteristics of local relief and percent gentle slope.

Problem 17

Examine the topographic map in Figure 6.11. For each landform listed below, choose the location on the map that best illustrates the following features:

- hill
- valley
- summit
- closed depression
- cliff
- gentle slope

Problem 18

Determine the slope of the hillside between the 50-foot and 350-foot contour lines along the line X–X′ in Figure 6.11. First calculate the slope in percent (elevation change per hundred feet of horizontal distance. Next use the measured vertical and horizontal distances and the table of tangents with instructions in Appendix 8 to determine the slope in degrees.

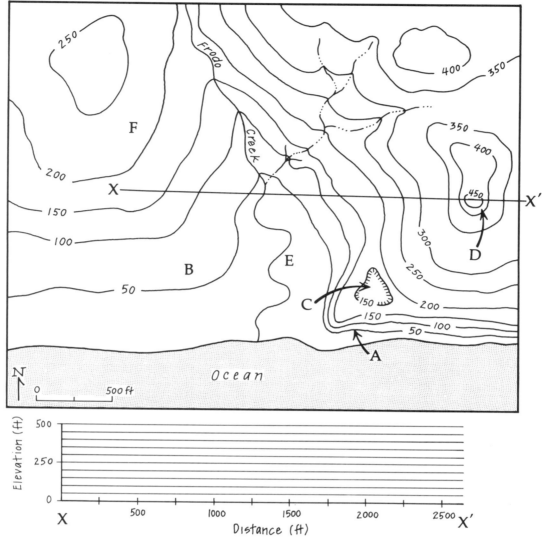

Figure 6.11
Topographic map of various landforms.

Problem 19

Draw the cross-profile of the stream valley in Figure 6.11 on the grid provided. Using your protractor, measure directly from the profile the average slope in degrees between the 50-foot and 350-foot contours. Which of your two methods for determining slope, the calculation using measured distances or the direct measurement with a protractor, do you think is the most accurate? Why?

Problem 20

Describe the type of topography—plains, hills, mountains, plains with hills or mountains, or tablelands—for each of the topographic maps in Figure 6.12.

A.

B.

C.

D.

E.

F.

Figure 6.12
Topographic maps of six different townships with their dominant landforms.

FIELD STUDY: TOPOGRAPHY AND GEOLOGY OF THE CONTERMINOUS UNITED STATES

In this field study you will use and compare three different maps of the conterminous United States— the maps of topography, geology, and physical divisions. The geologic map is a simplified map of the geologic units that appear at the surface. The map of physical divisions divides the United States into several regions with topographic and geographic affinities. The three maps are Color Maps 1, 2, and 5.

Problem 21

What general areas of the United States are below 500 feet in elevation?

Problem 22

Estimate the percentage of the total area that is below 500 feet.

Problem 23

In what physical divisions do these areas occur?

Problem 24

Where in the United States are there areas above 9000 feet in elevation?

Problem 25

In what physical divisions do these area occur?

Problem 26

What is the general shape and orientation of these areas?

Problem 27

If you were planning a transcontinental railroad to link the important population centers of Washington, D.C., New York, and Boston with Chicago, Denver, and San Francisco, where would you place your route to avoid areas of high elevation and relief? Consult both the topographic map and map of physical divisions. Draw your route on Color Map 1.

Compare the geologic and topographic maps with the map of physical divisions.

Problem 28

Find the line that separates the Great Plains from the Central Lowlands. To what topographic contour line does this division correspond? What is the general elevation east of the line? West of the line.?

Problem 29

To what major geologic division does this line generally correspond, especially from Kansas south? Compare topographic relief with the age of the rocks. Disregarding the Atlantic Plain (an area of very young sediments), are the rocks of the eastern United States generally older or younger than those of the West?

The Basin and Range area (Nevada and western Utah) is a geologic puzzle. North-to-south-trending mountain ranges made up of igneous, sedimentary, and metamorphic rocks are separated by basins. We shall study the region further in Exercises 7 and 18.

Problem 30

What is the surface geology of the valleys? Volcanic rocks of the Basin and Range are of mid-Tertiary age. Which is older, the mountain or the valley material?

Problem 31

Is the general elevation of the Basin and Range province higher or lower than most other physical divisions?

Structural Geology: Deformations in the Earth

Structural geology is the study of the architecture of the earth as it is expressed in the deformation of the earth's crust. Rock layers, including sedimentary deposits that must have been planar and horizontal when they were formed, are often found tilted, folded, or broken and displaced along fractures. Geologists interpret these deformations as evidence of forces that act within the solid earth to compress, stretch, or shear. The exact origin of these deformational forces is not clear, but the required energy is undoubtedly supplied by the earth's internal heat. This interior heat is transferred to outer cooler regions primarily by movement of material in the earth's inner layers, which causes stresses in the crust. The structural geologist is concerned with three major questions that we shall consider in this exercise: (1) What rock structure is present? (2) Under what conditions and by what mechanism did it form? (3) When did it form?

We shall first look briefly at forces operating within the earth, then at the results of these forces as expressed in deformation of crustal rocks. In the next exercise, we shall be concerned with representation of geologic structures on geologic maps. The data on many such maps will enable you to answer the three questions posed above, once you have learned the material in this exercise.

FORCES WITHIN THE EARTH

A *force* is a vector quantity (that is, it possesses both a direction and a magnitude) that tends to change the motion or shape of a body. The term *stress* is a quantity more often used to describe directed forces within the earth; it denotes a force acting over a unit area, for example, pounds per square foot.

Stresses are *compressive* if they push material together, *tensile* if they pull material apart, and *shear* if they distort material asymmetrically (as in the sliding of a deck of cards). Figure 7.1 illustrates these three types of stresses.

Strain is the deformation caused by stress. If rock is subjected to stress it may respond in three ways:

1. *Elastic strain:* If the stress is small and of short duration, most rock behaves in an *elastic* manner; when the stress is withdrawn, the rock body returns to its original size and shape. Earthquake waves passing through the earth generate elastic strain, which may be measured by seismographs. This topic will be covered in Exercise 16. The present exercise will focus on the other two kinds of strain.

2. *Plastic (ductile) strain:* Plastic deformation occurs when stress exceeds the limit of elastic strain or when small stresses act on solid rock for very long periods of time and the rock flows permanently. In this case the rock remains distorted from its original shape after the deforming stress is removed.

3. *Rupture (faulting):* When there is a continued increase in stress, fractures may develop in a rock body, and it may fail by *rupture*.

Substances that undergo a large plastic deformation before rupture are said to be *ductile*. Those that rupture before any significant plastic deformation are *brittle*. Whether a given rock body undergoes plastic deformation or rupture depends on several factors, notably the pressure, temperature, and time span over which the stress is applied.

Experiments done in the laboratory at high pressures and temperatures, like those within the earth's

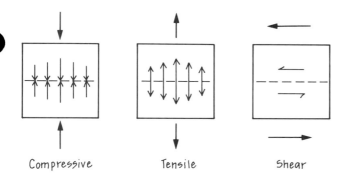

Figure 7.1
Tensile, compressive, and shear stresses.

crust, indicate that rocks behave quite differently at depths greater than 5 km than they do at the surface. (The elevated pressure mentioned here is *lithostatic pressure*, which is due to the weight of overlying rocks.) Because it is applied uniformly in all directions, it is not a directed force or stress. The stress that produces deformations is a directed stress (stress that is different in different directions), which tends to deform rock by tension, compression, or shear. At low pressure and temperature (surface conditions), most rocks exhibit brittle behavior; they deform by rupture after only small amounts of elastic strain (1 to 2 percent shortening under compressive stress) with almost no plastic deformation. In geologic terms, most rocks at the earth's surface react to stress by fracturing or faulting. (*Faults* are fractures along which significant movement of the rock has occurred.) At higher lithostatic pressures, many rocks demonstrate marked ductile behavior. Sandstone, shale, and limestone, for example, may undergo as much as 25 percent strain before rupture as long as they are under a lithostatic pressure of 2000 bars (where 1 bar very nearly equals atmospheric pressure).

Problem 1

Look back at Figure 4.1 in Exercise 4. To what depth of burial does a pressure of 2000 bars correspond? (Pressure in bars is approximately 100 times the average overlying rock density times the depth in kilometers. Average crustal density is 2.7.)

Elevated temperature at depth within the crust also increases the ductility of rock, meaning that less stress is needed to produce a given plastic deformation. The combined effects of elevated temperature and pressure at depth render rocks more ductile at depth and more likely to deform in a plastic manner instead of fracturing. Plastic deformation of rocks under compressive stress takes the form of folding, a process in which solid rock can actually flow, although very slowly. The bending of a sheet of warm wax is an everyday analog.

The third factor affecting deformation is the length of time over which the stress operates. The longer the time span, the more strain is produced. Over the great length of geologic time, comparatively small but continuous directed stresses may produce large deformations.

Problem 2

Arrange these four processes according to the depth at which each occurs, shallowest to deepest, realizing that there is some overlap in the depths:

• metamorphism

• deformation by folding (plastic deformation)

• deformation by faulting (rupture)

• melting of rock

DEFORMATIONS WITHIN THE EARTH

Tilted Strata: Strike and Dip

The simplest evidence of deformation in rocks is tilting of formerly horizontal rock layers. The layers remain approximately planar but at some orientation other than horizontal. Two simple measurements are used to describe the orientation of tilted surfaces, strike and dip. The *strike* is the compass direction of the line formed by the intersection of a horizontal plane with the tilted layer—a level line along the tilted surface. Another way to think of strike is to remember that only one horizontal direction is contained in any non-horizontal plane. This direction is the strike. The *dip* of the plane is the acute angle between the dipping

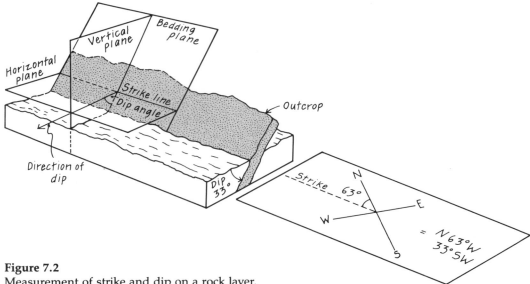

Figure 7.2
Measurement of strike and dip on a rock layer.

plane and the horizontal (see Figure 7.2). The direction of dip is always perpendicular to the strike and in the direction water would flow if it ran down the plane.

The strike is given as a compass direction relative to north. "North sixty-three degrees west" is a strike that is 63° west of geographic north (see Figure 7.2). For any given strike, two directions of dip at the same angle are possible, so the compass direction of the downward slope must be specified. "Thirty-three degrees southwest" is the dip of the rock layer in Figure 7.2. Strike and dip measurements for the example are written as follows: N63°W, 33°SW.

Because it is necessary to record the orientation of rock layers in a simple way on a map, geologists measure the strikes and dips of all rock layers they encounter while mapping an area in the field. The data are written directly onto a map at the point where the measurement is made. The symbols used are explained in Figure 7.3 and in more detail in Appendix 6. For the problems in this exercise, assume that north is at the top of the page, unless otherwise indicated.

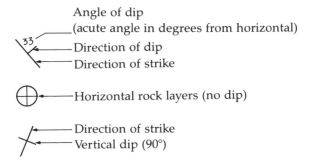

Figure 7.3
Symbols for strike and dip.

Problem 3

Use the written and symbolic form of the strike and dip measurement to express the following orientations. Use a protractor to determine the correct strike.

Example:

strike, north forty-five degrees east (N45°E)
dip, sixty-three degrees southeast (63°SE)

1. strike, north twenty-two degrees east
   dip, twenty-five degrees northwest

2. strike, north sixty-seven degrees west
   dip, eighty-three degrees northeast

3. strike, south two degrees east
   dip, ninety degrees

Problem 4

Determine the orientation of the strike and dip symbols in Figure 7.4 using your protractor.

A.

B.

C.

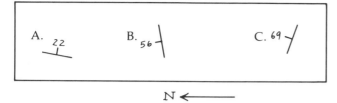

Figure 7.4
Strike and dip symbols to be measured.

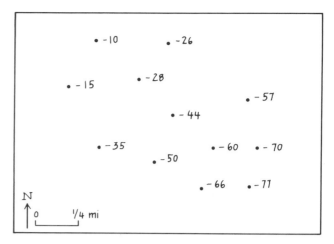

Figure 7.5
Elevations in feet of a sandstone layer, determined
by drill holes.

Problem 5

Figure 7.5 shows the elevations (in feet) at which drill holes contacted a quartz-rich sandstone. From these elevations draw contours on the sandstone bed and determine the strike and dip of the sandstone bed. Note that elevations are given as negative numbers, indicating that they are below sea level. Use a contour interval of 10 feet. Contours should be straight and parallel because the surface is planar, not irregular. Use the instructions and Partial Table of Tangents in Appendix 8 to determine the dip. Show your work.

Faults

A fault is a fracture in the crustal rocks along which has occurred significant displacement of one body of rock relative to another. Color Plate 61 shows small-scale faulting in bedded sandstone of the Dakota Formation. The surface on which this movement has taken place is the *fault plane*. The relative direction of the rock displacement defines three major types of faulting.

1. *Strike-slip faulting:* The direction of displacement in strike-slip faulting is horizontal; the two rock bodies slip past each other in a horizontal direction, with little or no vertical component to the motion. The movement is parallel to the strike of the fault plane, hence the term *strike-slip*. Strike-slip move-

ment usually occurs on a vertical or near-vertical fault plane. Strike-slip faults are called *left-lateral* if the block of crust on the opposite side of the fault appears to move to the left to an observer across the fault, and *right-lateral* if the opposite block appears to move to the right. Strike-slip faults often result from shear stress in the crust, but they may also occur in compression or extension.

2. *Dip-slip faulting:* The two rock bodies slip past each other on the fault plane in a direction parallel to the dip of the fault plane, that is, perpendicular to the strike. The fault plane is usually tilted 30° to 60° to the horizontal. Two traditional mining terms are used to refer to the sides of a dip-slip fault: the *hanging wall* is above the fault plane; the *footwall* is below it (see Figure 7.6).

   Dip-slip faults are divided into two subtypes. A *normal fault* is one in which the hanging wall moves downward relative to the footwall. Normal faults (which are no more "normal" than any other type) occur in extension or stretching of the crust. A *reverse fault* is one in which the hanging wall moves upward relative to the footwall; it is associated with crustal compression (see Figure 7.6). An important type of reverse fault is the *thrust fault*. In this type of fault, the fault plane is at a low angle (less than 20 degrees), the overthrust sheet is typically relatively thin, and long-distance horizontal movements of the crust are involved. Thrust faults imply very substantial compression of the crust.

3. *Oblique-slip faulting:* This type has components of both strike-slip and dip-slip movements and may result from extension, compression, or shear of the crust (see Figure 7.6).

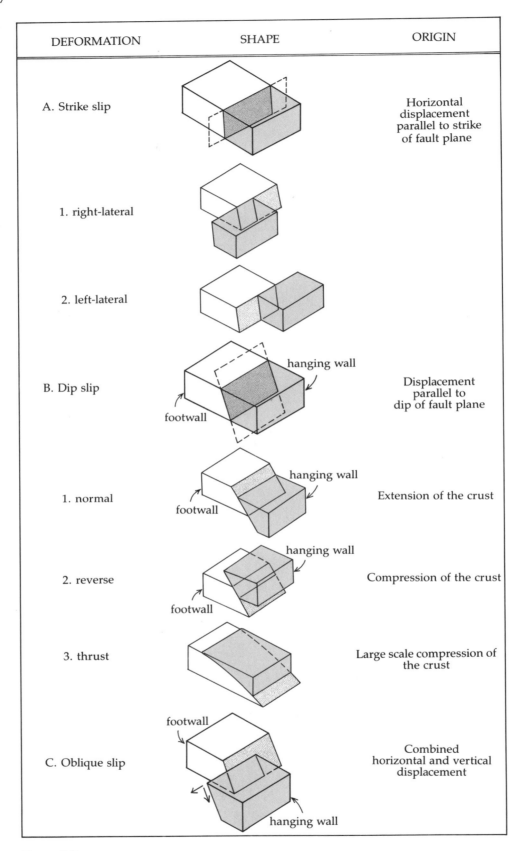

Figure 7.6
Fault nomenclature.

Folds

Folds are wavelike bends in rock layers resulting from compression of the crust. Originally horizontal strata move up or down in response to compression to accommodate shortening of the crust. Folds range in size from millimeters to hundreds of kilometers in length and have various shapes. There are three main types of folds, based on their shapes:

1. A *monocline* is a one-sided fold that typically occurs in flat-lying strata of stable continental areas.

2. An *anticline* is an upward-closing fold with the deepest (and therefore oldest) rock layers in the center.

3. A *syncline* is a downward-closing fold with the uppermost (and therefore youngest) rock layers in the center.

In anticlines and synclines the rock layers dip away from and toward, respectively, a central line of greatest curvature called the *fold hinge*. The hinge is at 90° to the direction of maximum compression. Anticlines and synclines are generally adjacent, forming alternating crests and troughs in the rock layers. Color Plate 62 shows small folds in phyllite. Monoclines, anticlines, and synclines are illustrated in Figure 7.7. The hinges are the straight lines along crests and troughs of the folds.

Basins, *domes*, and *plunging folds* are variations in the anticline and syncline fold shapes. Basins and domes are downward and upward bends that lack a linear hinge. They are oval or bowl-shaped, such that the strata dip away from or toward, respectively, a point rather than a line. Basins and domes, unlike most other folds, need not form solely from compression but may be due to a variety of factors. Plunging folds are anticlines and synclines that have a tilted hinge. These folds are illustrated in Figure 7.8.

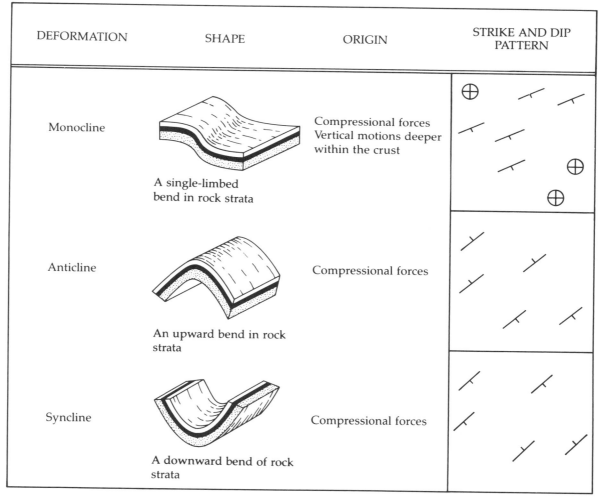

Figure 7.7
Types of folds.

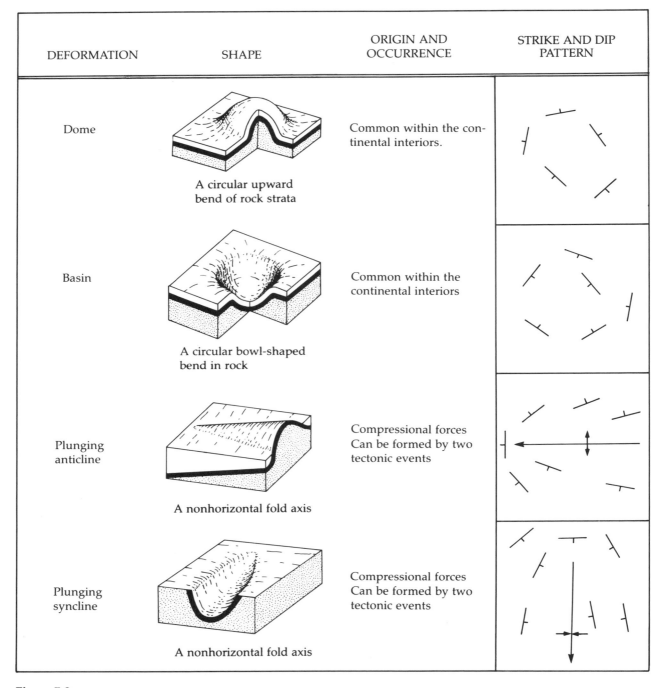

DEFORMATION	SHAPE	ORIGIN AND OCCURRENCE	STRIKE AND DIP PATTERN
Dome	A circular upward bend of rock strata	Common within the continental interiors.	
Basin	A circular bowl-shaped bend in rock	Common within the continental interiors	
Plunging anticline	A nonhorizontal fold axis	Compressional forces Can be formed by two tectonic events	
Plunging syncline	A nonhorizontal fold axis	Compressional forces Can be formed by two tectonic events	

Figure 7.8
Subsidiary fold types.

Figure 7.9
Instructions for making box models of structures.

Problem 6

Make the structural models on the pages immediately following, or use the models provided in the laboratory (see Figure 7.9).

1. Cut the model along its borders and discard the excess paper.

2. Fold the paper model along the marked lines, so that the illustrated side is facing out.

3. Fold the corners inward along the "fold lines," so that the edges of the box are aligned.

4. Tape the corners to the inside of the box.

5. Close the box by taping the top down. Make sure that the stripes on adjacent faces match.

6. Secure all loose areas with transparent tape.

7. Orient the box on your table so that the number is located in the lower right corner.

(Continued)

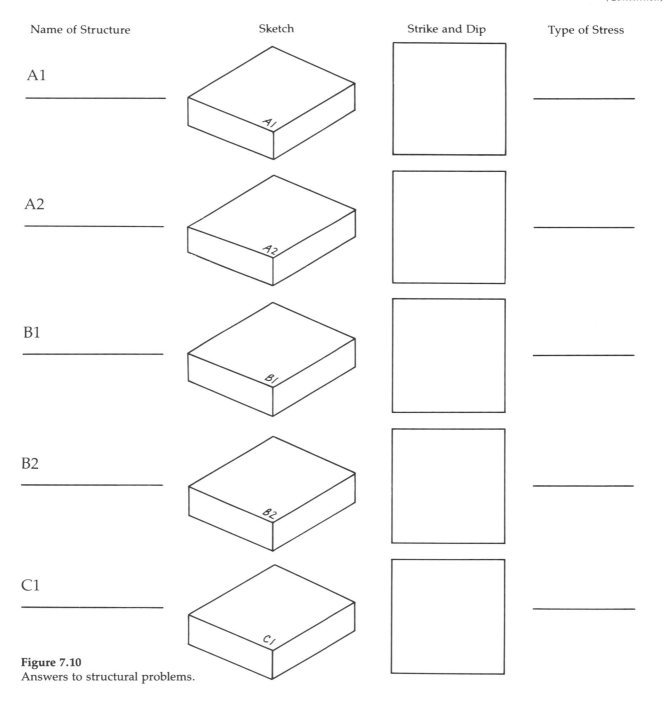

Name of Structure	Sketch	Strike and Dip	Type of Stress
A1 _____			_____
A2 _____			_____
B1 _____			_____
B2 _____			_____
C1 _____			_____

Figure 7.10
Answers to structural problems.

Name and sketch the structures illustrated in models A through E, using the spaces provided on Figure 7.10 for your answer. Be sure to inspect the sides as well as the top of each model to determine the type of structure. On that figure also record strike and dip directions of layers, and state the type of stress that produced the deformation (compression, extension, or unknown). Show the orientation of the stresses with arrows on your sketch.

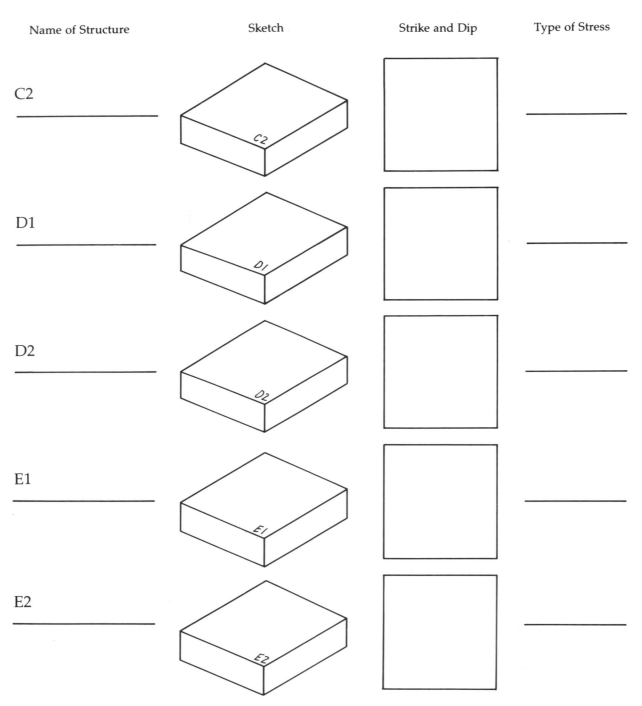

Name of Structure	Sketch	Strike and Dip	Type of Stress
C2			
D1			
D2			
E1			
E2			

Figure 7.10
Answers to structural problems.

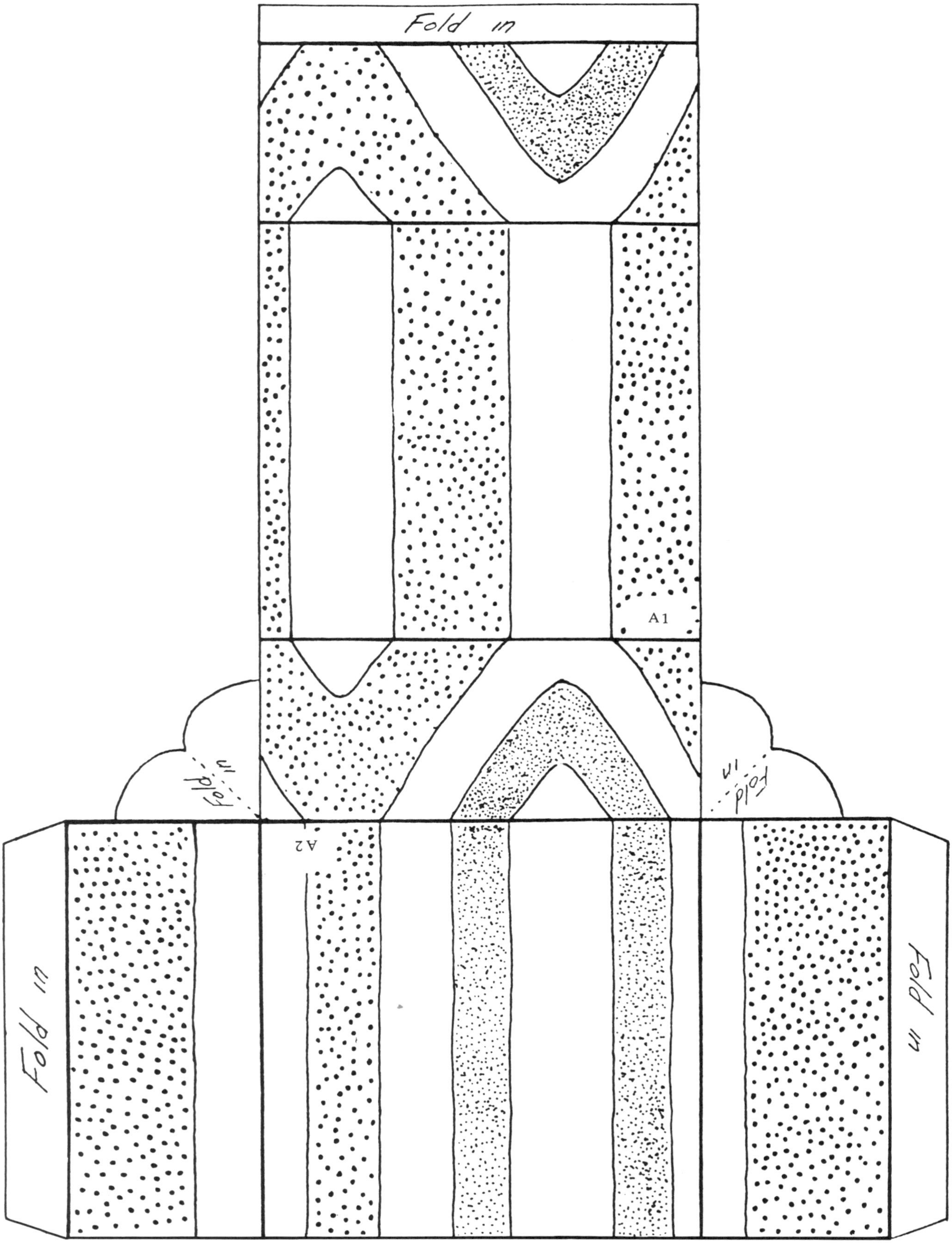

93

Fold in

Fold in

Fold in

Fold in

Fold in

Fold in

A1

A2

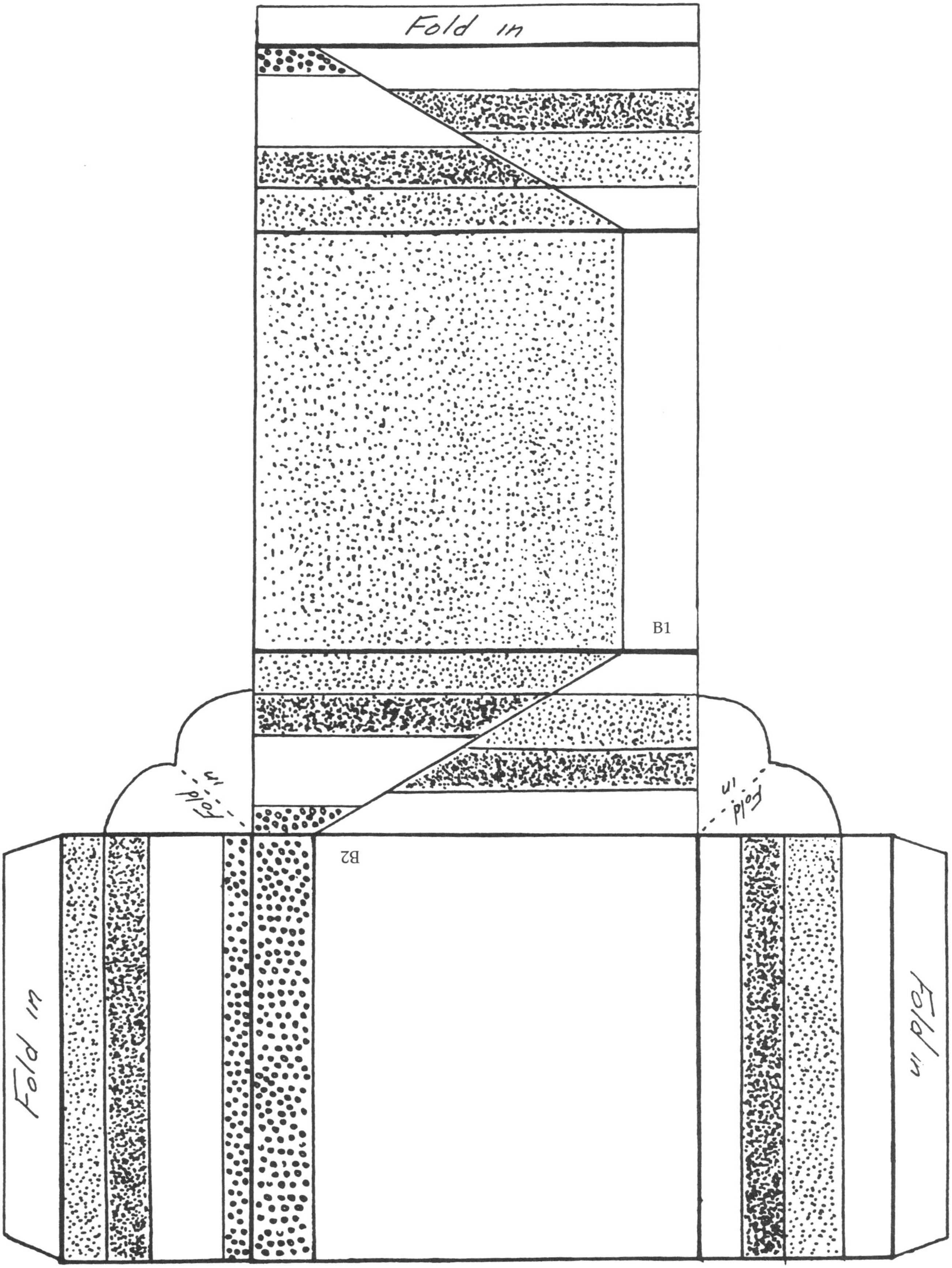

Fold in

B1

B2

Fold in

Fold in

Fold in

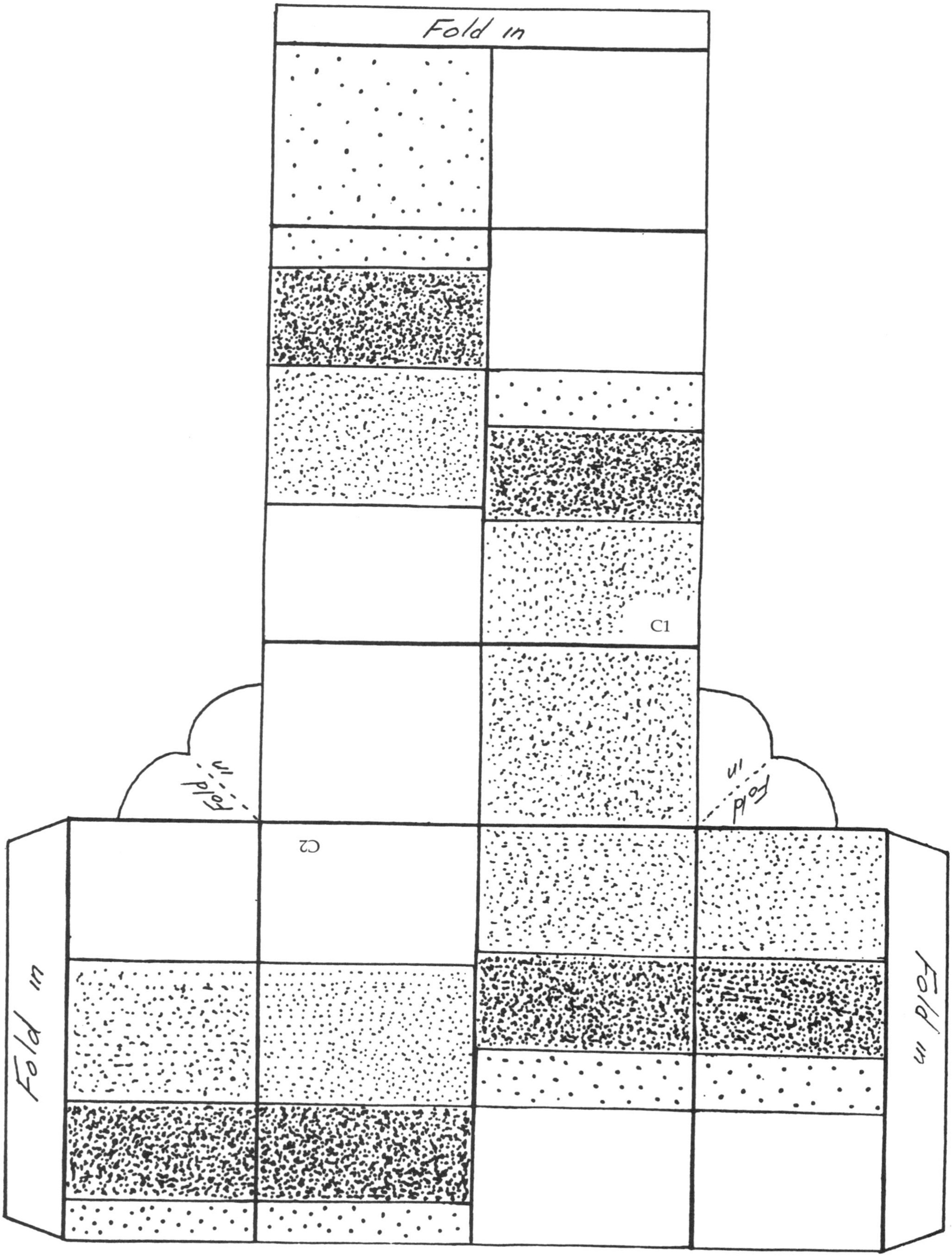

Fold in

Fold in

Fold in

Fold in

C1

C2

99

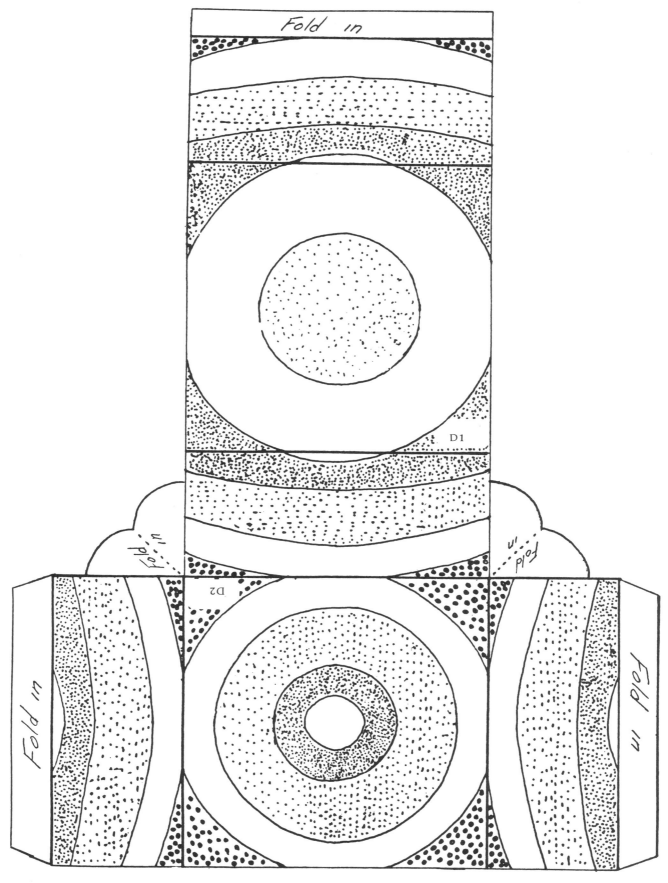

Fold in

Fold in

Fold in

Fold in

D1

D2

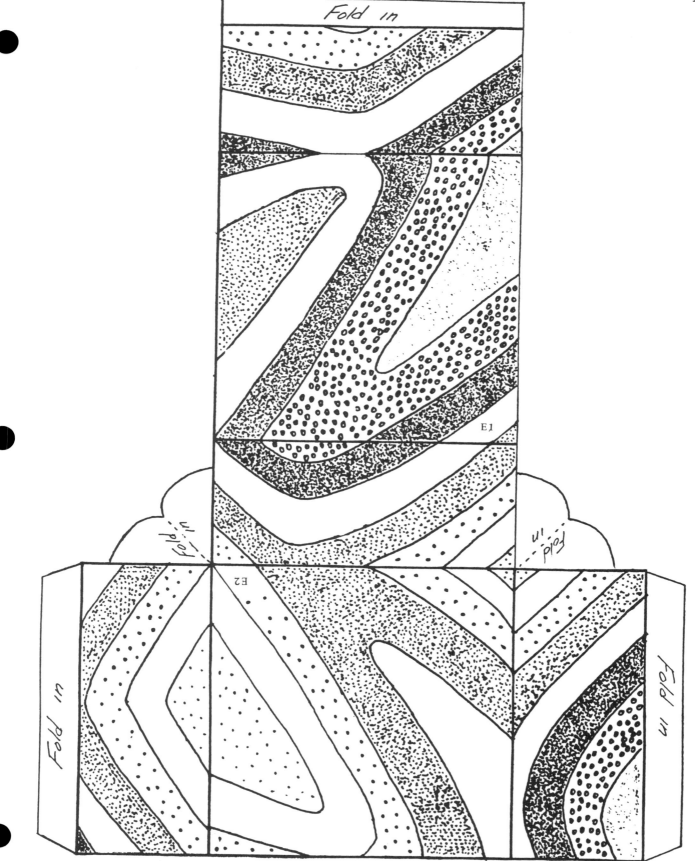

Unconformities

You should recall from Exercise 5 that an *unconformity* is a surface of erosion that separates younger strata from older rocks. Structural geology is concerned with unconformities, even though their origin involves erosion and deposition, because unconformities commonly imply vertical movement of the crust upward to above sea level, where erosion takes place. There are various kinds of unconformities, depending on the rocks involved, but in the interest of simplicity we shall use the general term *unconformity* to represent any time gap in the rock record.

Unconformities are difficult to distinguish in map view from certain types of faults unless careful field observations have been made. There is no specific map symbol for an unconformity.

Problem 7

In Exercise 5 you determined that the rock type often found directly above an unconformity was a conglomerate or a sandstone. Assume that an old erosional surface of foliated schist is partially covered by an encroaching ocean that deposits a coarse basal sand that is buried and lithified before sea level drops again (see Figure 7.11A) and erosion exposes it. Complete the map view (Figure 7.11B) of the schist and sandstone formations resulting from such an event.

Structural Study: Dakota Formation in Capitol Reef National Park

Return to the stratigraphic column of Capitol Reef National Park given in Exercise 5 to refresh your memory on the position of the Dakota Formation in the succession of rock units in the Colorado Plateau.

Just as topographic contours illustrate the topography of the earth's surface, so can lines of equal subsurface elevation be used to describe the form of a structure beneath the land surface. These *structure contours* are read in the same way as topographic contours; the numbers refer to the elevation of a feature, often a contact between two rock units, that is not completely exposed at the surface. Because much of the feature of interest is beneath the surface, structure contour maps are made from only a limited number of data, generally drill-hole or outcrop measurements. Hence structure contour maps only approximate the actual form of the feature, and they are not as accurate as topographic maps.

Figure 7.12 shows three maps of the same area in Capitol Reef National Park, Utah: (A) a strike and dip map of rock layers at the surface, (B) a structure contour map drawn on the base of the Dakota Formation, and (C) a map of the ages of rock units.

Figure 7.11
Erosional surface with encroaching ocean.

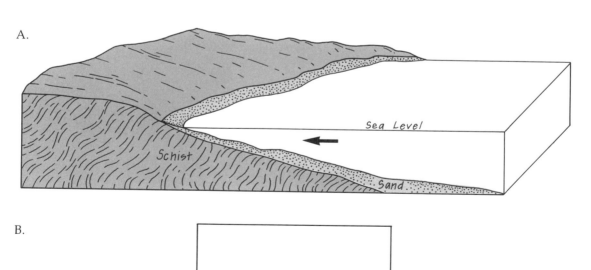

A. Strike and Dip Map

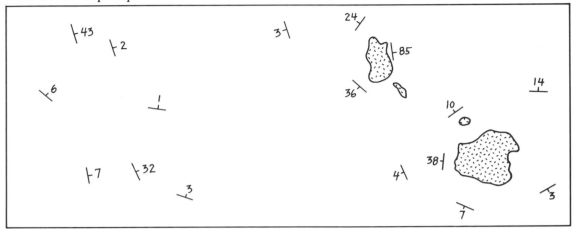

B. Structure Contour Map

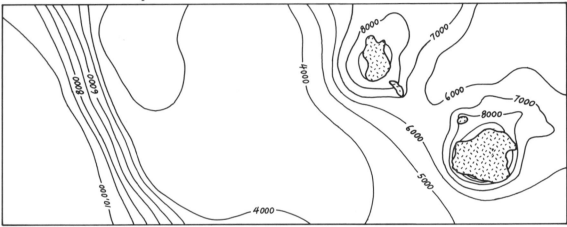

C. Age Map

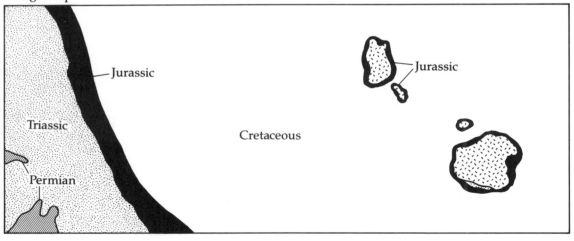

Figure 7.12
Maps of part of Capitol Reef National Park and vicinity.

Problem 8

From the strike and dip measurements, what structural deformations are present in this area, and where are they?

Problem 9

Examine the structure contour map of the base of the Dakota Formation. Does this map reflect the same deformations shown by the strikes and dips? Explain.

Problem 10

What might explain the localized deformation surrounding the igneous intrusions in this area?

Problem 11

The igneous intrusions are 44 million years old. During what geologic period did the deformations in this area occur? Cite evidence to support your answer.

Figure 7.13 shows subsurface elevations to the top of the same Dakota Formation in the Gramp's oil field in Colorado.

Problem 12

Draw structure contours on the top of the Dakota Formation.

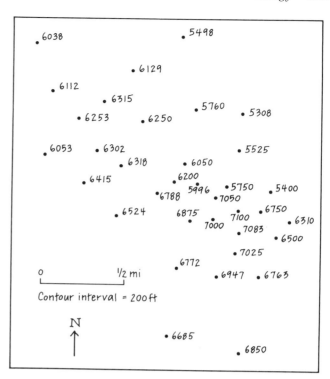

Figure 7.13
Subsurface elevations of the top of the Dakota Formation, Gramp's oil field, Colorado.

Problem 13

Notice that the average elevations of the north and south halves of this area are different. What structural feature might cause this difference? What is its orientation? Trace the trend of this feature on your map in red and label it A.

Problem 14

There is another structure present in this area. What is it? Trace the trend of this feature in blue. Label it B.

Problem 15

Which feature is older, A or B? How do you know?

REPRESENTATIONS OF ROCK STRUCTURES: CROSS SECTIONS

A geologic cross section shows the internal structure of the crust from a side view; it is the view of the crust if it were sliced open like a cake. You learned to construct topographic profiles (cross sections) from topographic maps in Exercise 6. Figure 7.14 shows how to construct a geologic cross section, representing the contacts of different formations accurately.

From a geologic cross section, one may often determine the relative geologic sequence of events in the formation and deformation of the crust, using the stratigraphic laws of superposition, original horizontality, and crosscutting relationships. Cross sections of layered rocks, for example, often show horizontal beds above tilted beds. This type of structural arrangement indicates at least four events: (1) the original deposition of strata, (2) tilting of the strata, (3) erosion of the tilted strata, and (4) deposition of the overlying horizontal strata.

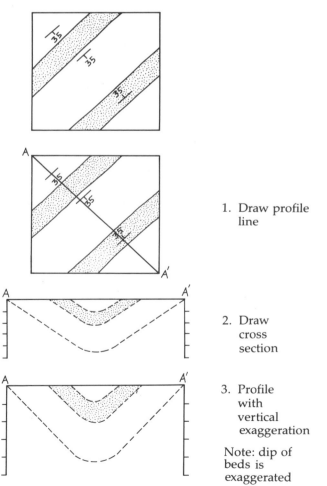

1. Draw profile line

2. Draw cross section

3. Profile with vertical exaggeration

Note: dip of beds is exaggerated

Figure 7.14
Construction of a geologic cross section.

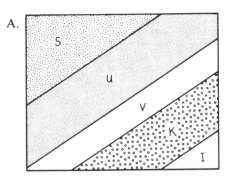

A.

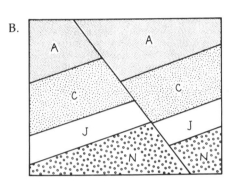

B.

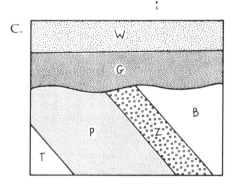

C.

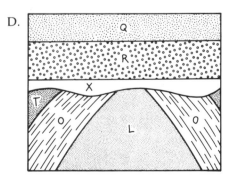

D.

Figure 7.15
Cross sections of deformed sedimentary beds.

Problem 16

Determine the sequence of geologic events (deposition and deformation) for each of the cross sections shown in Figure 7.15. Assume that these are sequences of sedimentary beds.

A.

B.

C.

Problem 17

From the maps and stratigraphic column given in Figure 7.16, describe the structures present in each area (folding, faulting, unconformity).

A.

B.

C.

Problem 18

What structures are suggested by the strike and dip patterns in Figure 7.17?

A.

B.

C.

Problem 19

Draw cross sections X–Y for the designated diagrams in Figures 7.16 and 7.17. Use the answer spaces provided in the figures.

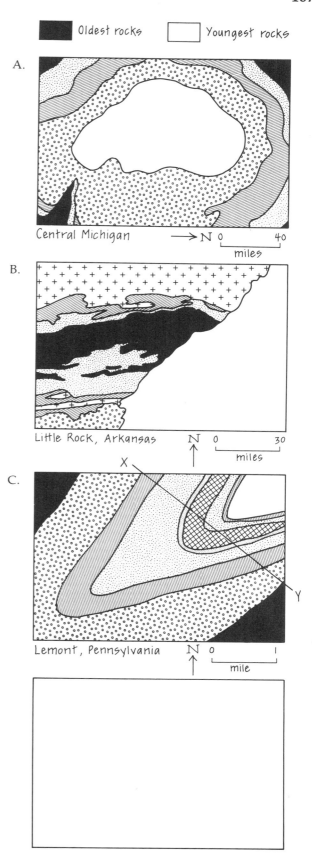

Oldest rocks   Youngest rocks

A.

Central Michigan   → N   0   40
miles

B.

Little Rock, Arkansas   N   0   30
miles

C.

X

Y

Lemont, Pennsylvania   N   0   1
mile

Figure 7.16
Geologic maps of areas that have undergone deformation.

A.

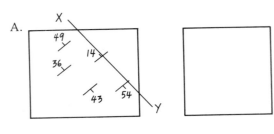

B.

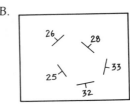

C.

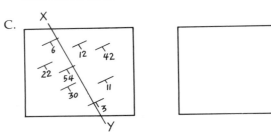

Figure 7.17
Strike and dip problems.

FIELD STUDY: STRUCTURES OF THE CONTINENTAL UNITED STATES

Examine the structure map of the conterminous United States, Color Map 3.

Problem 20
In what main regions is the crust folded?

Problem 21
Do the folds seem to occur singly or in belts?

Problem 22
What is the general orientation of the fold hinges?

Problem 23
Summarize the evidence supplied by fold belts. What type of stresses do they represent? In what direction were the stresses applied? Where in the continent have these stresses been active?

Two different fold types exist in the Central Lowland region. In this area, Paleozoic and Mesozoic sediments were deposited over a relatively flat Precambrian erosional surface.

Problem 24
Return to the geologic map, Color Map 2, and examine the shape and sequence of Paleozoic sedimentary outcrops between Lake Michigan and Lake Huron. What structure is this?

Problem 25
Now examine the structure contours (red lines) drawn on the Precambrian surface shown on Color Map 3. Is this structure in the Paleozoic layers reflected in the Precambrian rocks? Explain.

Problem 26
Look at the shape and sequence of Paleozoic sedimentary rocks around Cincinnati, Ohio, to Frankfort, Kentucky, shown on Color Map 2. What structure is this?

Problem 27
Is this structure reflected in the Precambrian rocks? Explain.

Problem 28

Locate on the geologic map the Black Hills, an oval area just west of Rapid City, South Dakota. Compare the geology with the structure map of the area, and explain what the Black Hills represent.

Problem 29

Where is the crust broken by thrust faults on Color Map 3?

Problem 30

What do the location and orientation of these faults suggest about the type and direction of stresses that have affected the crust in these areas?

Problem 31

Does this evidence support the evidence supplied by the fold belts?

Problem 32

Where is the crust broken by normal faults?

Problem 33

What do these faults imply about the type and direction of stresses in the crust in these regions?

Problem 34

Does this conclusion agree with your previous conclusions?

Problem 35

Look at the ages of the rocks offset by normal faults and thrust faults in the Basin and Range area. Which is older, the thrust faulting or the normal faulting? (*Hint:* The normal fault that runs north–south just east of Great Salt Lake displays the relationship well.) Explain your answer.

Problem 36

Examine the contact in Alabama between the Appalachian fold belt and the very large area of platform deposits. What type of contact is this, conformable or unconformable?

Problem 37

What sequence of geologic events does this contact suggest?

Problem 38

Return to the topographic and geologic maps of the United States (Color Maps 2 and 5). Briefly summarize the relationship of the structural elements you have discussed above to the topography and geology. (Pay particular attention to the Appalachian Mountains and to the Basin and Range area.)

Reading Structure from the Map

And some rin up hill and down dale knapping the chucky stanes to pieces wi' hammers, like sae many road makers run daft. They say it is to see how the world was made.

Sir Walter Scott

The first four exercises in this book dealt with the identification of rocks and knowledge of the processes that formed those rocks, whether igneous, sedimentary, or metamorphic. An understanding of what rocks are present in an area and how those rocks formed is crucial to deciphering the geologic history of the area. These data are most often presented in the form of a *geologic map*, which shows the surface distribution of rock units, much as conventional political maps show the areas of cities, states, or nations.

Most geologic maps are superimposed on the topographic map of the area: The topography is a powerful tool in the study of the other important element in the geologic history of an area, the structures that are present and the deformational history they represent. In Exercise 7 you studied geologic structures—dipping beds, folds, faults, and unconformities. You should be familiar with the surface outcrops produced by these structures on a flat and horizontal surface. In this exercise we shall consider the more realistic situation of surface outcrops on irregular surfaces. This knowledge is the key to formulating the geologic history—the deposition of the various rocks present and their subsequent deformation—of the Grand Canyon area from the information presented on the geologic map of the Bright Angel quadrangle.

The size and shape of the area where a formation intersects the earth's surface is the formation's *outcrop pattern.* Exercise 7 dealt with the simple case of outcrop patterns produced by rock layers of uniform thickness and a constant dip on a flat horizontal sur-

face. Now we shall introduce more complex examples. First, outcrop patterns depend not only on the structure but also on the thickness of the layer and its dip with respect to the topography. Second, the outcrop patterns on a map reflect irregularities in the topography. What appears as a straight line on a flat horizontal surface is distorted into characteristic patterns on irregular surfaces.

THE DIP OF THE LAYER

The dip of a rock layer affects the width of its outcrop at the surface of the earth, an effect best seen by assuming for the moment that the earth's surface is flat and horizontal. Inspect the geologic cross section shown in Figure 8.1.

Problem 1

Measure the true thickness in millimeters of rock unit A, that is, the distance between the upper and lower contact, measured at 90° to the contacts, at points X, Y, and Z.

X.

Y.

Z.

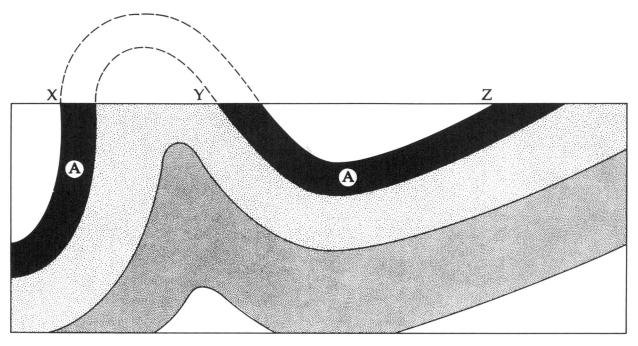

Figure 8.1
Geologic cross section of deformed layers.

Problem 2

Measure the outcrop width in millimeters of rock unit A in the three areas where it intersects the surface of the earth, X, Y, and Z.

X.

Y.

Z.

Problem 3

What is the relationship of the outcrop width to the dip of the layer?

IRREGULARITIES OF THE EARTH'S SURFACE

Thus far, the surface of the land has been assumed to be flat and horizontal in order to simplify the discussion. In this section we deal with the more realistic situation of an irregular land surface. The three-dimensional model base to be used in this section rep-resents a sloping stream gully crossing a flat surface. On this base you will superimpose several overlays representing outcropping rock layers that dip in different directions and then record the map view of the outcrop on a geologic map.

Make the model base according to instructions given in Exercise 7, Figure 7.9. Cut along the heavy black lines. There will be an open space running across the top of the model. Make the five different structural overlays according to the diagram shown in Figure 8.2.

1. Cut the overlay from the page.

2. Fold the overlay along the central axis so the printed side faces inward. Unfold the bend.

3. Bend the overlay along the marked fold lines so the printed side faces outward.

4. Bend the paper into the illustrated shape (step C in Figure 8.2).

5. Snip the paper along the dotted lines.

6. Fold the exterior flaps downward.

7. Place the overlay over the model base. Insert the exterior flaps behind the front of the base. Secure the overlay with rubber bands.

A.

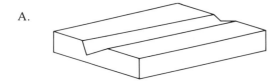

B.

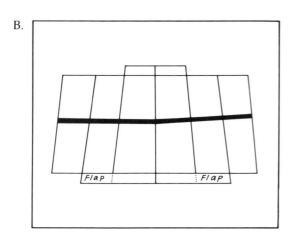

C.

D.

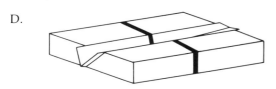

Figure 8.2
Models for structural overlays.

Problem 5

Draw the outcrop exposure as seen from directly above the model (in plan or map view) for each structural model in Figure 8.3, column B. To describe the orientation of the rock layer in the model, place strike and dip symbols on the map. Notice that, in all but one of the map views, the outcrops display a characteristic V-shaped pattern in the gully. The direction in which the V points depends on the direction in which the layer dips.

Problem 6

In column C, Figure 8.3, state whether the outcrop pattern points upstream, downstream, or does not have a V-shape in the map view.

Problem 7

From your diagrams, formulate a general rule as to how the dip direction of a layer may be determined from its outcrop pattern across a stream valley, as seen on a geologic map. In the case of overlay 5, the layer dips downstream but at a smaller angle than the slope of the floor of the gully. Be sure to distinguish this example from two cases that appear similar from the side view: the horizontal layer and the layer that dips downstream more steeply than the slope of the gully.

Problem 4

Sketch the rock layer illustrated in each structural overlay, 1 through 5, in Figure 8.3, column A, p. 121.

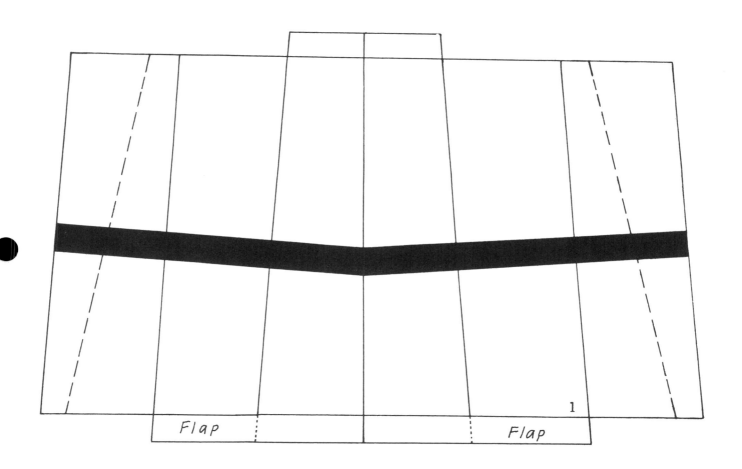

Flap

Flap

1

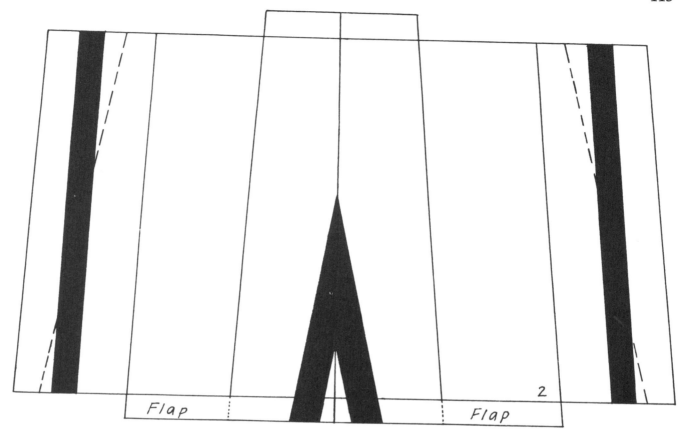

Flap   Flap   2

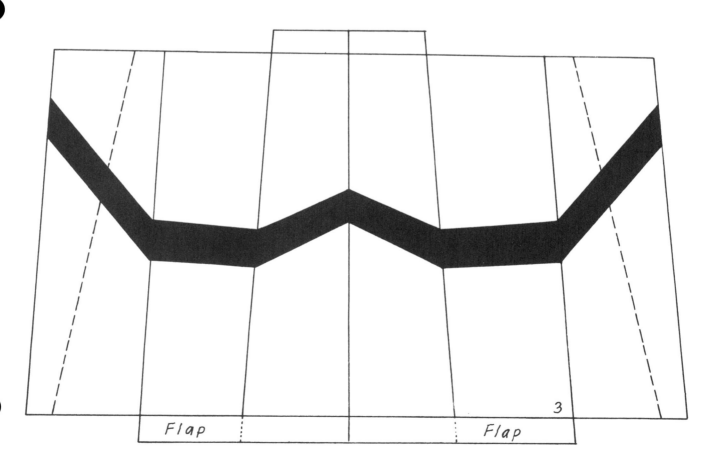

Flap   Flap   3

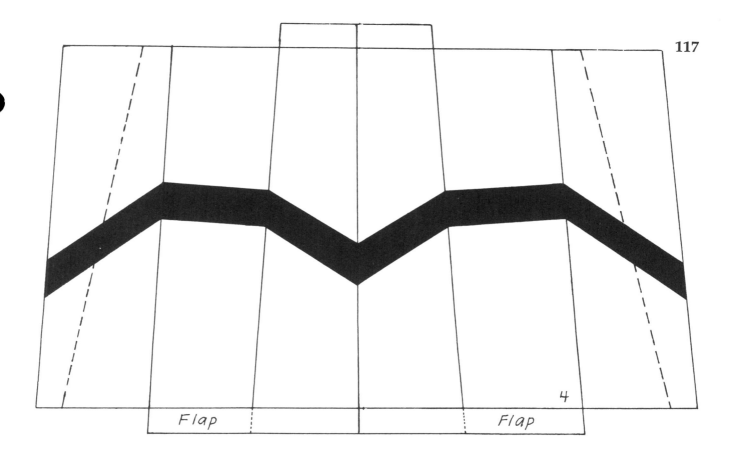

Flap   Flap

4

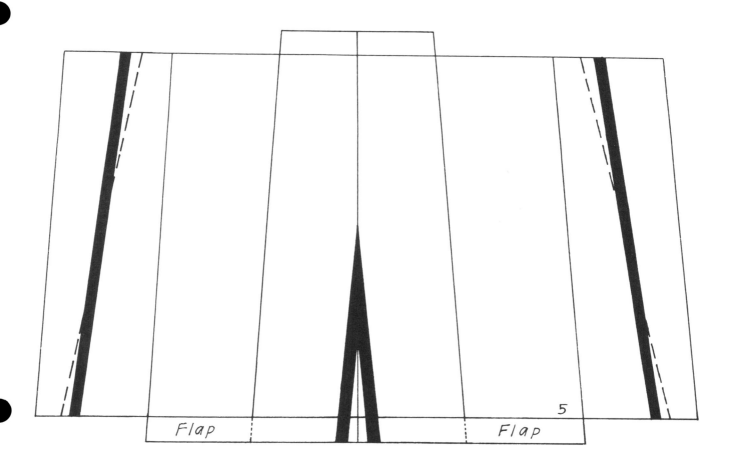

Flap   Flap

5

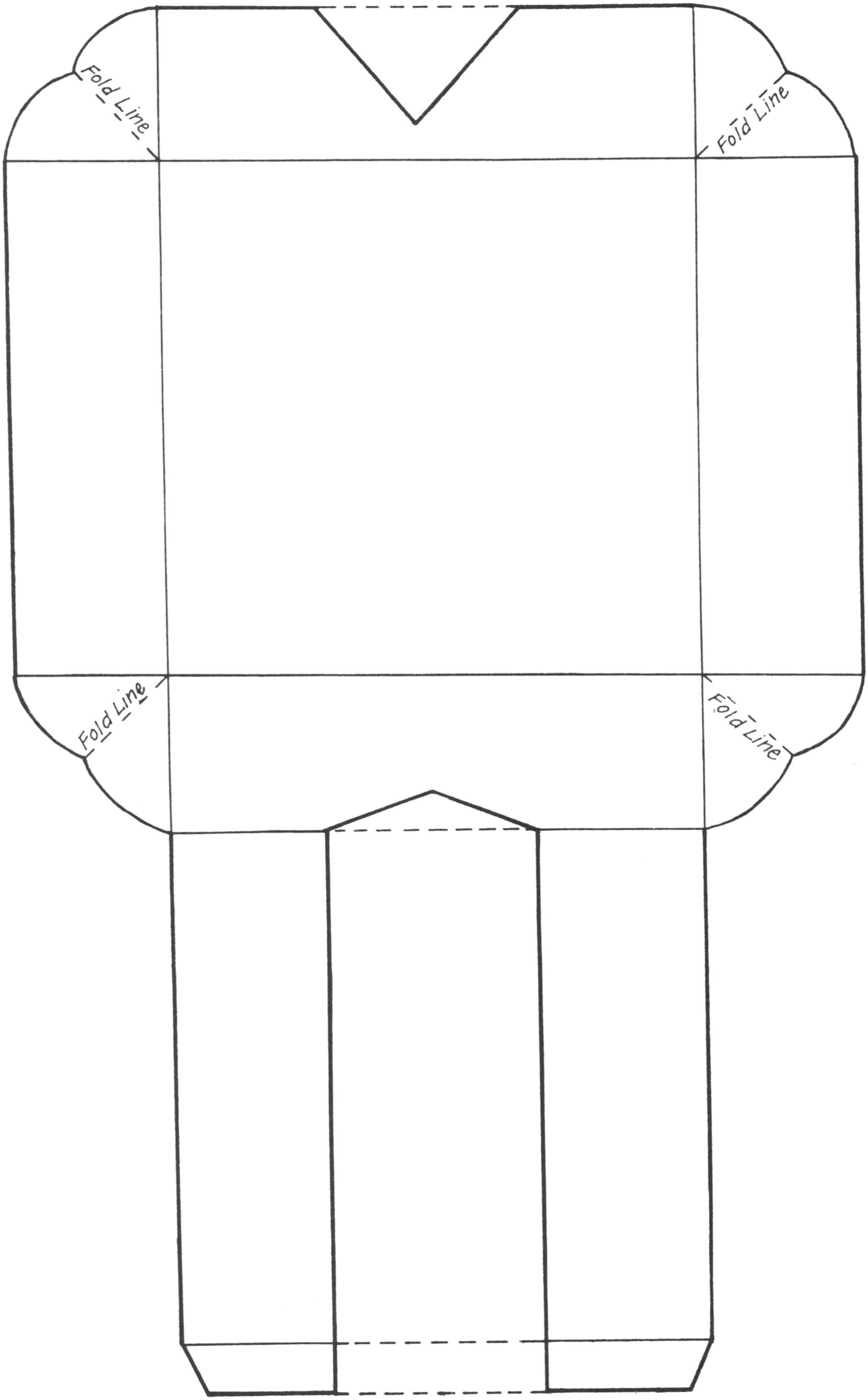

Fold Line

Fold Line

Fold Line

Fold Line

A. Sketch   B. Map view   C. Outcrop direction

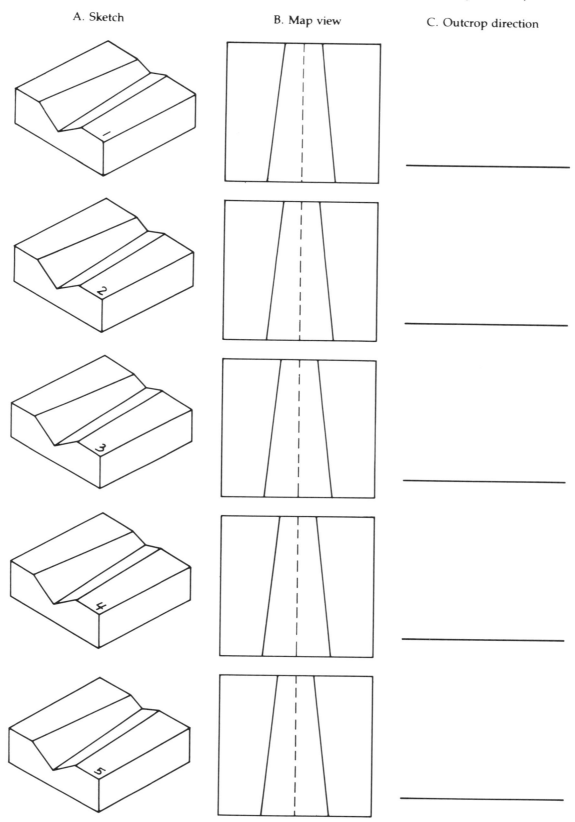

Figure 8.3
Answer sheet for gully problem.

Figure 8.4
Stereoscopic aerial photographs, Bright Angel Canyon.

A.

FIELD STUDY: GRAND CANYON NATIONAL PARK, ARIZONA

Use the stereoscopic photographs (Figure 8.4), the topographic map of the Bright Angel quadrangle (Figure 8.6), and the stratigraphic column of the Grand Canyon area you used in Exercise 5 to answer the following questions and ultimately to synthesize the geologic history of the Grand Canyon area. Symbols for faults, fold hinges, strike and dip of layers, and some other structures are given in Color Map 4, along with the names of rock formations. See also Appendix 6 for map symbols.

A Walk Through Time

It is a clear sunny morning on the north rim of the Grand Canyon as you prepare for your hike to Roaring Springs. Because you are an avid amateur geologist, you have come prepared with aerial photographs, a geologic map (Color Map 4), and the stratigraphic column of the Grand Canyon (Exercise 5). Orient yourself on the aerial photographs of Figure 8.4; the trailhead is at point 1.

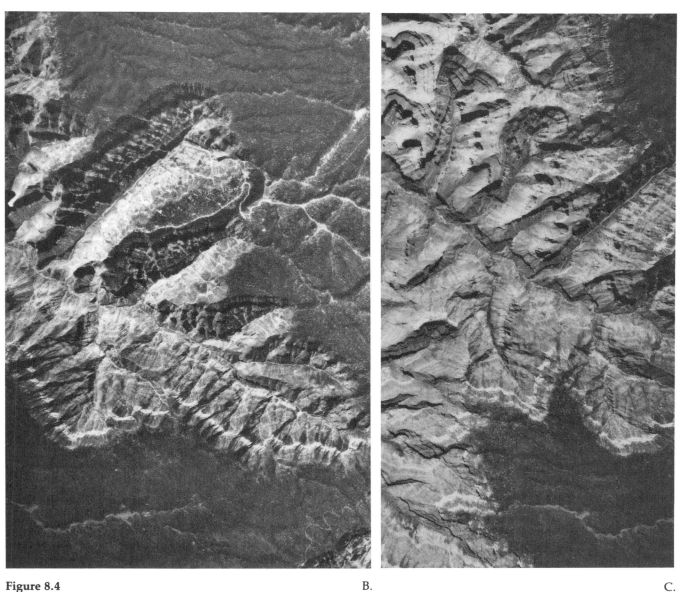

Figure 8.4
Stereoscopic aerial photographs, Bright Angel Canyon.

B.

C.

Problem 8

By matching features on the map with those on the photographs, determine which direction is north on the photographs. Place a north arrow on the photographs. Check your answer with your instructor before proceeding.

Problem 9

Transfer the location of point 1 to your geologic map of the Grand Canyon. What rock type do you see at the surface at the trailhead? What is the name of this formation?

Problem 10

Is this formation the oldest or the youngest rock you will see on your hike down into the canyon?

Problem 11

What depositional environment, marine or land, is suggested by this formation?

You walk down the Bright Angel Trail to point 2 on the aerial photographs. This conspicuous white cliff-forming unit is the Coconino Sandstone.

Problem 12

Transfer the location of point 2 to your topographic and geologic maps. Notice that a similar white cliff appears at the same elevation as you look southeast across Bright Angel Canyon. Is it likely that the Coconino Sandstone was originally deposited as isolated patches of sediment separated by a canyon, as it is at present, or was it deposited as a continuous layer of sediment and later eroded by the stream? Show the original distribution of the Coconino Sandstone on the cross profile in Figure 8.5. Formulate this rather commonsense idea into a statement that might be called the "law of original continuity." This law is the third of Steno's observations governing the deposition of sediments (the others are the law of superposition and the law of original horizontality that you formulated in Exercise 3).

Problem 13

What environment of deposition is suggested by the grains of well-rounded, well-sorted quartz sand and the tracks of terrestrial reptiles in the Coconino Sandstone? What change in depositional environment, therefore, occurred between the time when the Coconino sediments were deposited and the unit at point 1 was deposited?

Between units 1 and 2, you have covered a short distance horizontally and have also walked downhill through the third dimension. But because you have also descended vertically through part of the stratigraphic column at the Grand Canyon, you have also walked back through the fourth dimension, time. As we walk about in the three dimensions of space, we also move back and forth in time through earth history as it is expressed in rocks of various ages.

Problem 14

Find point 3 on the aerial photographs, and transfer its location to the geologic map. What is the topography here?

Problem 15

This formation is the Redwall Limestone. What has been the change in depositional environment from the time when the Redwall Limestone was deposited to the time when the Coconino sediments were deposited?

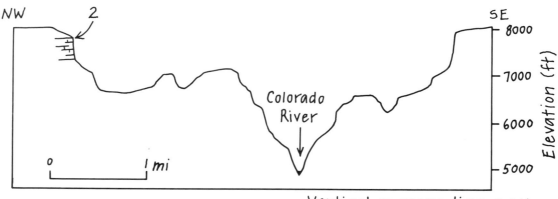

Figure 8.5
Cross-sectional view of side canyon.

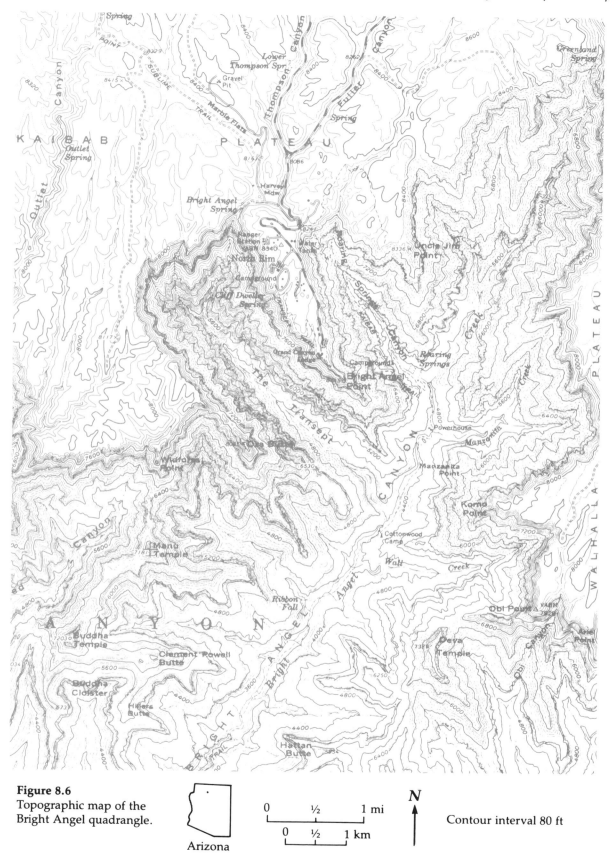

Figure 8.6
Topographic map of the
Bright Angel quadrangle.

Arizona

0   ½   1 mi

0   ½   1 km

N

Contour interval 80 ft

The contact of the Redwall Limestone with the unit below it is puzzling at first. It is not a gradational horizontal surface like the others you have seen on your hike down the canyon, so you make a quick sketch of the relationship of the Redwall to the rock units below as you walk along the trail (Figure 8.7).

Problem 16

Recall the law of original horizontality of sedimentary layers. Assuming, then, that the top of the Muav Limestone was approximately flat and horizontal when the sediments were deposited about 300 million years ago, how do you account for the irregular undulating top surface of the Muav Limestone revealed in Figure 8.7?

Problem 17

The upper surface of the Temple Butte Limestone is similar in shape to the top of the Muav Limestone. Is the contact between the Redwall Limestone and the Temple Butte Limestone conformable or unconformable?

Problem 18

Starting with the deposition of the Muav Limestone and ending with the deposition of the Redwall Limestone, list the sequence of depositional or erosional events that produced the two contacts shown in Figure 8.7. Indicate clearly the progression in age from oldest to youngest.

Problem 19

Transfer the location of point 4 to your geologic map. What is the name of the rock formation here? What change in the depositional environment is reflected by the transition from the Muav Limestone to your present position?

Problem 20

Point 5 marks your descent down through the cliff-forming Tapeats Sandstone: transfer its location to your geologic map. In what water depth might a coarse-grained sandstone such as the Tapeats typically be deposited?

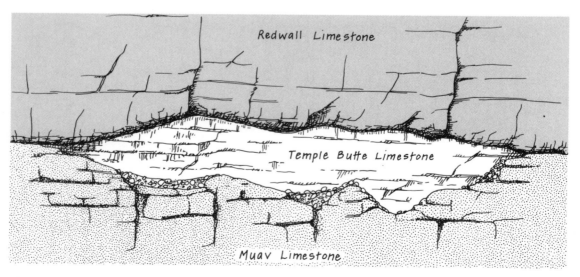

Figure 8.7
Sketch of Redwall Limestone and underlying Temple Butte and Muav Limestones.

Problem 21

Summarize your observations from your hike, the aerial photographs, the geologic map, and the stratigraphic column by outlining the geologic history of the Paleozoic section of the Grand Canyon. Start with the deposition of the Tapeats Sandstone, and work your way up through the section (forward in time to the present), including depositional and erosional events in their proper order. For deposition of sediments, briefly describe the depositional environment (deep marine, shallow marine, desert or beach, etc.) of each stratigraphic unit in the Paleozoic section. Fill in the right side of the block diagram given in Figure 8.8 by noting the environments of formation of the rock units named on the left side of the diagram.

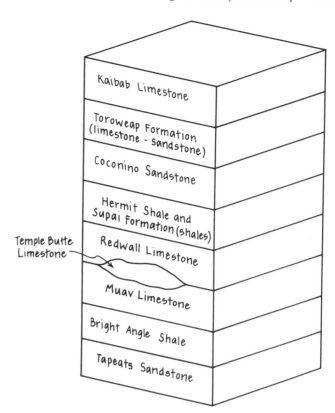

Figure 8.8
Block diagram of Paleozoic section, Grand Canyon.

Topography and Rock Type in a Dry Climate

Problem 22

What type of topography (mountains, plains, tablelands, etc.) is exemplified by the Grand Canyon region?

Problem 23

Notice the relationship of rock type to cliff or slope within the Paleozoic strata. Rock types of the Paleozoic strata are given below. Which rock types form cliffs, and which form gentler slopes?

Formation	Rock Type
Kaibab	limestone
Coconino	sandstone
Supai	sandstone, shale
Muav	limestone
Tapeats	sandstone
Toroweap	limestone
Hermit	shale
Redwall	limestone
Bright Angel	shale, sandstone

Horizontal and Dipping Planes

Problem 24

The Paleozoic strata of the Grand Canyon are horizontal. Compare the contact between the Muav Limestone and the Bright Angel Shale with the contour lines in the vicinity of the contact, and state how you know the contacts are horizontal.

Problem 25

Study the contact between the two units of the Vishnu Group in the inner gorge near Clear Creek. This contact has no similarity to those of the horizontal Paleozoic formations on either side of the inner gorge. What is the orientation of the contact between the schists? Explain how you can determine the orientation from the form of these two contacts on the map.

Consider the Bright Angel fault, and recall the surface outcrops of horizontal and dipping planes as they cross an irregular land surface.

Problem 26

What is the dip of the fault plane in the north part of Bright Angel Canyon? How do you know?

Problem 27

Find the short northeast-trending fault just northwest of Phantom Ranch. What is its dip? How do you know?

Conformable and Unconformable Contacts

Problem 28

Sketch an approximate geologic cross section across Clear Creek near the outcrop of the Zoroaster complex on the profile given in Figure 8.9, showing the angular relationships between the units of the Vishnu Group, the Zoroaster Granite, and the Tapeats Sandstone. Is the lower contact of the Tapeats Sandstone (the basal contact) a conformable contact, or is it unconformable?

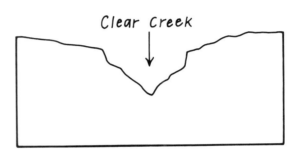

Figure 8.9
Cross section across Clear Creek.

Problem 29

Follow the basal contact of the Tapeats Sandstone in the map area. The Tapeats Sandstone was deposited on an erosional surface developed on the schist and granite. Erosional surfaces of this type are often smooth surfaces with very little topographic relief. Examine the elevation of the upper surface of the schist (the old erosional surface) where it is exposed along the Colorado River and Bright Angel Creek. What is the approximate relief of the erosional surface (the ancient landscape) developed on the schist and upper Precambrian rocks?

Problem 30

What topographic features of this ancient landscape might account for the absence of Tapeats deposits in areas where the Bright Angel Shale lies directly on the schist?

Folds

Problem 31

Do you see any evidence of folding in any of the Grand Canyon rocks? If so, what rocks appear to be folded?

Faults

Problem 32

Which fault is older, the Bright Angel or the Phantom fault? Which fault is older, the Bright Angel or the Tipoff fault? Cite your evidence.

The approximate age of the fault labeled 400, a northwest-trending fault about 2 miles west of Phantom Ranch, can be estimated by considering the relationship of rock formations along the fault north of the Colorado River. Study the relationship of the rock units along the fault near Trinity and Ninetyone Mile Creeks and answer the following questions.

Problem 33

What formations have been displaced by the fault? Which have not been displaced?

Formations Displaced   **Age of Formation**

Formations Not Displaced   **Age of Formation**

Problem 34

From your answers to Problem 33, and recalling the principle of crosscutting relationships, during what geologic period was the fault active?

Locate the major outcrops of the upper Precambrian series of rocks (Bass Formation through the Diabase intrusives) in the map area. This group of rock units is sometimes called the Grand Canyon Group. They were deposited unconformably on an erosional surface developed on the older Precambrian schist and granite.

Problem 35

The original thickness of these rocks has been established at up to 25,000 feet. What is the explanation for the present thickness of only a few thousand feet? Is the upper contact of this series of rocks with the overlying Tapeats Sandstone conformable or unconformable?

Geologic History

Problem 36

Construct a geologic structure section of the rock beds at and beneath the surface along the line B–B' on the Bright Angel map (Color Map 4) from Sumner Point to a point just northeast of Newton Butte. Use the profile provided in Figure 8.10. Show the orientations and contacts of the underlying Precambrian rocks as well as those of the Paleozoic rocks above.

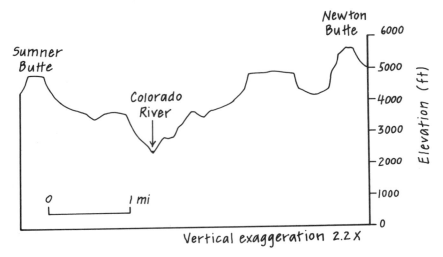

Figure 8.10
Profile of Grand Canyon, B–B', from Sumner Butte to Newton Butte.

Problem 37

From the geologic cross section you have just drawn and from your answers to all the preceding questions, synthesize the geologic history of the Grand Canyon area. Outline the geologic events up to the deposition of the Tapeats Sandstone. State the events in the correct sequence, from oldest to youngest, beginning with the deposition of the parent rocks of the schists. Remember to include in your geologic history the two unconformities you have discovered and the faulting events as well as deposition of the various rocks.

REFERENCES FOR PART B

Books

Berry, W.B.N., *Growth of a Prehistoric Time Scale*. Blackwell Scientific Publications, Palo Alto, Calif., 1987.

Billings, M.P. *Structural Geology*. Prentice-Hall, Englewood Cliffs, N.J., 1973.

Davis. G.H., *Structural Geology of Rocks and Regions*. John Wiley & Sons, New York, 1984.

Dott, R.H., Jr., and R.L. Batten, *Evolution of the Earth*. McGraw-Hill, New York, 1980.

Eicher, Donald L., *Geologic Time*. Prentice-Hall, Englewood Cliffs, N.J., 1968.

Harland, W.B., A.V. Cox, P.G. Llewellyn, C.A.G. Pickton, A.G. Smith, and R. Walters, *A Geologic Time Scale*. Cambridge University Press, New York, 1983.

Hills, E.S., *Elements of Structural Geology*. John Wiley & Sons, New York, 1972.

Rowland, S., *Structural Analysis and Synthesis*. Blackwell Scientific Publications, Palo Alto, Calif., 1987.

Simpson, G. G., *Fossils and the History of Life*. Scientific American Books, New York, 1983.

Stanley, S., *Earth and Life Through Time*. W.H. Freeman, New York, 1986.

U.S. Government Publications

U.S. Geological Survey Information Pamphlets, available free from

Distribution Branch
U.S. Geological Survey
Box 25286, Federal Center
Denver, CO 80225

Elevations and Distances in the United States, 1980.
Geologic Maps: Portraits of the Earth, 1982.
Geologic Time, 1978.
Land Use and Land Cover and Associated Maps, 1982.
Landforms of the United States, 1984.
Map, Line and Sinker, 1977.
Naming (and Misnaming) of America, 1981.
Quiet Revolution in Mapping.
Tree Rings: Timekeepers of the Past.
Topographic Maps: Silent Guides for Outdoorsmen, 1978.
Topographic Maps: Tools for Planning, 1980.
Types of Maps Published by Government Agencies, 1978.

Other

Geological Highway Maps, published by The American Association of Petroleum Geologists, P.O. Box 979, Tulsa, OK 74101.

Roadside Geology Guides, available from

Mountain Press Publishing
P.O. Box 2399
Missoula, MT 59806

for Colorado, Texas, Oregon, Northern California, Southern California, Northern Rockies.

Coastline and surf near Santa Cruz, California. The Pacific Ocean continues to erode and weather the western coastline of North America, as it has for hundreds of millions of years.

Surface Processes

Almost all earth processes may be understood as the interaction of two heat machines that transform heat energy into motion. The earth's *internal energy* drives such processes as volcanism, plutonism, metamorphism, and deformation of crustal rocks. Internal energy is derived from the heat of materials in the earth's interior and is generated largely by radioactive decay.

The earth's *external energy*, in contrast, comes from the sun. The sun supplies the heat energy that evaporates water from land and sea and drives the circulation of the earth's atmosphere, carrying water vapor over the land, where it may condense and fall as rain or snow. Agents that wear down and erode the earth's surface—running water, glaciers, and wind—all run on energy from the sun. Weathering of rock material to soil is a process on which all life on earth depends; weathering, too, is ultimately powered by the sun's energy.

An easy way to understand the ways in which the earth's internal and external heat machines interact is to consider the interrelationships among the three major rock types. Igneous rocks are in a sense the primary rocks of the earth's crust, having solidified from the originally molten state of the primordial earth. They weather at the surface in response to the conditions there, producing soil and sediments. These loose materials are usually transported from their place of formation by slope processes, running water, or, less commonly, by glaciers or the wind. When they are deposited, they may undergo diagenesis and lithification to sedimentary rocks at shallow depths in the crust. Under conditions of deep burial, sedimentary or igneous rocks undergo heating at high pressure that transforms them to metamorphic rocks. If temperatures are high enough, the rock material melts to form magma, completing the cycle. Refer to Figure 9.1.

In Part C you will study the surface processes of the earth, which are sustained by external energy from the sun. Exercise 9 concerns the interaction of the sun's energy, water, and the atmosphere to physically and chemically weather igneous, sedimentary, and metamorphic rocks. In some cases the resultant material is soil, the subject of a Field Study in Exercise 9. Exercise 10 deals with slope processes, the movement of loose material downslope under the influence of gravity. Exercises 11, 12, and 13 explore the hydrosphere, including the water that flows beneath the surface (Exercise 11), the system of surface drainage (Exercise 12), and water in oceans and other open bodies of water (Exercise 13). Exercise 14 concerns the action on the landscape of frozen water in the form of glaciers. Exercise 15 deals with the generation and action of the wind on the surface of the earth.

Weathering in the Surface Environment

Rust never sleeps.

NEIL YOUNG

The earth's surface processes and products result from the interaction of four principal features: the lithosphere, the hydrosphere, the atmosphere, and the biosphere (see Figure 9.1). Powered by the sun's energy, rocks, water, and gases interact to produce the surface environment. The *lithosphere* is the solid part of the earth. Most of the earth's crust is composed of rocks that have formed at high temperatures and pressures.

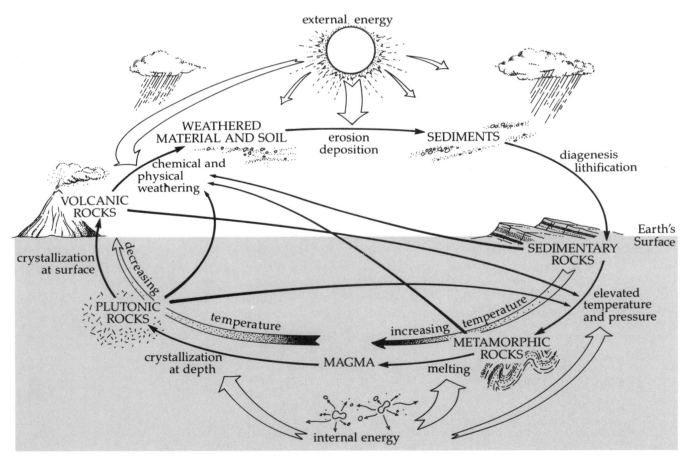

Figure 9.1
Internal and external energy are responsible for all geologic processes.

The minerals of igneous rocks, which crystallize at relatively high temperatures and sometimes at high pressures, are not necessarily stable under prolonged exposure to the low temperatures and pressures of the earth's surface. They are susceptible to change by the various agents active at the surface, the most important of which is water.

The *hydrosphere* includes all of the earth's free water, that is, water that is not chemically bound. (In gypsum, for example, water is chemically bound within the mineral structure, $CaSO_4 \cdot 2H_2O$.) The oceans contain the vast bulk (97.3 percent) of the earth's free water. Also included in the hydrosphere are glaciers (2.1 percent of the free water), groundwater (0.6 percent), and rivers and lakes (0.01 percent). The erosional effects of these forms of water will be discussed in Exercises 11 through 14. This exercise is concerned solely with the weathering or disintegration of rocks at the earth's surface, a process that usually must occur before rock material can be eroded or transported. Water plays a vital role in all weathering processes.

The *atmosphere* is the familiar gaseous envelope surrounding the earth. Ninety-nine percent of the earth's atmosphere consists of two gases, nitrogen (78 percent) and oxygen (21 percent). Nitrogen was almost certainly produced early in earth history by volcanism. Oxygen is a relative newcomer, the product of photosynthesis by green plants. Of the two gases, oxygen is the most active chemically and is vitally important in the weathering of rocks as well as in biological activity.

The *biosphere* consists of all living organisms. Life processes derive energy from the sun and nourishment from the atmosphere, lithosphere, and hydrosphere.

THE EARTH'S EXTERNAL ENERGY

Energy from the sun drives the cycle of surface processes that modify the earth's landscape, namely, the weathering of rocks and the transport of the weathered material. The earth intercepts an enormous amount of energy radiated by the sun. The average incoming solar energy (the earth's *external energy*) is about 6000 times its *internal energy* (energy derived from hot material in the earth's interior). If the earth were to retain all of the incoming solar energy, the granitic crust would melt to a depth of 2500 feet in 500 years. Obviously, the sun's energy is not entirely retained; over 50 percent is reradiated back into space. The incoming and outgoing radiation maintain a balance that accounts for the relatively narrow surface temperature range from $-120°F$ ($-84°C$) to $120°F$ ($49°C$).

Problem 1

Examine the global distribution of average sea-level temperatures in July and January illustrated in Figures 9.2A and B. In what direction do the isotherms (lines of equal temperature) trend?

Problem 2

What is the reason for this trend?

Problem 3

Look at the 30°N latitude line in central Asia or North America. Estimate the average range of sea-level temperatures from January to July. Now look at the 30°N latitude line in the Atlantic Ocean; estimate the temperature difference from January to July over the ocean. How does the temperature range on the continents differ from the temperature range over the oceans?

WATER AND THE HYDROLOGIC CYCLE

Water is the only substance that can exist in three states within the narrow temperature range of the earth's surface—as a solid (ice), liquid (water), and gas (water vapor). Above $0°C$ ($32°F$), water as a liquid is stable; below $0°C$, ice crystallizes. Water vapor forms at any temperature, the rate increasing at higher temperatures. Upon cooling, water vapor may condense to a liquid or crystallize directly to a solid. The three states of water control to a great extent the weathering of the solid crust near the surface.

Problem 4

How does the density of water vary in its three states (ice, water, and water vapor)? Since these substances are of the same chemical compound, H_2O, how do you explain the density differences?

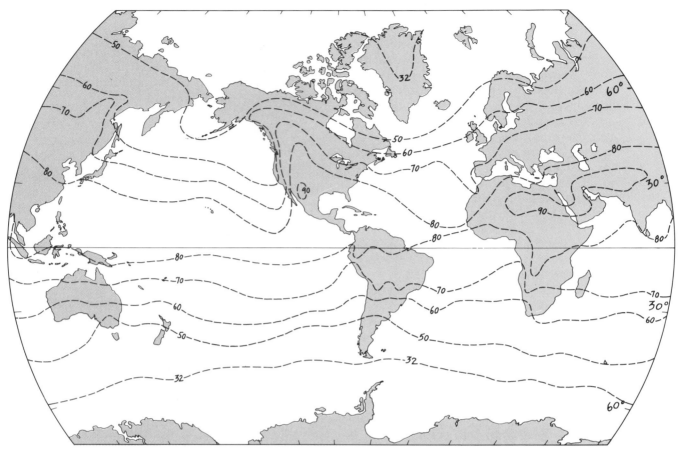

Figure 9.2
A. Distribution of average sea level temperature (°F) in July.

The *hydrologic cycle* encompasses the processes that transfer water between solid, liquid, and gaseous states in the surface environment. This cycle may be regarded as the history of a water molecule in the surface environment. Although the cycle is continuous and never ending, it can be visualized as beginning in the largest water source, the oceans. Water is transferred by evaporation to the atmosphere as water vapor, which winds then circulate over the earth. The water returns to the surface through precipitation as rain or snow. Of the total precipitation over the earth, approximately three-quarters fall into the oceans, and one-quarter falls on land. The hydrologic cycle is completed with the return of the water to the oceans. This cycle is represented in Figure 9.3.

The water that falls on land may proceed in three possible paths to the ocean. First, water may return directly to the atmosphere through evaporation or transpiration by plants and then precipitate back into the ocean. Second, the water may flow directly into streams and then back to the ocean. Finally, water may be stored below the surface as groundwater or on the

surface as ice in glaciers. This storage, though measured in years, is only temporary in geologic terms, and eventually this water also returns to the ocean. In the United States approximately 70 percent of the total precipitation evaporates, 30 percent flows into streams, and less than 1 percent accumulates as stored water.

Problem 5

Compare the global distribution of precipitation (Figure 9.4) with that of temperature. What temperature conditions seem to be associated with high precipitation?

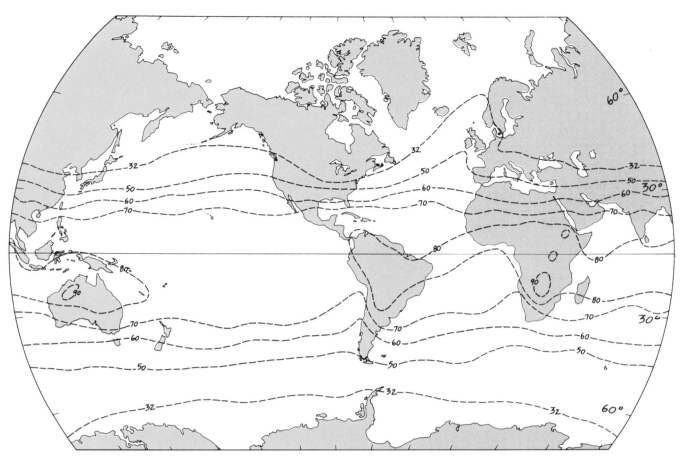

Figure 9.2
B. Distribution of average sea level temperature (°F) in January.

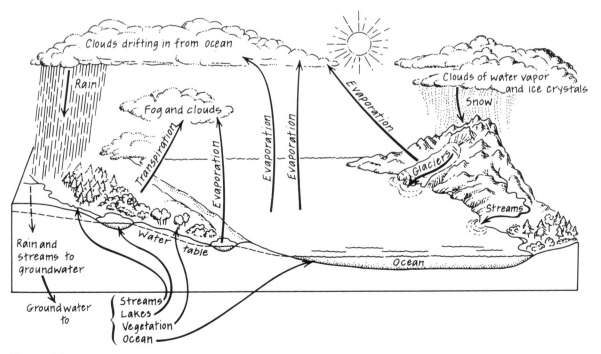

Figure 9.3
The hydrologic cycle.

138

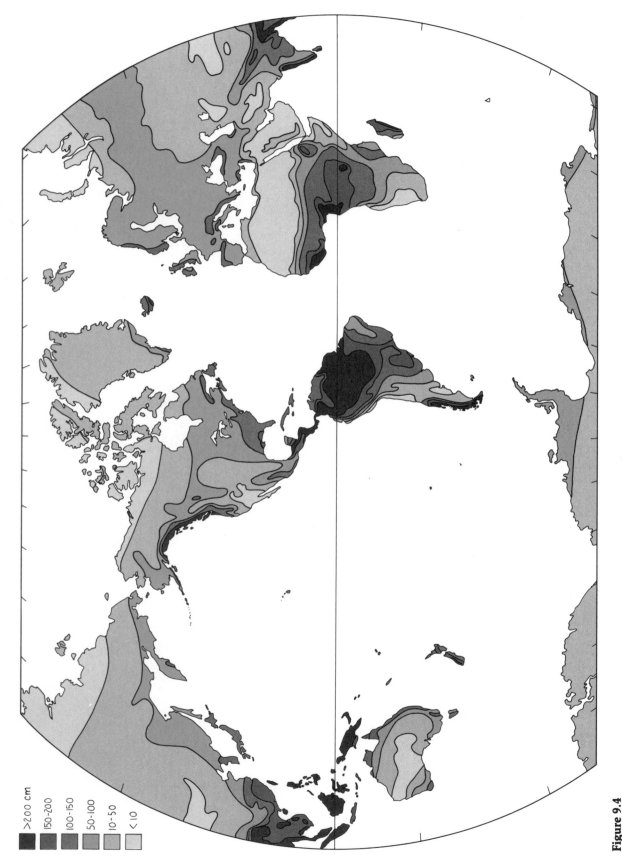

Figure 9.4
Global map of average annual precipitation.

Problem 6

The continental United States has an average annual precipitation of approximately 30 cm (12 inches) of water. The total area of these 48 states is about 200 million acres. One inch of rain covering 1 acre of ground weighs about 200,000 pounds. Estimate how many pounds of water fall on the continental United States each year. Show your work.

Problem 7

What energy source lifts this enormous amount of water from the surface?

WEATHERING

Weathering is the disintegration of rocks and minerals in the surface environment. It is divided, for simplicity's sake, into two types, *chemical weathering* and *physical weathering,* although in nature the two types are inseparable. Chemical weathering involves dissolution of minerals by water or the chemical reaction of minerals with water (hydrolysis), oxygen (oxidation), or other chemical agents present near the earth's surface. Physical weathering is the breakup of rock into smaller fragments. The expansive action of water in wedging open rock fractures, by freeze–thaw cycles or by swelling of some minerals when exposed to water, is the most important element in physical weathering.

Balance Between Physical and Chemical Weathering

Although the two types of weathering occur together in nature, in certain climates chemical weathering is more important than physical weathering, and vice versa. Because chemical weathering changes minerals chemically and forms new substances, whereas physical weathering merely breaks rock apart into smaller particles without changing them chemically, careful examination of sediments or sedimentary rocks may shed some light on the climatic environment that prevailed when the source rocks of the sediments were weathered.

Problem 8

Refer to the global maps of temperature and precipitation. Active chemical weathering requires both abundant precipitation and relatively high temperatures. In what regions would you expect to see chemical weathering predominate over physical weathering?

Problem 9

When cars are driven in cold climates, care must be exercised to prevent the water or coolant from freezing in the engine block since this often results in cracking of the block due to the expansion in volume when water changes to ice. Relate this phenomenon to physical weathering of rocks in the natural environment.

Problem 10

From the data in Figures 9.2 and 9.4 or from temperature and rainfall data supplied by your instructor, use the graph in Figure 9.5 to estimate the balance between physical and chemical weathering for your area.

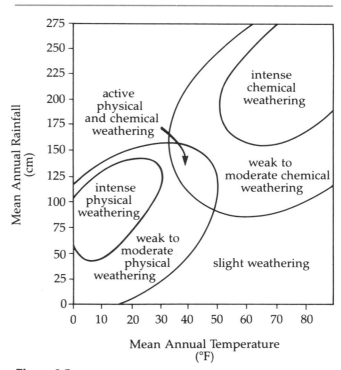

Figure 9.5
Relation of weathering type to temperature and rainfall conditions.

Problem 11

What is the balance between physical and chemical weathering for the following locations?

- New York
- Florida
- Los Angeles
- Central America
- Southwestern Australia

Problem 12

Summarize your predictions of the balance between chemical and physical weathering by placing on Figure 9.2A a P in regions where physical weathering predominates and a C where chemical weathering predominates.

Problem 13

Examine Color Plate 63, which shows a weathering rind in a diorite cobble. Is this an example of physical or chemical weathering? How do you know?

Differential Weathering

Chemical weathering attacks some minerals more strongly than others, a process called *differential weathering*. Its effects may be seen by studying individual minerals and by looking at rocks containing different minerals.

Problem 14

Examine and describe the mineral samples provided.

	Mineral Sample 1	Mineral Sample 2
• Color		
• Shape		
• Size		

Problem 15

Combine the samples and place them in a container with water. Stir the liquid and mineral mixture for approximately 2 minutes. Separate the liquid from the solid by straining or decanting. Examine and describe the remaining solid.

	Mineral Sample 1	Mineral Sample 2
• Color		
• Shape		
• Size		

Problem 16

What has happened to each mineral sample?

Problem 17

Based on your observations above, what type of sedimentary rocks, chemical or detrital, would be most affected by chemical weathering?

Problem 18

Color Plate 64 shows a weathered boulder of sandstone (tan) and mudstone (gray). Which rock is more resistant to weathering? Considering the substances that cement detrital sedimentary rocks, why might this be so?

Examine the rock samples provided by your instructor, or refer to Color Plate 63. Notice that at least one side of the sample is altered by weathering.

Problem 19

How does the rock's weathered surface differ from a fresh or unweathered surface?

	Fresh Surface	Weathered Surface
• Color		
• Luster		
• Other properties		

Problem 20

How thick is the weathered surface?

Problem 21

Are certain minerals preferentially weathered? Which minerals are severely weathered? Which are only slightly weathered?

Sediments Produced by Weathering

Detrital sedimentary rocks are the ultimate product of the weathering of preexisting rocks. Since igneous rocks are the primary source for clastic sediments, the difference in the composition of igneous and sedimentary rocks is a result of the chemical weathering process. Figure 9.6 illustrates the distribution of minerals and rock fragments by size in detrital sedimentary rocks. To determine what minerals are most common in a given size range, find the size on the horizontal axis, and read the percentage of minerals vertically. For example, a medium sand is composed, on the average, of 65 percent quartz clasts, 10 percent feldspar clasts, and 25 percent lithic fragments.

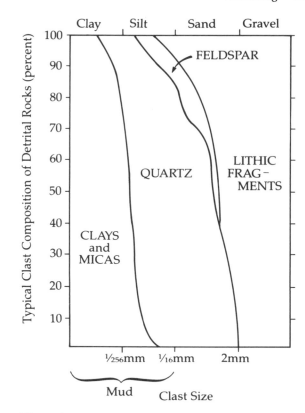

Figure 9.6
Distribution of minerals and rock fragments by size in clastic sedimentary rocks.

Problem 22

What is the most abundant clast composition in the following rocks?

- Conglomerate
- Sandstone
- Mudstone

Problem 23

Why are single mineral fragments not common in pebble-sized clasts?

Problem 24

Compare the mineral composition of sedimentary rocks with that of igneous rocks by stating in the chart below whether the mineral or mineral group is absent, rare, common, or abundant in sedimentary and igneous rocks. The mineral composition of igneous rocks is given in Exercise 2, Figure 2.5.

Mineral	Igneous Rocks	Sedimentary Rocks
• Olivine		
• Pyroxene		
• Hornblende		
• Mica		
• Feldspar		
• Quartz		
• Clay		

Problem 25

What igneous minerals are altered in the process of chemical weathering?

Problem 26

What igneous minerals are resistant to chemical weathering?

Problem 27

How do relative mineral stabilities relate to the temperature at which they crystallize? Refer to the Bowen reaction series, Exercise 2, Figure 2.2

Problem 28

What new mineral forms in the chemical weathering process?

Problem 29

Considering mineral stability and type of weathering, what sedimentary rocks are likely to be formed by moderate physical and chemical weathering of granite?

Physical Weathering Predominates	**Chemical Weathering Predominates**

FIELD STUDY: SOILS

Weathering of the earth's lithosphere over an extended period of time under appropriate conditions produces the unconsolidated material called *soil*. Soil is truly the bridge between the inanimate and the living world. Soil is rock that has been attacked chemically so that part of it is dissolved or altered to new substances, such as clay minerals. Other components may remain unaltered, such as any crystalline quartz in the parent rock. Still other materials have been added to the soil, typically organic material from decay of vegetation and microorganisms. Typical soils have a layered profile that reflects the transition from solid rock at depth to a complex mixture of residual, altered, and accumulated material near and at the surface.

In its idealized form, a mature *soil profile* consists of three main layers, usually designated as the A, B, and C *soil horizons*. The A horizon is a dark layer nearest the surface, where organic matter is concentrated and where leaching has removed some of the mobile ionic components, such as Na^+, K^+, Al^{3+}, Fe^{3+}, and sometimes Ca^{2+}. The B horizon is the zone where material leached from the A horizon accumulates— iron and aluminum oxides and clay, for example. The C horizon is slightly weathered parent material. Below the C horizon is the R horizon—bedrock. There are various subdivisions in the three main soil horizons; soils are very complex and varied entities. There may be an O horizon—largely organic materials—above the A horizon. An E horizon is sometimes present as an intensely leached zone between A and B horizons. Soil horizons are shown in Figure 9.7.

Problem 30

Examine with a hand lens the soil sequence provided by your instructor. (See also Color Plate 65.) These samples were taken at various intervals from the surface (1) to the specified depth in a well-developed soil of your area. What trends do you note in the properties listed below?

- Color

- Organic content

- Clay content

- Weathering of mineral grains

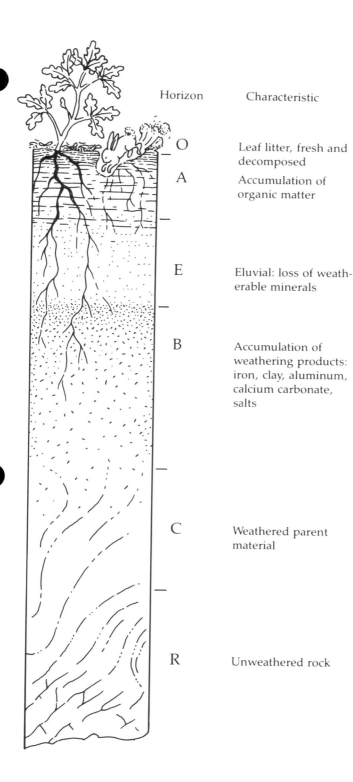

Horizon	Characteristic
O	Leaf litter, fresh and decomposed
A	Accumulation of organic matter
E	Eluvial: loss of weatherable minerals
B	Accumulation of weathering products: iron, clay, aluminum, calcium carbonate, salts
C	Weathered parent material
R	Unweathered rock

Figure 9.7
Idealized soil profile.

Problem 31

How do you explain these trends in terms of soil horizons and the processes that produce them?

The formation of soil involves the complex action and interaction of many factors. The climate, parent rock material, topography, vegetation, and time all influence the properties of the soil formed in an area. The relative effect of each of these factors can be studied by analyzing sites where four of the factors have been essentially constant and only one has varied.

Climate is perhaps the most important factor controlling the soil-forming process. Areas of high rainfall, for example, are susceptible to more leaching (dissolving) than arid areas. Especially if rain is slightly acidic from absorbing carbon dioxide from the soil and air, it attacks ionic bonds in minerals, dissolving the more soluble components from materials in the upper layers. Sodium, potassium, calcium, and sometimes iron and aluminum are removed from the A horizon in this manner, leaving behind the more stable minerals, notably crystalline quartz. The upper layer of the soil is generally the leached zone, except for the organic material (leaf litter, microorganisms, etc.) that has been added to it by life on or near the surface. Very soluble materials leached from the A horizon may be carried away completely by high rainfall, as is commonly the case with sodium, potassium, and calcium, unless they are stabilized in the A horizon by vegetation that metabolizes them. Less soluble components, such as iron and aluminum, may precipitate in lower layers (B horizon) if rainfall is insufficient to flush them from the soil completely. White calcium carbonate accumulates in the lower B or upper C horizon if the rainfall is quite low. This layer is called *caliche* and is sometimes designated as the K horizon. If the average temperature is high, chemical weathering reactions proceed more rapidly, and soils develop more quickly and to a greater depth than in cold regions.

Problem 32

To visualize the effects of climate (rainfall plus temperature) on soil formation, examine the two soil descriptions below. All the factors of soil formation, except climate, are determined to have been the same for both. The parent rock material is granite, the topography is flat to gently sloping, and the time interval of soil formation is 100,000 years.

Soil Sample 1

Horizon	Depth (cm)	Description
O	Surface	Partly decomposed leaves.
A	0–6	Dark brown sandy loam[1], friable[2], common fine roots and insect burrows.
E	6–10	Yellow-brown sandy clay loam, friable, few fine roots.
B1	10–27	Light reddish-brown sandy clay loam, few large roots.
B2	27–82	Reddish-brown silty clay, very firm, few large roots.
B3	82–135	Light reddish-brown clay loam, firm.
C	135+	Mottled reddish-brown, yellow, black, and gray weathered granite. Crushes easily to sandy clay loam.

Soil Sample 2

Horizon	Depth (cm)	Description
A	0–5	Light gray-brown loamy sand, few fine roots.
B1	5–11	Gray-brown sandy loam, few fine roots.
B2	11–21	Gray-brown sandy clay loam with 50 percent rock fragments, traces of $CaCO_3$.
B3	21–52	Fractured rock with thin clay coatings and traces of $CaCO_3$ on fracture surfaces.
R	52+	Jointed granite bedrock.

[1]Loam: a soil material composed of roughly equal parts of sand, silt, and clay.
[2]Friable: soft, porous, and crumbly.

The climates that produced these two soils differ mainly in the amount of rainfall. One soil formed in New Mexico, with a mean annual temperature of 55°F and precipitation of 12 inches a year. The vegetation is sagebrush. The other soil formed in North Carolina under deciduous forest; the mean annual temperature is 60°; rainfall is 50 inches a year. First compile your evidence from the various soil horizons, and then state which soil formed in New Mexico and which formed in North Carolina. Consider differences in the depth of the soil, the organic content, the color, the texture, and the composition, and give four statements of evidence for your answer.

1.

2.

3.

4.

Parent rock material is the raw geologic material from which the soil develops. This material may be bedrock, alluvium, glacial deposits, or wind deposits. It may be of any lithologic composition. For example, bedrock could be granite, basalt, limestone, schist, or shale; alluvium could be composed of a variety of rock types or only a few. Because the parent rock provides most of the chemicals that the soil is made of, it exerts an important influence on the properties of the soil.

Problem 33

Consider two soils formed side by side over a time period of 100,000 years in a moist temperature climate—one on granite, one on gabbro. The mineral composition of each parent material is given below:

Granite		Olivine Gabbro	
Quartz	25%	Plagioclase	50%
Orthoclase	40	Augite	30
Na-plagioclase	25	Olivine	20
Biotite	10		

From what you know about the chemical compositions of these minerals, their relative stabilities, and their weathering products, contrast the following properties of the two soils:

- Clay content (mafic minerals weather almost completely to clays)

- Redness (red color in soil is usually due to Fe^{3+})

- Sand content (consider quartz sand only)

- Fertility (plants need Fe, Ca, K, Mg, and P, among other things, from rock weathering)

Topography—the slope of the land surface—affects the development of soil in a rather obvious fashion.

Problem 34

Thick soils can develop over flat or gently sloping bedrock. Would you expect soils over a steep bedrock surface to be thick and well developed? Why or why not?

The *vegetation* of an area, although related to climate, exerts an influence of its own. Grasses help stabilize alkali elements (sodium, potassium, and calcium) in soils because they utilize them in metabolism; thus grassland soils tend to be very fertile and rich in organics. Deciduous forests, however, produce more acidic soils because they do not use alkalis to such an extent, thus allowing them to be leached from the soil. Coniferous forests produce especially acidic soils of low fertility.

The influence of *time* on soil development is best seen by examining the two soils described below. In both examples, the parent rock material is gravelly alluvium derived from granite. The topography is flat to gently sloping, and the climate is semiarid (20 inches of precipitation a year, falling in winter), with a mean annual temperature of 60°F. The vegetation is grassland with scattered oaks, located in the western foothills of the Sierra Nevada in California. Only the time available for soil formation varies. In one case the soil has been forming for 20,000 years; the other has been forming for only 1000 years.

Soil Sample 1

Horizon	Depth (cm)	Description
O	Surface	Partly decomposed grass.
A1	0–2	Dark gray-brown loamy sand, friable, abundant fine roots and insect burrows.
A2	2–8	Gray-brown gravelly loamy sand, friable, common fine roots.
C1	8–20	Gray, gravelly granitic sand, loose.
C2	20–24	Gray laminated sand, loose.
C3	24–60	Gray gravelly, loamy sand, loose.

Soil Sample 2

Horizon	Depth (cm)	Description
O	Surface	Partly decomposed grass.
A1	0–10	Very dark brown sandy loam, very friable, abundant fine roots and insect burrows.
A2	10–30	Dark gray-brown sandy clay loam, friable, common fine roots and earthworm holes.
B1	30–45	Brown clay loam, firm.
B2	45–70	Gray-brown sandy clay loam, firm.
C	70–90	Gray-brown gravelly loamy granitic sand.

Problem 35

Considering the evidence of soil depth, color, texture, organic content, and composition, state which is the younger and which is the older soil. State clearly four lines of evidence for your conclusions.

1.

3.

2.

4.

Slope Processes

Not only sands and gravels
Were once more on their travels,
But gulping muddy gallons
Great boulders off their balance
Bumped heads together dully
And started down the gully.

. . . A world torn loose went by me.

ROBERT FROST

Slope processes are processes that move rock and soil downhill. Often called *mass wasting*, the downslope movement of the weathered rock material may occur as individual particles or as enormous masses weighing many thousands of tons. The rate of downslope movement ranges from millimeters per year to hundreds of feet per second. Because any material on an incline is susceptible to downslope motion due to gravity, mass wasting is the most widespread surface process. It bridges the gap between weathering, which occurs in place, and stream erosion, in which material is carried away by running water. Mass wasting occurs above or below sea level and with or without the physical elements of water, wind, or ice.

Rock and soil materials deposited by slope processes are called *colluvium*. It may occur as a thin layer that covers the bedrock or as a thick wedge of sediment filling a hollow or at the base of a cliff. The size of individual fragments in colluvium ranges from mud to boulders. Layering within the unit, if present at all, is indistinct. Because colluvium is derived by weathering from bedrock, the composition of the fragments is usually similar to the bedrock. Although colluvium is widespread over the earth's surface, it is rarely preserved in the geologic rock record. Figure 10.1 shows landforms associated with slope processes.

DISTRIBUTION OF LANDSLIDES

Examine Figure 10.2, which shows the major landslide areas in the United States.

Problem 1

Recall the topography (Color Map 5), land surface forms (Color Map 6), and precipitation (Exercise 9) of the continental United States. What is the correlation between these factors and the prevalence of landslides?

Problem 2

Is there any correlation between landslides and the bedrock geology (Color Map 2)? What physical regions (Color Map 1) have the highest landslide potential?

Two types of forces act on material along a slope, resisting forces and downhill forces, examples of which are shown in Figure 10.3. Resisting forces hold material in place on the slope. One of these is friction, which is proportional to the roughness of the slope surface and to the component of gravity that is directed perpendicular to the slope. Cohesion of particles within the colluvium is another important resisting force. Other resisting forces are roots of plants and any manmade supports. The downhill force is gravity.

The slope, rock or soil type, geologic structure, and precipitation of an area are factors that influence the potential for mass wasting. Movement most often

147

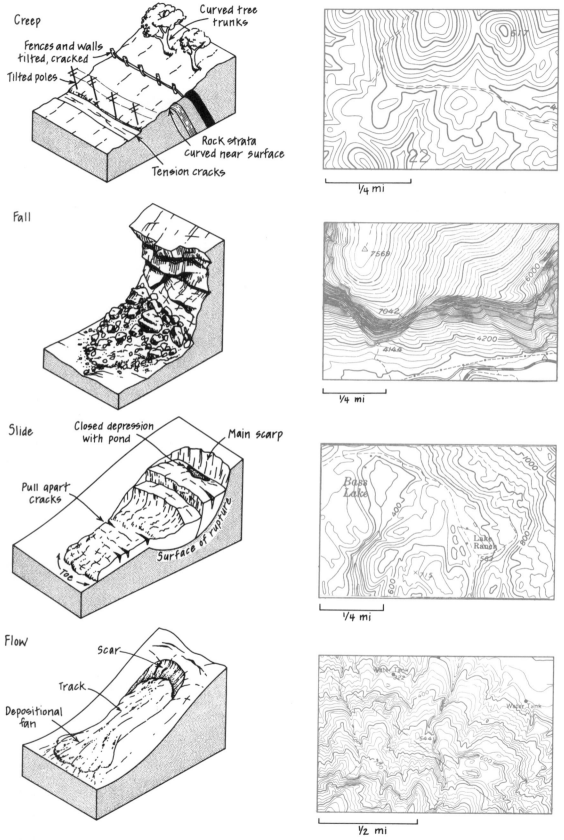

Figure 10.1
Landforms created by slope processes.

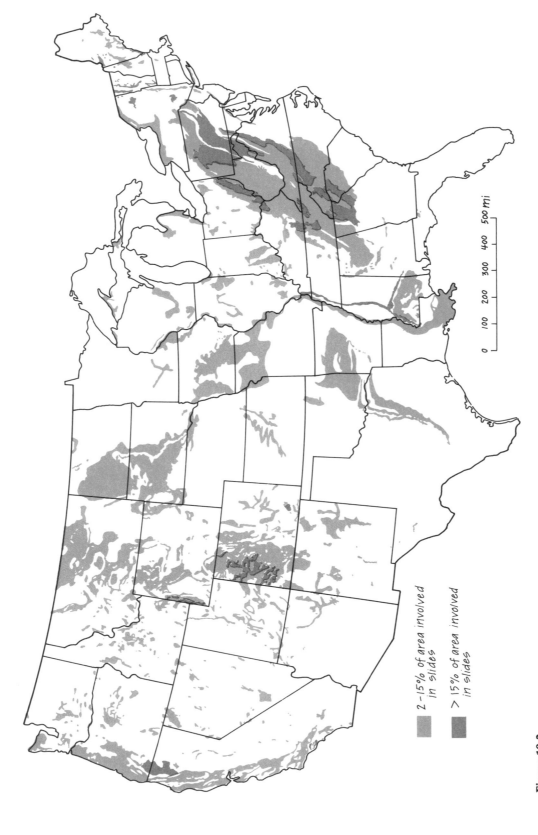

Figure 10.2
Distribution of landslides in the conterminous
United States.

2 – 15% of area involved
in slides

> 15% of area involved
in slides

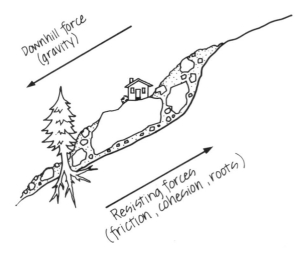

Figure 10.3
Resisting and downhill forces.

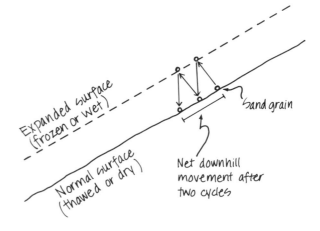

Figure 10.4
Shrink-swell creep caused by freeze–thaw or wet–
dry cycles.

occurs on steep slopes because of the large downhill component of gravity, but movement may also develop on gentle slopes where there are thick masses of weak or unconsolidated rock material, particularly during heavy rainstorms or earthquakes. Tilted rock layers are also prone to sliding. The inclined bedding planes tend to act as shear or sliding surfaces over which movement occurs. A very important factor in most landslides is water. Precipitation, especially long heavy downpours, promotes landslide movement in several ways. Water that saturates the ground greatly increases the weight of the loose material on a slope. In addition, water between fragments and in cracks acts to buoy up the rock mass and counteract the frictional force holding it to the slope.

SLOPE PROCESSES

There are two broad classes of slope processes, *creep* and *landslides.* Landslides may be further subdivided into *falls, slides,* and *flows.*

Creep

Creep, the imperceptible movement of rock material downslope by the motion of individual grains and fragments, occurs within a meter or two of the surface. Its motion is observable only over a period of years. Some signs that creep is occurring include tilted fence posts or utility poles, curved tree trunks, and pavement pulled apart by tension. Types of creep include rainsplash creep, shrink-swell creep, and biogenic transport. Figure 10.4 illustrates how shrink-

swell creep may be caused by freeze–thaw or wet–dry cycles. In many areas creep is the dominant form of mass wasting, but its effects are subtle and often overlooked.

To observe rainsplash creep in loose sediments, use the following materials provided by your instructor:

- Shallow dish or lid
- Salt or clean light-colored sand
- Eyedropper
- Bottle of colored water

1. Fill the dish with salt or sand. The surface should be level with the rim and smooth. (Variations in this procedure are encouraged after the initial experiment.)

2. Tilt the board at an angle of about 30°, and place the salt-filled dish in the center, making sure it does not spill.

3. Using the eyedropper, extract some colored water from the bottle. Hold the eyedropper 15 to 20 inches over the lid containing the salt, and carefully squeeze a drop from the dropper so that it falls on the surface.

Problem 3

Examine the impact site. Record and sketch your observations. Include a size scale.

Problem 4

Continue to drop water on the salt surface until most of the surface is moist. What other observations can you make?

Making a model of a natural system, as you have done with the drop experiment, is a common procedure in geology and in other sciences as well. An important step in utilizing a model is to relate this model to the natural system you wish to study and to take into account any differences between the simpler model and the (usually) much more complex natural system. The questions that follow are designed to help you do that.

Problem 5

What must be the texture of the material for raindrop impact to be an important surface process? (Compare bedrock with your sand model.)

Problem 6

How would vegetation affect creep induced by raindrop impact?

Problem 7

Describe a natural setting (include the nature of the surface, the slope angle, the type of vegetation, and whatever you think would be relevant) where raindrop erosion would be an important surface process.

Problem 8

Describe a natural setting where raindrop erosion would be minimal.

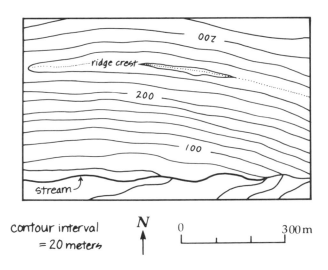

contour interval = 20 meters

N

0   300 m

Figure 10.5
Calculation of hillside creep due to earthworms.

Another approach to studying complex natural systems is to observe carefully a small portion of the system as quantitatively as possible and then make numerical calculations that can be extended to the whole system. An excellent example of this method was provided in 1881 by Charles Darwin in his last (and underestimated) major work, "The Formation of Vegetable Mould, Through the Action of Worms, with Observations on Their Habits," in which he studied, among other things, biological transport of material downhill by earthworms, one of the processes called biogenic creep.

Problem 9

Darwin estimated that earthworms displace 0.25 cm^3 of soil downslope per year for each centimeter of stream bank. Using his calculations as an estimate for the creep rate, determine the volume of soil that enters the creek at the base of the slope shown in Figure 10.5 in one year due to the action of earthworms. Show your calculations.

Figure 10.6
Stereopair of aerial photographs of mass movements in the Rio Grande Gorge near Taos, New Mexico.

N ⟶   A.

Landslides

The movement of rock material in a landslide occurs by falling, sliding, or flowing along a discrete shear surface.

Falls.   Falls are the free fall through the air of rock or debris from a cliff or overhang. Fragments collect at the base of the slope, forming an unconsolidated rock pile called *talus*.

Problem 10

What type of weathering, physical or chemical, leads to the type of clifflike topography in which falls are important?

Slides.   In contrast to the very slow movement of creep, *slides* are perceptible downslope movements of a unit of rock material that slides along a shear surface. Slides are especially common in steep mountain and plateau regions, but they can also occur on gentler slopes where there are thick masses of weak or unconsolidated material. Types of slides include *slumps*—movement of material along a spoon-shaped, concave-upward shear surface—and *glides*—movement of material along an inclined planar surface, such as a bedding plane. Slumps are illustrated in Figure 10.1. Notice that, in the case of slumps, the transported units are tilted backward, creating closed depressions in which water may accumulate to form lakes or ponds.

Figure 10.6 is a stereopair of aerial photographs and Figure 10.8, p. 156, is a topographic map of the Rio Grande Gorge south of Taos, New Mexico. Here the river canyon is cut into weak sediments lying beneath a cap of basaltic lava that forms the plateau in the left side of the photographs.

Figure 10.6
Stereopair of aerial photographs of mass movements in the Rio Grande Gorge near Taos, New Mexico.

B.

Problem 11

Trace in red pencil on Figures 10.6 and 10.8 the main scarp line of the mass movements shown. Trace in blue pencil some of the minor scarp lines. Place an arrow on the photographs showing the direction of movement. What kind of mass movement is this?

Problem 12

Why would the geology of this region, described above, lead to movements of this sort?

Problem 13

Why is there no discernible toe at the base of these movements? (See illustration of toe in the slide diagram in Figure 10.1)

Flows. Flows are movements of material of any size that flow like a viscous fluid along a discrete shear zone. Unlike the rock and colluvium in a slide, which moves more or less as a coherent unit, the material in flows is churned around in a turbulent fashion. Although dry flows of sand and silt are known, most flows contain water. Flows may move slowly or rapidly, depending mainly on the fluid content of the material.

In many cases, mass movement that begins as a rockfall or slide at higher elevations becomes quite fluid due to air and water entrapped in the mass and may travel great distances at lower elevations as a flow of debris. An *avalanche* (the most rapid of these compound movements) in May, 1970, that originated as a rock and ice fall from Huascarán peak high in the Peruvian Andes evolved into a devastating flow that traveled over 20 km at an average velocity of 300 km/hr, buried the town of Yungay, and killed 18,000 people.

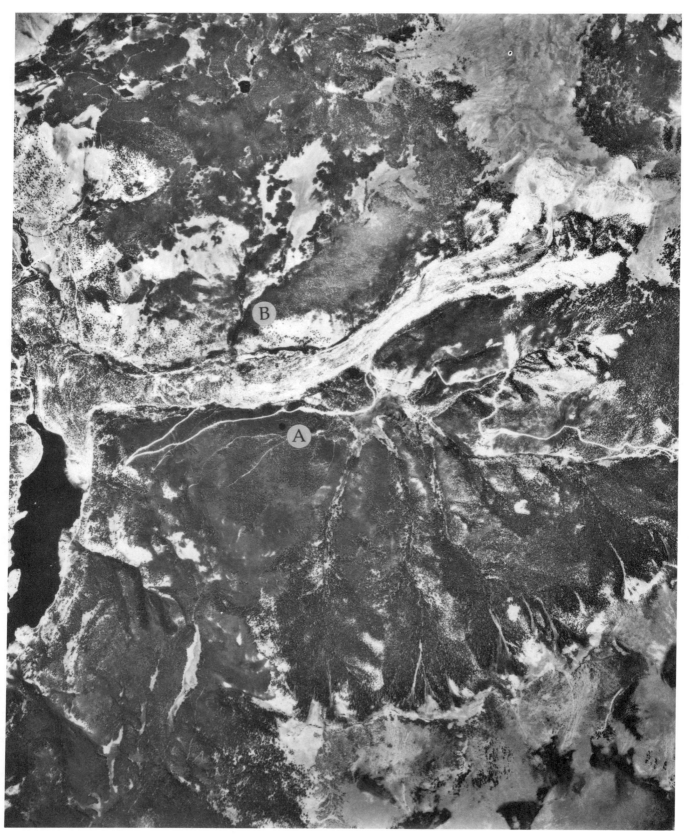

Figure 10.7
Photographs of part of the Lake San Cristobal quadrangle, Colorado.

0 ⊢———————————⊣ 1 mi

N

A.

Figure 10.7
Photographs of part of the Lake San Cristobal quadrangle, Colorado.

B.

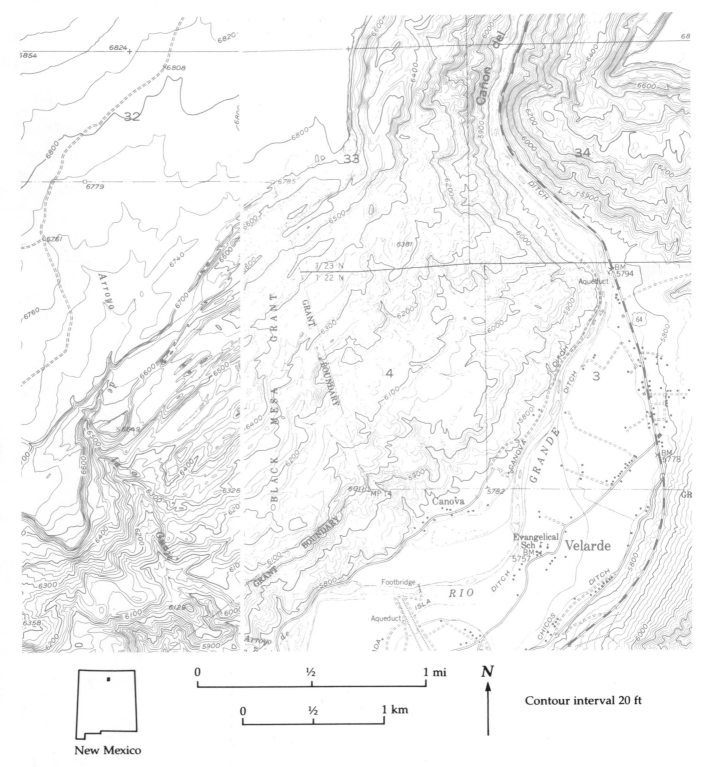

New Mexico

N

Contour interval 20 ft

Figure 10.8
Topographic map of part of Velarde quadrangle, New Mexico, showing area in
photographs in Figure 10.6.

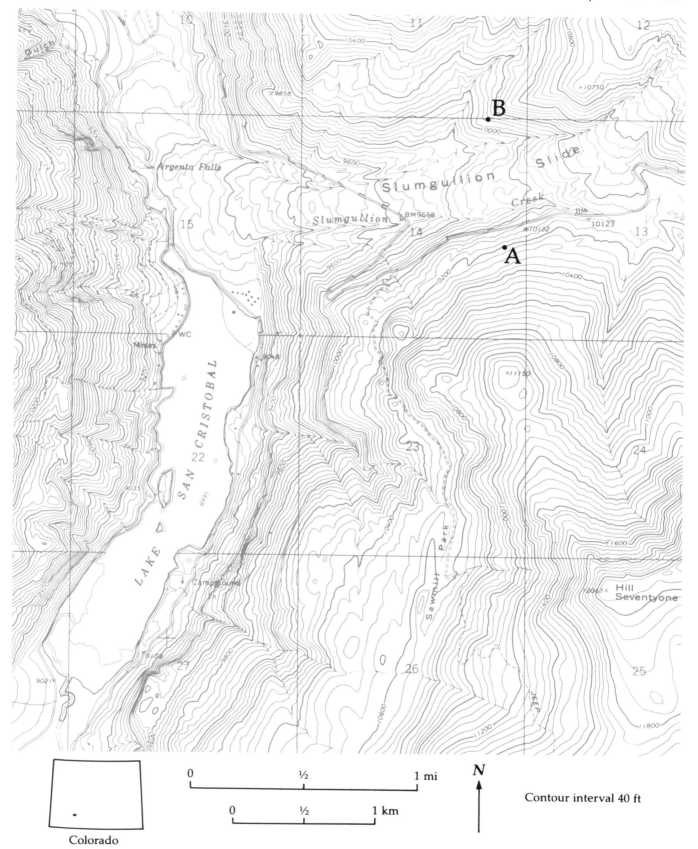

Figure 10.9
Topographic map of part of the Lake San Cristobal quadrangle, Colorado,
showing area in photographs in Figure 10.7.

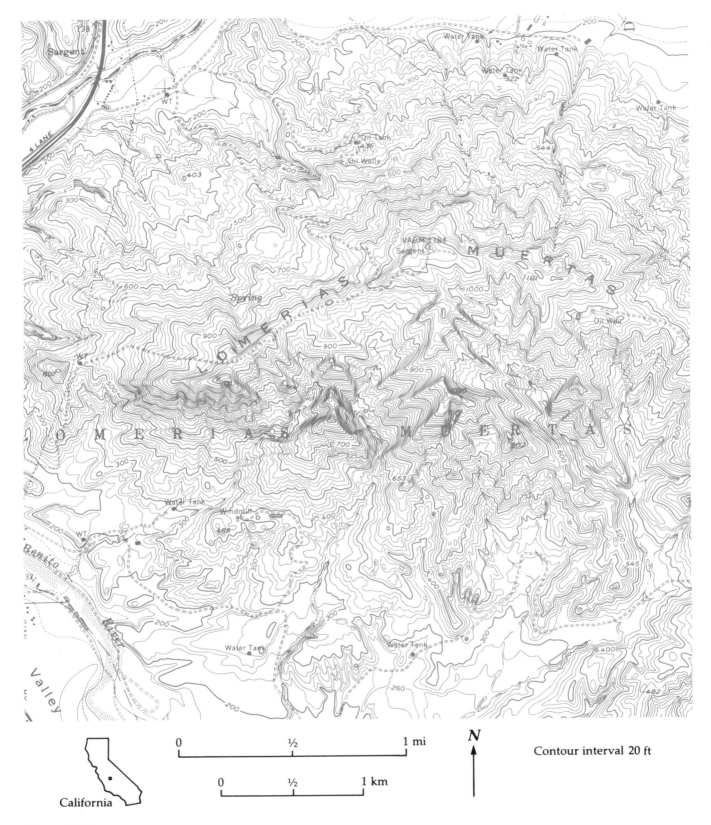

0   ½   1 mi

0   ½   1 km

N

California

Contour interval 20 ft

Figure 10.10
Topographic map of part of the Chittenden quadrangle, California, showing area in photographs in Figure 10.12.

Problem 14

Examine the photographs and topographic map of Slumgullion Slide in the Lake San Cristobal quadrangle, Colorado (Figures 10.7, pp. 154–155, and 10.9, p. 157). This "slide" is actually a compound motion of slide at upper elevations and flow at lower elevations. How do the topographic contours over the slide area differ from contours typical of stream valleys?

Problem 15

From the topographic contour line pattern, outline the slide area with a red pencil. Dash the contact line where you are unsure of the boundary. What is the general shape of the slide?

Problem 16

Draw a profile across Slumgullion Slide from A to B. Use the grid provided below. Describe the shape of the profile.

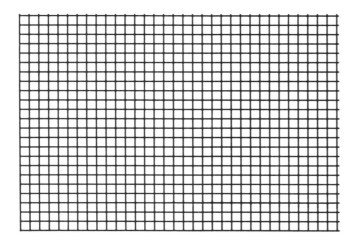

Problem 17

What is the origin of Lake San Cristobal?

Problem 18

Reexamining the entire map and the aerial photographs, do you see any other areas where landslides have occurred? Outline these areas in purple.

LANDFORMS ASSOCIATED WITH SLOPE PROCESSES

The balance between erosion by running water and the mass wasting processes of creep and landslides largely determines the shapes of hillsides. Steep concave-up hillsides occur where flow-type landslides dominate the slope processes. Slump-dominated topography is jumbled and chaotic, with many isolated knobs and closed depressions. Rockfall terrain is characterized by steep cliffs with straight slopes at their bases (talus). Creep forms distinctive hillslopes with rounded ridge tops. The lower part of creep-dominated hillslopes is often concave-up due to erosion by running water. Review the maps in Figure 10.1.

Problem 19

For each of the topographic maps of Figure 10.11, state whether the dominant slope process is creep, flows, slumps, or falls.

A.

B.

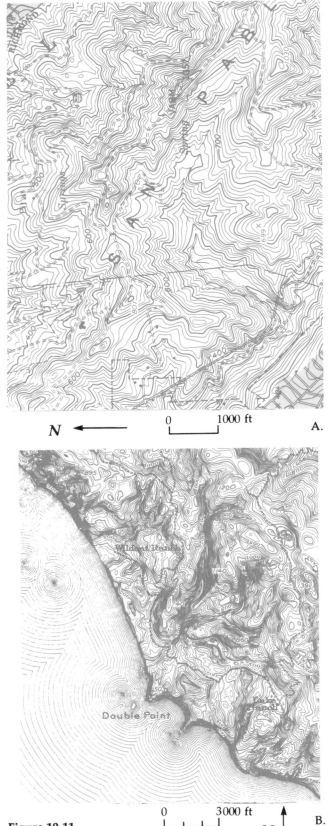

N ← | 0   1000 ft | A.

Figure 10.11
Topographic maps of slope
process landforms.

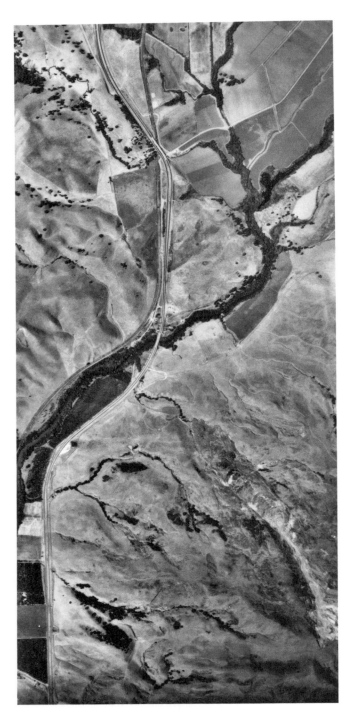

Figure 10.12   A.
Stereo photos of Chittenden quadrangle, California.

FIELD STUDY: LOMERIAS MUERTAS, CHITTENDEN QUADRANGLE, CALIFORNIA

Lomerias Muertas, south of San Francisco, is an area of extensive mass movements. In this exercise you will map the extent and relative ages of these movements onto a topographic base map, Figure 10.10, p. 158, using the aerial photographs in Figure 10.12.

Your first step in transferring data from aerial photographs to a topographic map is to orient yourself on the photographs. Locate a prominent feature on the photographs, and find the same feature on the topographic map. Label the photographs with a north arrow, and check with your instructor before proceeding.

Problem 20

On your topographic map sketch the outlines of all mass movements you can recognize in the area for which the photographs give you stereoscopic relief.

Problem 21

Some of the movements are recently active; color these red. Others are older and have been eroded by streams; color them blue. Label the most recent earth movement you can see as 1.

Problem 22

The style of movement and its extent are different on the north and south sides of the ridge marked by the words *Lomerias Muertas*. To isolate the various factors contributing to these mass movements, complete the following chart:

North Slope   South Slope

1. The slope with the most extensive area of movement:

2. Factors that may affect extent of movement:

   a. *Dip of rock strata:* Which slope is parallel to the dip of the strata?

   b. *Steepness of slope:* Which slope is steeper?

   c. *Moisture content of slope:* Which slope would contain more moisture?

Problem 23

Which of the three factors seems to be most important in causing mass movement? List the three factors in their order of importance.

1.

2.

3.

Problem 24

Is there a difference in the style of movement (slump, slide, flow, creep) from the north to the south slope?

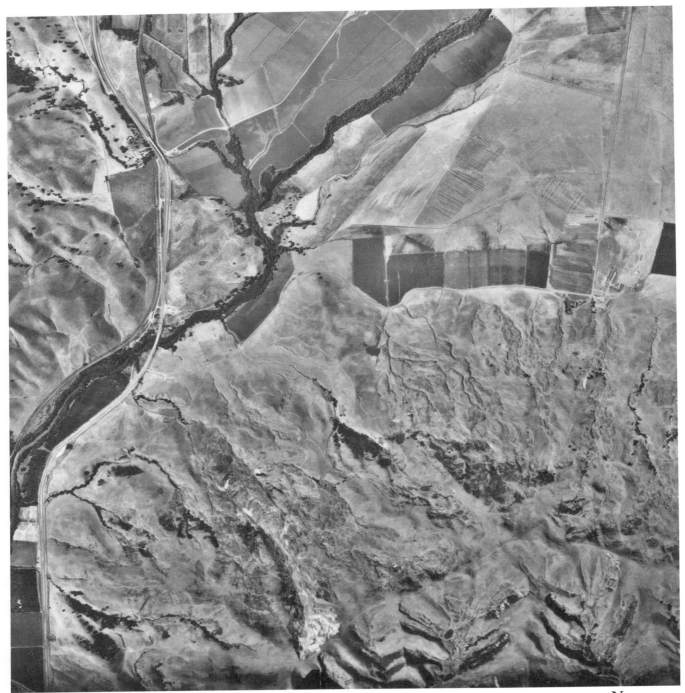

Figure 10.12
Stereo photos of Chittenden quadrangle, California.

0   1 mi

N

B.

Figure 10.12
Stereo photos of Chittenden quadrangle, California.

C.

Exercise 11

Groundwater

Water that fills the open spaces in rocks underground is called *groundwater*. Water from rain or melting ice may run off the surface and join the system of rivers and streams, or it may sink directly into the ground to become part of the groundwater system. Once beneath the surface, water percolates downward until it reaches the *water table*, the boundary between unsaturated and saturated rock material. Below the water table, all the spaces in cracks and between fragments of rock are completely filled with water; this is the zone of saturation (the *phreatic zone*). Above the water table is the zone of aeration (the *vadose zone*). Figure 11.1 shows a profile of the groundwater system.

Where precipitation is plentiful, the water table is usually near the surface, but in arid areas the water table is as much as several hundred feet down. In general the water table reflects the shape of the overlying topography (Figure 11.2); it is higher under hills and lower beneath valleys. The water table is a smoother surface, however, lacking the minor irregularities of the land.

DISTRIBUTION OF GROUNDWATER

Reservoirs of groundwater, called *aquifers*, are present in porous rocks and in beds of loose sand and gravel. Figure 11.3 shows the locations of major aquifers in the United States.

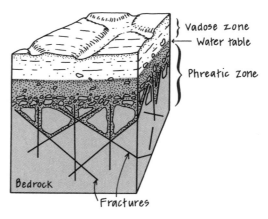

Figure 11.1
Vertical profile of the groundwater system.

Problem 1

Refer to the geologic, structural, and topographic color maps of the United States. What correlation exists between large continuous aquifers and rock type (igneous, sedimentary, or metamorphic)? Topography? Folded and faulted crust?

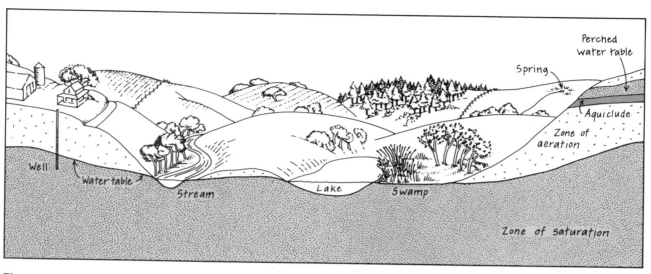

Figure 11.2
The water table with respect to ground surface and some groundwater features.

Problem 2
What factors might explain the lack of large aquifers in the western United States?

POROSITY AND PERMEABILITY

The quantity of water contained in a given volume of material depends on the amount of void (open) space in the rock or sediment. *Porosity* is defined as the volume of void space relative to the volume of the solid. Unconsolidated sediment and highly fractured rocks have high porosities, usually greater than 25 percent. In contrast, unfractured crystalline rocks have low porosities, generally less than 5 percent.

Permeable rocks are those through which fluids can move. Movement of fluid through a solid occurs when the void spaces are interconnected and are sufficiently large that the effect of the surface tension of water, which tends to hold water to the surfaces of the rock, is negligible. Rocks having high porosities are not necessarily permeable. Mudstones can have porosities of 60 percent, yet have negligible permeability because the void spaces are too small to permit flow. Even in

permeable rock, groundwater movement is relatively slow, generally less than 6 inches per day.

In some areas an impermeable rock layer hinders the downward percolation of groundwater, creating a secondary water table above the primary water table. As illustrated in Figure 11.2, water accumulates above the impermeable layer, forming a *perched water table*. The impermeable layer is called an *aquiclude* and is commonly a mudstone or shale.

The porosity and permeability of rocks and sediments are important factors in the study of groundwater (and in the recovery of oil from its underground reservoirs). Determination of rock and sediment porosity and permeability is done routinely in the laboratory.

Problem 3
Before beginning the determinations of porosity and permeability, examine your sample with a hand lens, and provide a complete description of it. Note the grain size, sorting, rounding, clay content, closeness of packing, and whatever other features you think are important.

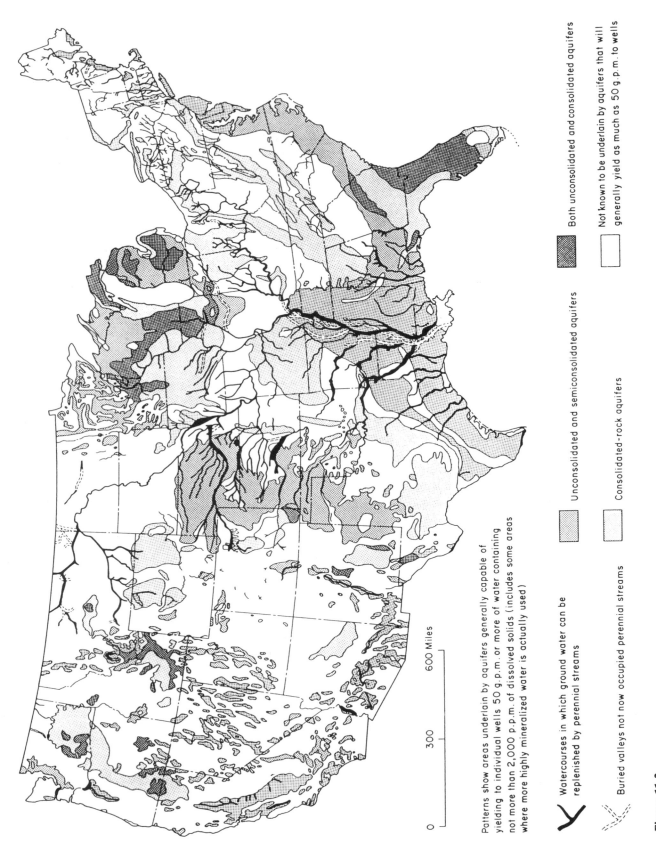

Patterns show areas underlain by aquifers generally capable of yielding to individual wells 50 g.p.m. or more of water containing not more than 2,000 p.p.m. of dissolved solids (includes some areas where more highly mineralized water is actually used)

Watercourses in which ground water can be replenished by perennial streams

Buried valleys not now occupied perennial streams

Unconsolidated and semiconsolidated aquifers

Consolidated-rock aquifers

Both unconsolidated and consolidated aquifers

Not known to be underlain by aquifers that will generally yield as much as 50 g.p.m. to wells

0   300   600 Miles

Figure 11.3
Distribution of major aquifers in the United States.

Problem 4

Measure the porosity of the solid furnished by your instructor. Follow the procedures described below, and record your descriptions and measurements in the table. Work in teams.

1. Fill a beaker about half full of water, and mark this level.

2. Pour this water into a graduated cylinder, and record its exact volume in row A.

3. Now fill the beaker up to the level of the water you marked in step 1 with the solid to be measured. Record this volume in row B.

4. Slowly pour the water from the graduated cylinder back into the beaker until the marked line is reached. Record the amount of water remaining in the cylinder in row C.

5. What is the volume of water in the beaker of solid and water? Record this volume in row D.

6. Porosity equals the volume of liquid in the container divided by the volume of solid, expressed in percent. Record the porosity in row E.

Porosity

A. Original
volume

B. Solid
volume

C. Remaining
liquid

D. Liquid
volume

E. Porosity

Problem 5

The permeability of solids is usually measured by their *hydraulic conductivity*. The hydraulic conductivity together with the hydraulic gradient (the driving force) determines the rate of flow of water through solids. The relationship is described mathematically by *Darcy's Law:*

$$\frac{g}{A} = \frac{Kdh}{dL}$$

where

g = flow, in cm^3/s
A = area of cross section, in cm^2
K = hydraulic conductivity, in cm/s
dL = length of flow, in cm
dh = hydraulic gradient, in cm

You will be able to measure g, A, dL, and dh and calculate K, the hydraulic conductivity, for the solid you used in the porosity determination. Work in teams. Figure 11.4 shows the experimental setup.

- dh, the hydraulic head, is the height of the water above the bottom of the cylinder.

- dL, the length of flow, is the thickness of the layer of solid.

- A, the area of cross section, is measured across the inside of the cylinder that contains the sample.

- g, the flow, is measured experimentally:

1. Set up your equipment according to Figure 11.4. If you are using sand, you need a 50-mL graduated cylinder; finer soils require a 10-mL graduated cylinder.

2. Place filter paper on the bottom of the sample cylinder, and add your sample up to the lower mark. Level the top of the sample.

3. Place the beaker under the sample cylinder to catch the drips, and add water very carefully up to the upper mark. Use a pipette to avoid stirring up the sample.

4. When the water level reaches the upper mark, replace the beaker with the funnel and graduated cylinder, and start timing. Take readings at regular time intervals. Record the elapsed time in row A of the chart on page 168 and record the volume of water in the graduated cylinder in row B.

5. Present the results in the form of a chart similar to that on page 167.

6. Determine g, and record its value in row C. Calculate K, and record it in row D.

When you have measured the porosity and permeability of your sample, provide the results plus the sample description to your instructor.

Sample description:

(Continued)

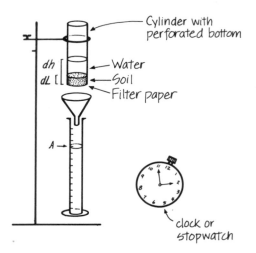

Figure 11.4
Experimental set-up for
permeability experiment.

Permeability

A. Elapsed
time

B. Volume of
water

C. g

D. K

Problem 6

Review the results obtained by the class. What factors
(grain size, shape, rounding, etc.) contribute to high
porosity? High permeability?

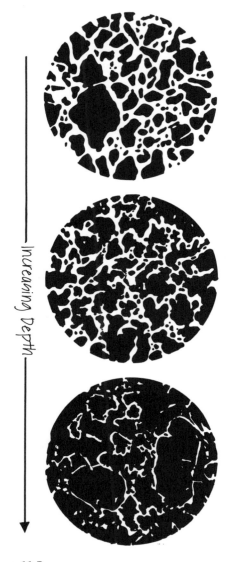

Figure 11.5
Tracings of grains in rock samples from different
depths in the Ventura Basin, California.

Figure 11.5 shows tracings from photomicrographs
of sandstones drilled at different depths in the Ventura Basin, California.

Problem 7

How does the porosity and permeability of the sandstones change with depth?

Problem 8

What causes this change?

Problem 9

What do these observations suggest about the amount of groundwater stored at depth in the earth's crust?

GROUNDWATER PROCESSES

Infiltration is the primary process by which water percolates underground to accumulate as groundwater. One may observe infiltration during a rainstorm by comparing a grass-covered surface with a paved one. The pavement always has rainwater flowing over its surface (*runoff*), but the natural surface rarely does. The difference in these two conditions is that water infiltrates the grass-covered soil but does not penetrate the pavement.

Problem 10

How might porosity and permeability affect infiltration rates?

Problem 11

Given two hillslopes—one underlaid by material of high hydraulic conductivity and the other by material of low hydraulic conductivity—which one will have more gullies eroded into it? Why?

Problem 12

Return to the determinations of permeability (Problem 5). The rainfall intensity of a typical thunderstorm is about 2.5 cm/hr. Using permeability values determined by the class, state which of your samples will have runoff.

Problem 13

Perhaps you have noticed while driving along paved roads in the country that the most lush growth of wildflowers and other vegetation is in a strip within 20 feet of the paved area. This pattern is especially evident in semi-arid to arid regions. Explain this observation in terms of runoff and infiltration.

The infiltration rate is a measure of how quickly water seeps into the ground. Normal infiltration rates are about 1 inch per hour, with variations depending on the surface material, vegetation, topography, and precipitation. In general, coarse-grained soils or sediments have faster infiltration rates than fine-grained material. Vegetation, by impeding runoff and increasing permeability, can double or triple the infiltration rate of a bare soil. The slope of the land also influences the infiltration rate; steep slopes favor runoff, and gentle slopes favor infiltration. Finally, infiltration decreases through time during a rainstorm as the cracks and open spaces in the ground become filled with water.

Groundwater moves through permeable rock toward areas of lower elevation. It moves around and between impermeable layers, usually traveling less than a few inches in a day. It may travel for several miles before emerging at the surface as a spring or seeping unseen into a stream, swamp, or lake. These groundwater features are illustrated in Figure 11.2. Oscillating lake levels and varying rates of flow from springs and rivers are fluctuations that arise from changes in the elevation of the water table. Lakes form where the land surface dips below the water table. Where the water table and the land surface are at approximately the same elevation, swamps and marshes form. Springs result from the discharge of groundwater at the surface.

Figure 11.6 shows infiltration rate measurements on three different surfaces. The experiment was controlled using artificial rainfall sprinkled over a clay-rich soil in the New Mexico desert. Three plots of land were tested—barren soil, soil with 18 percent vegetation cover, and soil with 37 percent vegetation cover.

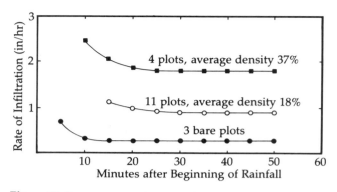

Figure 11.6
Infiltration rate changes under
different vegetation cover.

Problem 14

How does the infiltration rate change with time during the experiment? What factor might explain this pattern?

Problem 15

How does vegetation affect the infiltration rate? Why?

Hot Springs National Park, Arkansas, is an area of thermal springs. Water emerges from 41 springs at an average temperature of 145°F (63°C).

Problem 16

Recall that the geothermal gradient is the rise in temperature with depth in the crust. Assuming the average value for the geothermal gradient, 25°C/km, how far below the earth's surface would water have to sink before reaching a temperature of 63°C? Show your calculations.

Problem 17

If the geothermal gradient were higher, would the depth required for groundwater to reach 63°C be deeper or shallower?

Problem 18

If the groundwater traveled 5 feet per day, a very high rate, how long would it take for water to rise from the depth suggested by the average geothermal gradient?

Problem 19

Realizing that the groundwater cools as it nears the colder rocks at the surface, is the geothermal gradient in Hot Springs National Park higher or lower than average?

LANDFORMS ASSOCIATED WITH THE ACTION OF GROUNDWATER

Water moving through the ground dissolves minerals from the surrounding rock and soil. Given sufficient time and water migration, large openings may be eroded in the bedrock. Because calcite, gypsum, and halite are especially susceptible to dissolution, groundwater erosional landforms typically occur in limestone and evaporite bedrock.

In humid areas where thick limestone deposits are interbedded with other permeable layers such as sandstone, or where massive limestone is of variable permeability or is cut by joints, a peculiar and distinctive topography called *karst* may develop. Large underground *caverns* may form near the water table. Overlying layers can collapse into a cavern to form *sinkholes*, perhaps the most distinctive feature of karst landscape. Depending on the elevation of the water table, some sinkholes are dry, and others are filled with water. Another common characteristic of karst topography is the *lack of surface drainage*. Surface water tends to drain vertically through solution cracks in the limestone instead of flowing horizontally over the surface. Normal stream drainage may persist over sandstone or shale, only to flow abruptly underground where limestone is exposed at the surface, forming a *disappearing stream*. Conversely, under some condi-

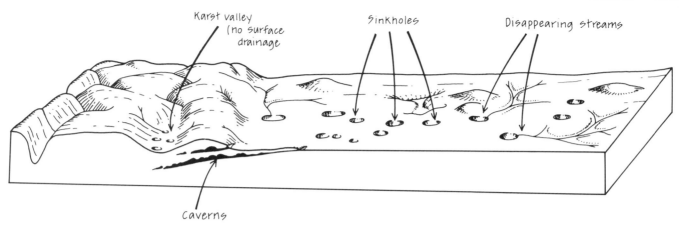

Figure 11.7
Karst landforms.

tions water may appear at the surface as a *spring* or *seep*. Figure 11.7 illustrates some common groundwater or karst landforms.

Once a solid such as limestone is dissolved, ions from it migrate with the groundwater to streams or adjacent rock layers. The chemical composition of streams that originate from subsurface water reflects the composition of the groundwater. Calcium, magnesium, and carbonate ions characterize water from limestone bedrock; sodium, potassium, and sulfate ions are common in waters from evaporite bedrock. If the mineral-laden groundwater seeps into open cavities formed previously, precipitation may occur, forming stalactites, stalagmites, and the other interesting and beautiful deposits of caves.

Examine the distribution of karst terrain across the continental United States, shown in Figure 11.8.

Problem 20

Compare the distribution of karst terrain with that of major aquifers.

Problem 21

What factors might explain the lack of karst terrain in the western United States?

Examine Figure 11.9, the topographic map of a section of St. Louis, Missouri (the Cave State).

Problem 22

What groundwater features are present in this area? Give an example of each.

Problem 23

What dissolved chemical elements would be present in the springwater in St. Louis?

Problem 24

Mark the direction of groundwater flow on the map. How do you know the direction?

Examine the stereoscopic photos (Figure 11.10) and the topographic map (Figure 11.12) of the area near Interlachen, Florida, an area of many small sinkholes.

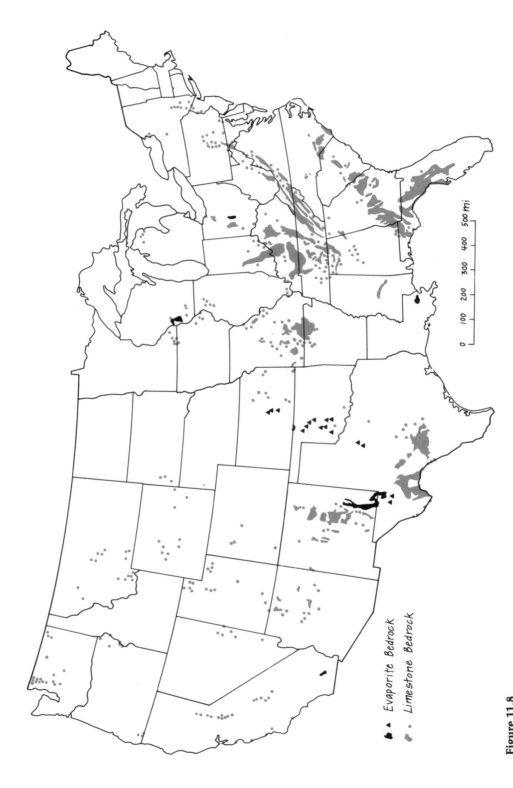

Figure 11.8
Distribution of karst terrain in the United States.

● ▲ Evaporite Bedrock

● Limestone Bedrock

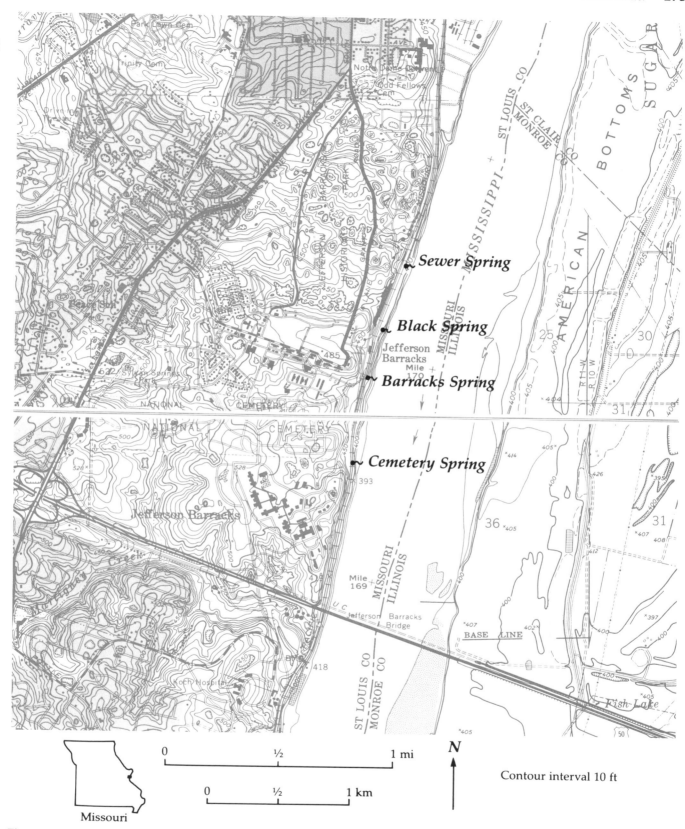

Figure 11.9
Topographic map of a section of St. Louis, Missouri.

A.

Figure 11.10
Stereopair of aerial photographs of Interlachen, Florida.
(Rotate these photos 90° clockwise for stereo viewing.)

Figure 11.10
Stereopair of aerial photographs of Interlachen, Florida.

0   1 mi

N

B.

Problem 25

Using the elevations of the small lakes as data points, contour the water table. Use a 5-foot contour interval.

Problem 26

How is the water table related to the topography?

Problem 27

If toxic waste were dumped into the lake near Oak Grove school and you were instructed to warn residents of possible well-water contamination, where would you concentrate your efforts?

Problem 28

There are no major rivers in Florida, despite its high average rainfall. What factor might account for this fact?

Figure 11.11   A.
Stereo photographs of Mammoth Cave area.

FIELD STUDY: MAMMOTH CAVE NATIONAL PARK, KENTUCKY

Mammoth Cave and the several surrounding cave systems consist of a vast network of underground passages beneath Mammoth and Flint Ridges. In 1972 explorers found the long-sought link between the Flint Ridge Cave Systems and Mammoth Cave, making the Flint–Mammoth cave system the longest in the world. At present there are over 250 miles of mapped passages. This underground marvel lies concealed beneath modest surroundings—a low plateau carved into hills and valleys by the Green River of central Kentucky.

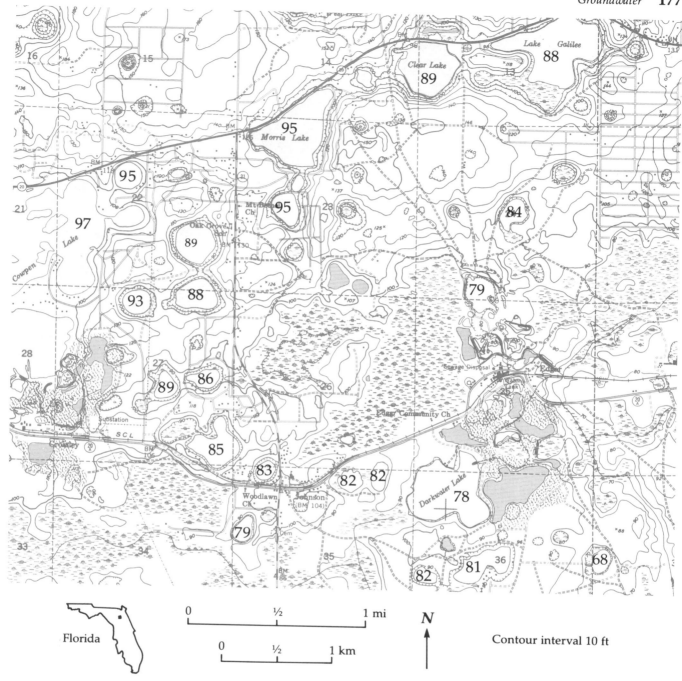

Figure 11.12
Topographic map of area surrounding Interlachen, Florida, showing the area in the photographs in Figure 11.10.

The only clues that something remarkable lies underground are subtle features of landscape that sometimes develop over thick layers of limestone weathered in a humid climate. Use the topographic map (Figure 11.14, p. 181) and the stereoscopic photographs of the Mammoth Cave quadrangle (Figure 11.11, pp. 176, 178, 179) to complete the following problems.

Problem 29

In what physical region is Mammoth Cave? What is the major rock type (igneous, sedimentary, metamorphic) of this region? What is the dominant land surface form? (See Color Maps 1, 2 and 6.)

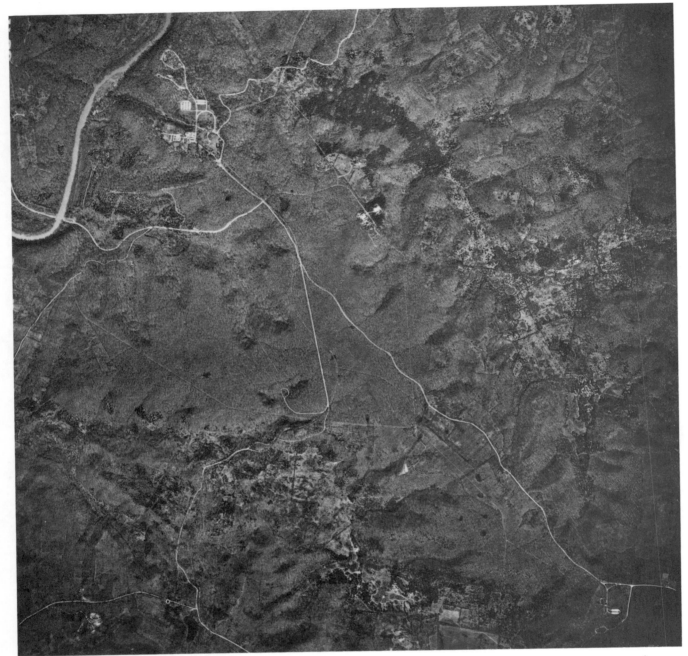

Figure 11.11
Stereo photographs of Mammoth Cave area.

0 ⊢——————————————————————————⊣ 1 mi   *N*↑   B.

Problem 30

List the groundwater features you see in this area, and cite an example of each.

Problem 31

Eaton Valley appears to be a normal valley eroded by surface streams, but one vital element is missing. What is it?

Figure 11.11
Stereo photographs of Mammoth Cave area.

C.

Figure 11.13 is a map of the passages of the Mammoth Cave system under Mammoth Ridge. Very few of the places where passages cross one another on the map represent connections. In most cases the passages are on different levels and do not connect.

Problem 33

The major cave passages (heavy dark lines) do not appear to be randomly oriented, but display a distinct directional trend. What is that trend?

Problem 34

There is a secondary trend shown best in the direction of shorter cave passages, especially those in the southeast half of the system. What is this secondary direction?

Problem 35

The limestone layers in which the caves develop generally strike northeast–southwest. What is the correlation between strike and dip and the orientation of cave passages?

The underground passages of the cave system are not at random depths but develop only at certain elevations. Although the directions of cave passages may be controlled by regional strikes and dips, the elevations of cave passages do not appear to be governed by structural and lithologic characteristics of the limestone. Details of the surface topography, along with the elevations of cave entrances, offer some clues.

Problem 32

This area of Kentucky averages about 35 inches of precipitation per year. When it rains, where does the water that falls in Eaton Valley go?

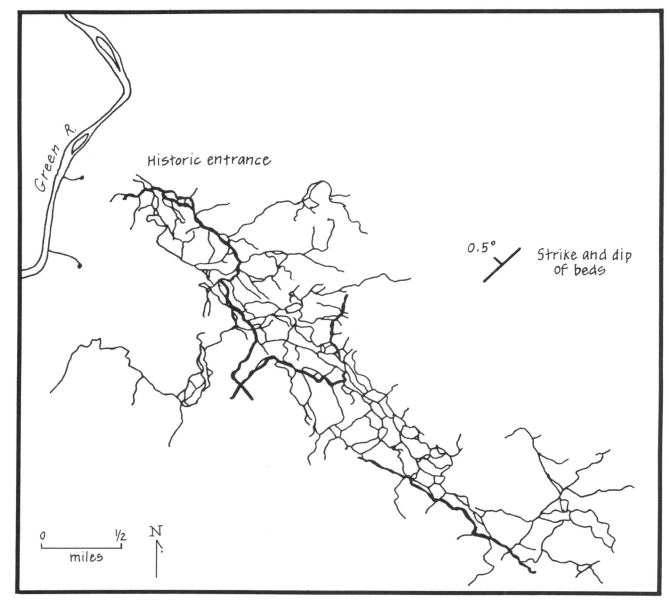

Figure 11.13
Map of cave passages under Mammoth Ridge.

Problem 36

The elevations of cave passages are at or just below the elevations of cave entrances. Cave entrances are indicated on Figure 11.14 by a Y symbol. The actual entrance is at the fork of the Y. Determine the elevations of the cave entrances listed below.

- Echo River
- "Cave" southwest of Colossal Cave
- Carmichael entrance

- Colossal Cave

- Cathedral Domes
- Styx River
- Whites Caves
- Dixon Cave
- Historic entrance to Mammoth Cave
- New Discovery entrance
- Violet City
- Dossey Cave (west bank of Green River)

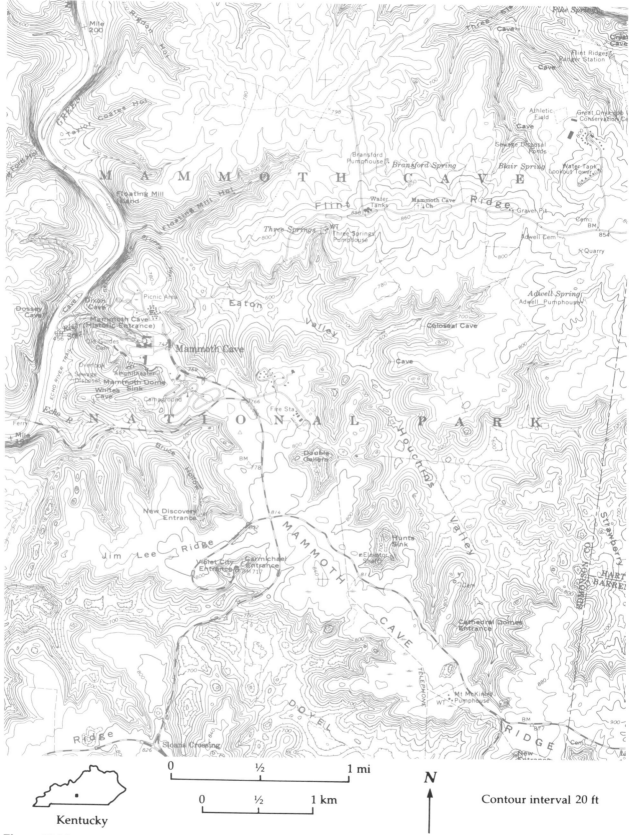

Figure 11.14
Topographic map of part of the Mammoth Cave quadrangle, Kentucky.

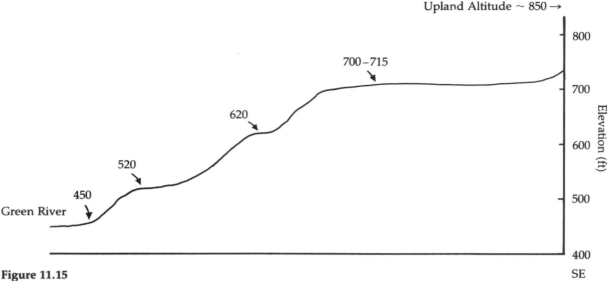

Figure 11.15
Cross-sectional sketch of river terraces, east side of Green River Valley, Kentucky.

Problem 37

Notice that the elevations of cave entrances (and thus the approximate elevations of cave passages) are not random but cluster at certain elevations. Around what four elevations do these cave entrances cluster?

Figure 11.15 shows in fine detail the topography of the valley side east of the Green River. Flat areas in the profile are river terraces formed in the past when the Green River was at higher elevations than at present. These terraces form when the river remains at the same elevation for long periods of time.

Problem 38

On Figure 11.15 draw horizontal lines at the four principal elevations of cave entrances. What is the correlation between the elevations of cave passages and elevations of old river terraces?

Problem 39

Since the terraces represent old river elevations, they also represent (approximately) old levels of the water table. What conclusions can you draw about where, with respect to the water table, most of the dissolution of limestone takes place to form the cave systems of the Mammoth area?

Problem 40

Summarize your knowledge of cave formation in the Mammoth Cave area by describing where dissolution takes place to form caverns and what factors govern the locations of cave passages in three dimensions.

Rivers and Streams

Nothing under heaven is softer or more yielding than water; but when it attacks things hard and resistant there is not one of them that can prevail.

LAO TZU

When water that has evaporated from the oceans by the sun's energy falls over the land as rain or snow, most of it flows over the surface of the land down to the sea under the influence of gravity. Flowing water is the dominant agent of *erosion,* the wearing down of the landscape.

We might tend to think of rivers as primarily erosional agents that wear down the landforms constructed by the earth's internal energy, but in doing so they become agents of transport and deposition as well. Especially in their lower reaches close to sea level, rivers can deposit vast thicknesses of sediment in the valleys through which they flow. In general (though there are exceptions), rivers erode the landscape at higher elevations near their sources and deposit material in the sea or in low-lying valleys near the sea. This exercise concerns the erosional and depositional characteristics of rivers and the landforms they produce as well as the ways in which rivers transport material. The adjective *fluvial* (from the Latin *fluvius,* river) refers to rivers and streams.

During a rainstorm, water seeps into the ground to join the system of groundwater, but as soon as the surface layer becomes saturated, the *overland flow* of water begins. Overland flow begins as a layer of water a few millimeters thick, called *sheetflow,* with only limited ability to erode and transport material. Surface irregularities and internal turbulence in overland flow, however, soon generate ripples and other concentrations of energy, which in turn result in dominant threads of current with erosive capability. *Rills* are small channels eroded by these currents; they may or may not persist from one storm to the next. More permanent and somewhat larger features are *gullies,* the first stage in the fluvial erosion of the landscape. Figure 12.1 depicts major landforms associated with fluvial processes.

DISTRIBUTION AND DIMENSIONS

Examine Figure 12.2, which shows the major rivers of the world.

Problem 1

Do all the continents have major rivers draining them?

Problem 2

A *divide* is the high land, usually a ridge, that separates one river system from another. The continental divide of North America separates streams and rivers that flow into the Atlantic Ocean from those that flow into the Pacific. Estimate the location of the continental divide in North America. Draw your estimate on the map in pencil. In what physical region does the continental divide occur? (See Color Map 1.) What is the predominant land surface form? (See Color Map 6.)

Problem 3

Does South America have an obvious continental divide that separates major river systems? Why or why not?

Problem 4

Explain the absence of major river systems over large areas of Australia, Africa, and Asia.

DEPOSITIONAL LANDFORMS

LANDFORM	CHARACTERISTICS	ORIGIN	TOPOGRAPHIC MAP
Point bar 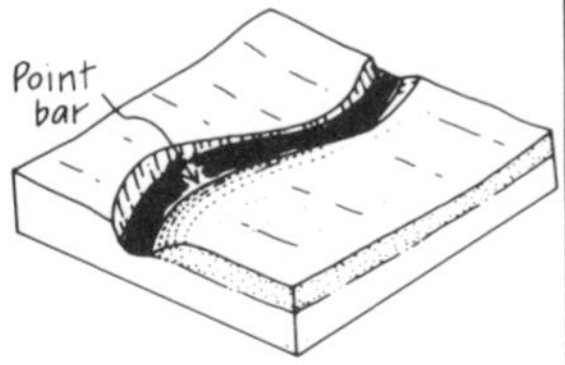	• The alluvial deposit on the inside of a meander bend • Crescent in shape • Composed of unconsolidated gravel, sand, and/or silt	Deposition by upward circulating currents moving around a meander bend	
Braid bar 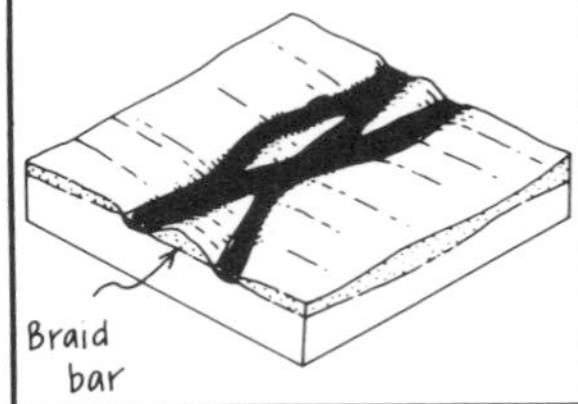	• An alluvial deposit within the stream channel • Generally elongated in shape • Composed of unconsolidated gravel, sand, and/or silt	A deposit by flood waters with subsequent enlargement	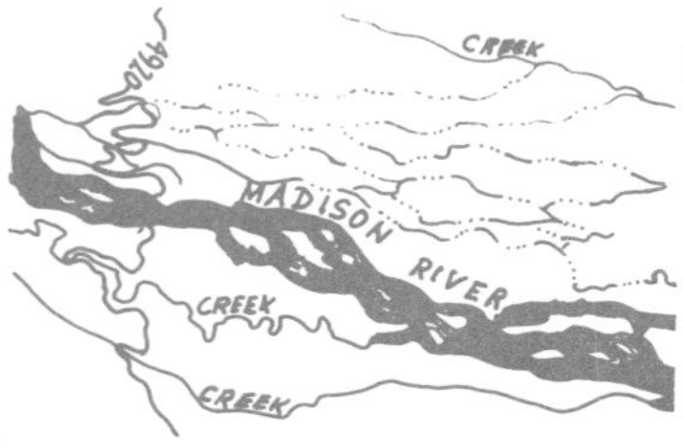
Levee 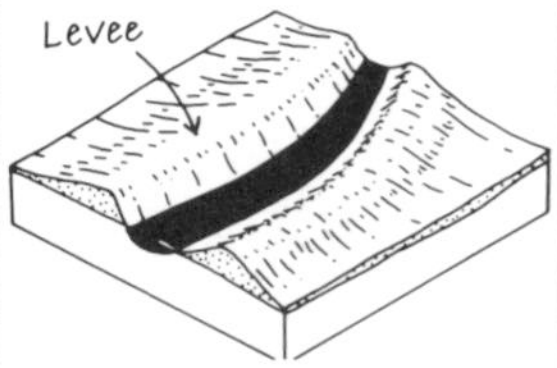	• An alluvial deposit on the perimeter of the channel • An elongated ridge parallel to the channel • Composed of sand and/or silt	A deposit by flood waters due to a sudden loss of water velocity along the channel borders	
Floodplain 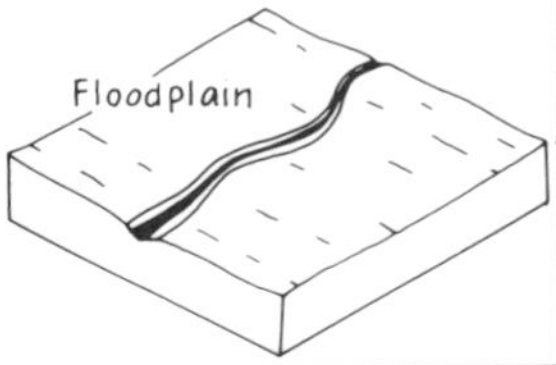	• The area adjacent to the channel covered in times of flood • Variable in size, topographically flat • Composed of sand and mud	Deposition by flood waters outside the channel	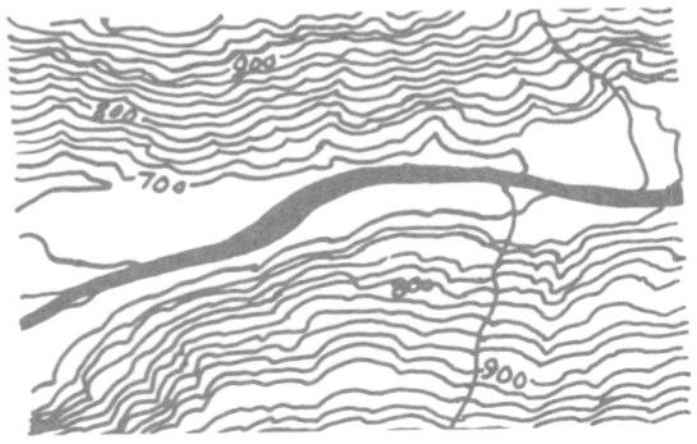
Alluvial fan 	• A large alluvial deposit at the base of mountains or hills • Fan-shaped • Composed of gravel, sand, and mud	Deposition by a river at a major break in slope where river flows from confined valley onto open plain	
Alluvial plain 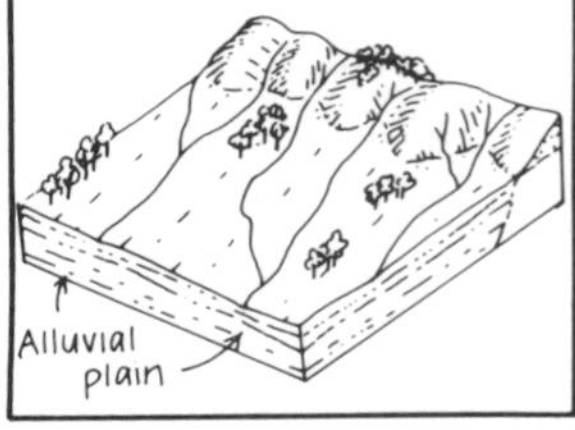	• A large alluvial deposit covering hundreds to millions of square miles • Topographically a smooth, flat to gently sloping surface • Composed of sand, silt, and clay	Deposition by flood waters from many streams	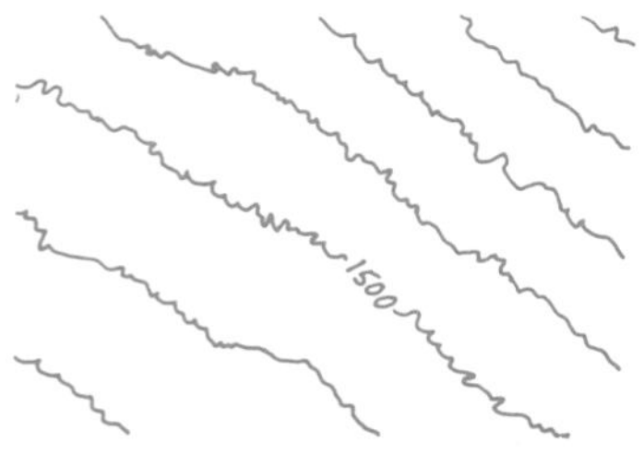

Figure 12.1
Landforms associated with fluvial processes.

EROSIONAL LANDFORMS

LANDFORM	CHARACTERISTICS	ORIGIN	TOPOGRAPHIC MAP
Cutbank 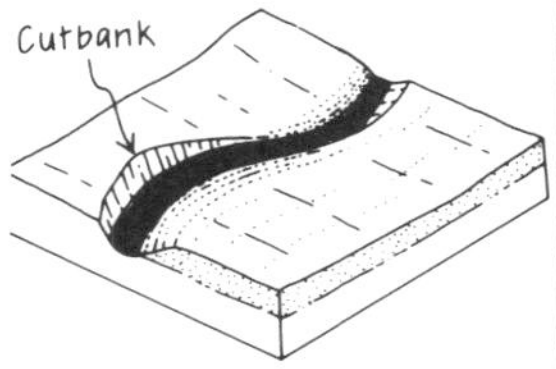	• The channel boundary along the outer edge of a meander band • Semicircular in shape, steep in slope • Composed of bedrock or alluvium	Erosion by downward circulating current moving around a meander band	
Meander cutoff 	• A semicircular depression adjacent to the channel. • Shallow and narrow. • Commonly filled by water to form an oxbow lake	Lateral erosion by a migrating channel.	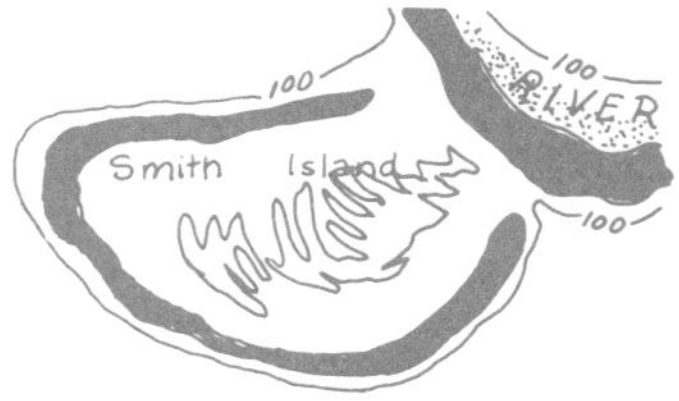
Stream valley 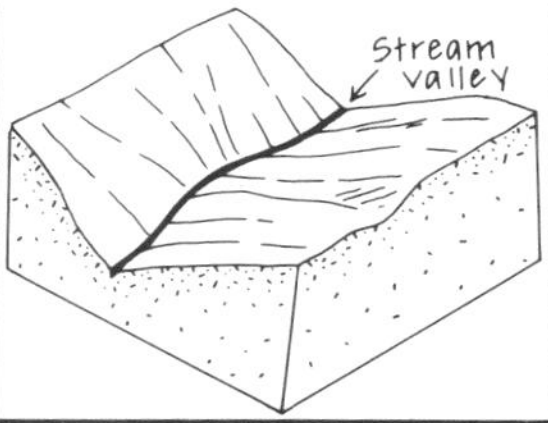	• A generally v-shaped cross profile • Size and slope vary among individual channels	Vertical erosion by a stream channel.	
Entrenched meander 	• A sinuous stream channel cut deeply into bedrock.	Vertical erosion by a meandering channel.	

Figure 12.1
Landforms associated with fluvial processes.

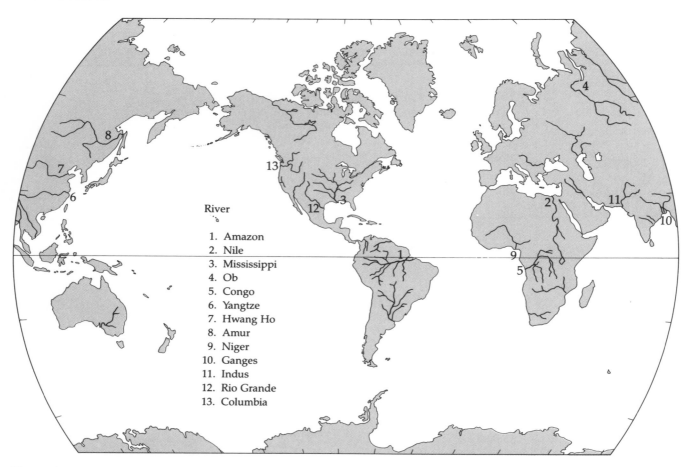

River

1. Amazon
2. Nile
3. Mississippi
4. Ob
5. Congo
6. Yangtze
7. Hwang Ho
8. Amur
9. Niger
10. Ganges
11. Indus
12. Rio Grande
13. Columbia

Figure 12.2
Major rivers of the world.

WATER MOVING IN A CHANNEL

The Drainage Basin

Gullies, creeks, streams, and rivers are channels of different sizes that transport rainwater from the land to the ocean. These channels interconnect to form vast drainage networks. A fundamental unit in the stream network is the *drainage basin,* the area that contributes water to the main channel. For small creeks the drainage basin may be less than a square mile in area, but for large rivers the drainage basin may exceed a million square miles. The boundary separating adjacent basins is the *drainage divide.* Topographically, the divide is a ridge between nearby lowlands.

Within the drainage basin, small channels called tributaries coalesce to form larger channels, which coalesce to form one large stream in an arrangement called the *drainage pattern.* Tributaries of highest elevation near the drainage divide are the headwater channels of the basin. The lowest elevation of the drainage basin is the stream *mouth.* Drainage patterns may take sev-

eral forms, depending on the nature of the underlying rock, the slope of the drainage basin, and the amount of rainfall. Figure 12.3 illustrates several kinds of drainage patterns.

Figure 12.4 is the drainage map of part of North America.

Problem 5

In green, draw the approximate location of the drainage divides that separate channels flowing into the Atlantic Ocean, the Pacific Ocean, the Gulf of Mexico, the Great Lakes, and Hudson Bay. What areas on the continent do not drain into these surrounding water bodies?

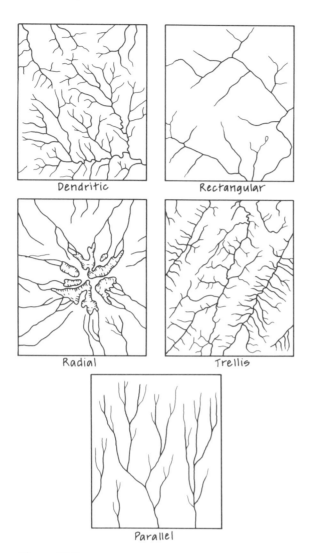

Figure 12.3
Common drainage patterns.

Problem 6

In red, draw the drainage divide between drainage basins of the Mississippi and Colorado Rivers. The mouths of these rivers are labeled 1 and 2, respectively. The area of the drainage basin for the Mississippi River is 1,300,000 square miles; that of the Colorado River is 243,000 square miles. Which of the two rivers would you expect to have the higher discharge (volume of water flowing past a point in a given time interval) at its mouth? Why?

Problem 7

In orange, draw the drainage divide for the Missouri and Yellowstone Rivers. The mouths of these rivers are labeled 3 and 4, respectively. What is the drainage pattern of the Yellowstone River in the Missouri drainage basin? Which river would have a higher discharge? Why?

Problem 8

From your observations of the Mississippi, Missouri, and Yellowstone drainage basins, what do you think would be the trend in discharge as you go downstream in a channel? How might the size of the channel change?

Geometry of the Stream Channel

The channel geometry describes the slope, width, and depth of a stream. Although these features vary among channels, the geometry of most streams is quite similar. For example, all streams increase in width and depth downstream. The channel geometry controls the rate of flow of the water in the channel.

The slope of a stream channel is the *gradient*, the change in elevation over a given horizontal distance. The gradient varies along the channel, typically ranging from less than 1 foot per mile near the mouth to over 100 feet per mile at the headwaters.

A *longitudinal profile*, a plot of distance versus elevation along the stream path, illustrates a river's gradient. Water flowing over a dipping surface will flow down the path of maximum gradient unless deflected by an obstacle.

Problem 9

Figure 12.5 shows the longitudinal profile for the entire Yellowstone River. How does the gradient change between the headwaters and the mouth?

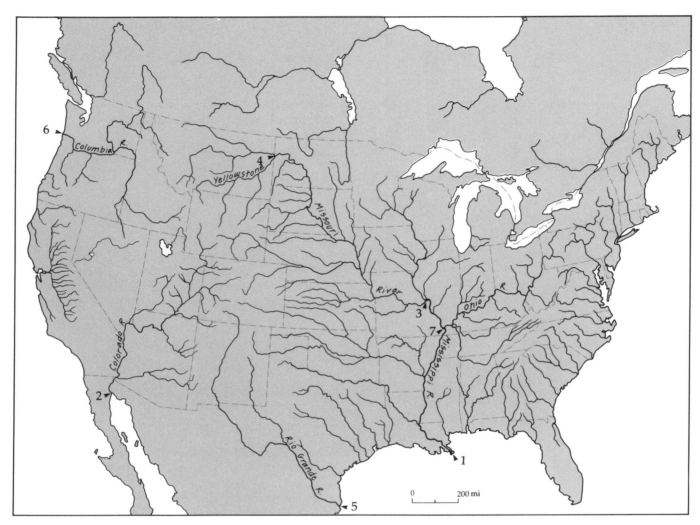

Figure 12.4
Drainage map of part of North America.

Problem 10

Figure 12.5 also lists the channel gradient and average discharge at three locations along the Yellowstone River in Montana: Sidney, Miles City, and Billings. Plot these points on the graph in Figure 12.5B. How do the channel gradient and discharge change as you go downstream?

Channel Depth

Channels vary in size from inches to miles in width and up to hundreds of feet in depth. Despite this variation, however, the cross-sectional shape of the channel is generally rectangular to semicircular. The amount of water in the channel depends on the size of the cross-sectional area, which is the product of the bank-to-bank width and average depth. The width, depth, and discharge increase downstream in almost every case. At any given time the channel depth oscillates between deep and shallow extremes called *pools* and *riffles*, respectively. These features are illustrated in Figure 12.6 for Seneca Creek, Maryland. The *thalweg* is the line connecting points of maximum depth along the length of the stream channel. It is the path that most of the water takes when flowing downstream.

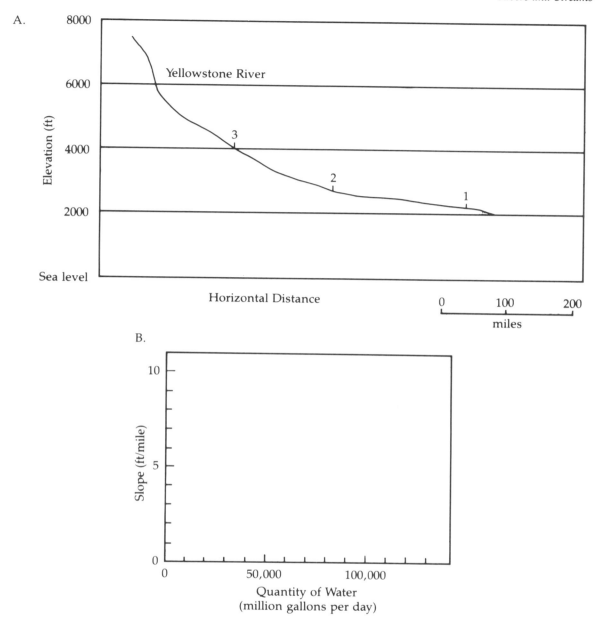

Figure 12.5
Longitudinal profile, channel gradient, and discharge of the Yellowstone River, Montana.

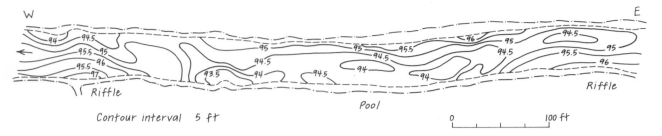

Figure 12.6
Topographic map of the channel bed of Seneca Creek, Maryland, showing alternation of pools and riffles.

Problem 11

With a red pencil, trace the thalweg in Figure 12.6. This stretch of stream seems to be rather straight if judged by the orientation of its banks. From the path of the thalweg, is the actual flow of water as straight as the banks seem to indicate?

Meandering Channels

One of the most widespread but puzzling features of stream channels is their universal tendency to bend or curve. This bending, or *meandering*, is common to large or small channels in all areas of the world, but it is more common at lower gradients. Meandering is measured as the *sinuosity*, the stream length divided by the valley length (Figure 12.7A). In straight channels, the stream length and valley length are equal, and the sinuosity is 1. As meandering increases, the stream length increases while the valley length remains the same, and the sinuosity becomes greater than 1. Any channel having a sinuosity greater than 1.5 is called meandering.

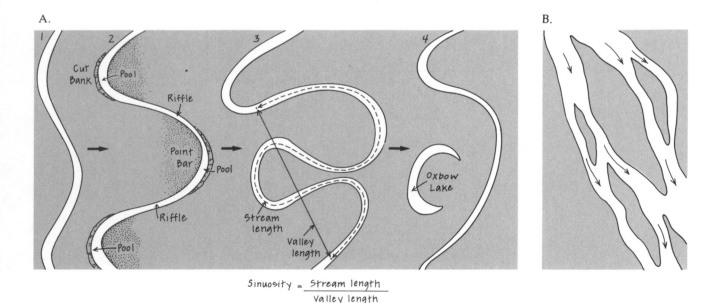

$$\text{Sinuosity} = \frac{\text{Stream length}}{\text{Valley length}}$$

Figure 12.7
Stream meanders. A. Four separate stages of cut-off in meandering channel.
B. Braided channel.

Problem 12

Measure the sinuosity of the thalweg in Seneca Creek.

Problem 13

Measure the sinuosity of the stream in Figure 12.7.

The bend in a meandering channel is further described as having a convex and a concave bank. The shore of the convex bank is the *point,* and is the site of sediment deposition. This accumulation of sediment is called the *point bar.* Conversely, the concave bank is eroded by the stream. This bank is the *cutbank.* In meandering channels, the pools are adjacent to cutbanks, and riffles occur where the thalweg crosses the center of the channel.

Most meandering streams do not flow along the same course for long. Rather, the meander loops migrate from side to side, eroding and enlarging their valleys. Because of the stream's gradient, the loops also tend to migrate in a downstream direction. Occasionally in times of flood a *meander cutoff* occurs. In this process the river cuts through the low neck of land between adjacent loops, straightening and shortening its course and leaving behind an *oxbow lake* in the abandoned meander loop. Migration of meanders and a meander cutoff are illustrated in Figure 12.7A.

Braided Channels

Braided streams are characterized by multiple channels that divide, meet, redivide and meet again and again. Braided channels are typical of streams with a relatively high and typically coarse sediment load, so that sediment is continually being deposited, blocking the channel, then carried away as the stream breaks through the barrier. The sediment load is usually gravel or sand, the gradient is often steep, and the stream banks are easily eroded. A braided channel is shown in Figure 12.7B.

The three types of stream channels—straight, meandering, and braided—often blend from one to another without sharp divisions, although the meandering stream is the most general form.

RIVER AND STREAM PROCESSES

Transport of Material

The movement of water in a channel causes particles of loose rock and soil to be picked up and transported with the water downstream. It has been estimated that over 90 percent of the potential energy of a river flowing to the sea is dissipated as internal turbulence within the water itself, leaving less than 10 percent of the potential energy to do the work of erosion and transport of materials. This small fraction of the river's total energy is nevertheless enough to transport enormous quantities of material. Running water transports material in three ways—as dissolved load, as suspended load, and as bed load. Solid material transported and deposited by running water is called *alluvium.*

Dissolved load consists of various ions dissolved from the rock units through which the river and its tributaries flow.

Problem 14

Consider the ways that sedimentary rocks (which cover 75 percent of the continental surface) form. Which types—detrital, chemical, or biogenic—would be most susceptible to dissolution by water? Why? What would be some common ions in dissolved load?

The *suspended load* of a river consists of fine-grained particles (silt and clay) that are kept aloft in the water by turbulence and eddies. These fine particles act much like sandpaper by abrading additional material from bedrock surfaces and sometimes polishing the bedrock to a smooth sheen.

Bed load is the material—usually larger grains, such as coarse sand, pebbles, and cobbles—that is pushed or rolled along the river bed by the moving current. The distinction between suspended and bed loads is illustrated in Figure 12.8.

Alluvium is the clastic rock material transported as suspended load and bed load. Throughout the drainage basin, all but the finest clay particles are eroded and deposited many times. In general, smaller grains travel farther and more frequently than the larger grains. Hence alluvium is naturally sorted by size through the channel network. Smaller particles are more abundant downstream.

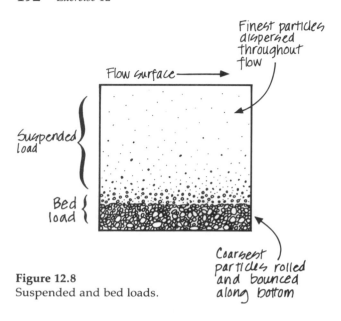

Figure 12.8
Suspended and bed loads.

Problem 15

Examine Figure 12.9, which illustrates the size of alluvium along the Platte River in Colorado and Nebraska. What is the largest average grain size? Where does it occur along the stream?

Problem 16

Plot the average grain size measurements on the graph below the map. How does the grain size change downstream?

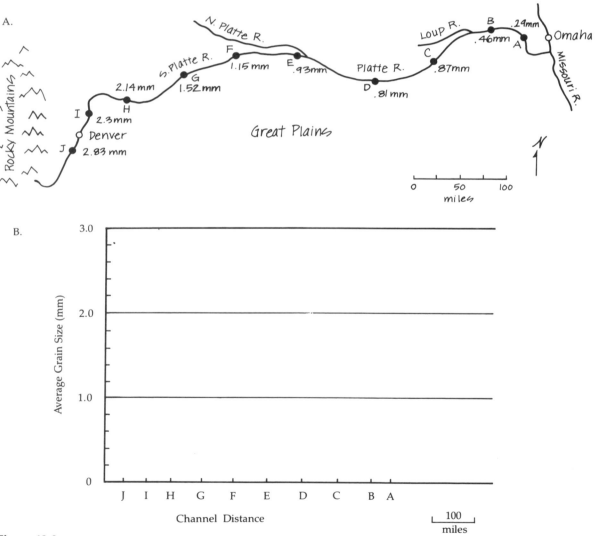

Figure 12.9
Alluvium size along the Platte River, Colorado and Nebraska.

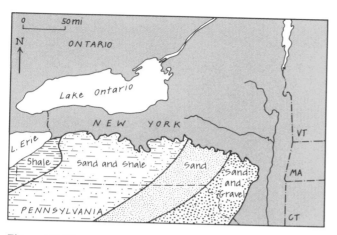

Figure 12.10
Particle sizes in Devonian sediments, southern New York.

Problem 17

Figure 12.10 shows the particle sizes in Devonian sediments deposited in southern New York. What was the direction of the currents that deposited these sediments? Suggest a likely source area for them.

Erosion

Stream erosion takes place along the bed and banks of the channel. Erosion of the bed lowers the elevation of the channel. Vertical downcutting erodes the bedrock to form a steep narrow valley with a V-shaped cross profile. Vertical downcutting is common in the upper portion of the drainage basin, especially near the drainage divide. Lateral erosion of the channel banks is caused primarily by meandering of the thalweg; as shown in Figure 12.7A, it is the means by which rivers widen the valleys through which they flow.

Problem 18

Figures 12.11 and 12.13 show the topography near the Mississippi River at Vicksburg, Mississippi. Contrast the topography of the eastern third of the map with the western two-thirds.

Problem 19

In what areas is vertical stream erosion the dominant process? Where is lateral stream erosion the dominant process?

Problem 20

What is the origin of Eagle lake in the north-central part of Figure 12.11?

Deposition

Deposition of alluvium in the drainage network occurs whenever the velocity and turbulence of the water decrease, such that erosion ceases and deposition begins. Alluvial deposits can be divided into channel deposits and overbank deposits.

Channel deposits occur within the banks of the channel and range in particle size from gravel to mud. Bars are the principal features of channel deposits. *Point bars* form along the inside bank of a meander curve, the point. These crescent-shaped bars form because of the pattern of water circulation around the meander, allowing sediment to drop out of suspension. At the same time erosion takes place on the outside of the meander bend (the *cutbank*), and the total sediment load of the river remains approximately constant. Figure 12.12A illustrates this process.

Overbank deposition occurs during floods when sediment-laden water spills over the banks onto the *floodplain*, the flat area adjacent to the channel. Once over the banks, the speed and turbulence of the water decrease, and the suspended sediment settles out. On many floodplains, accumulations of sediment from frequent floods form natural *levees*, low ridges that parallel the river course (Figure 12.12B). They may be a mile or more in width, are highest near the river, and slope gradually away from it. Once formed, natural levees tend to confine the river between its banks in high-water periods. In some areas natural levees have raised the surface of the river to elevations above that of the surrounding floodplain, preventing tributaries from entering the river. These smaller channels flowing parallel to the main river in the floodplain are called *yazoo streams*.

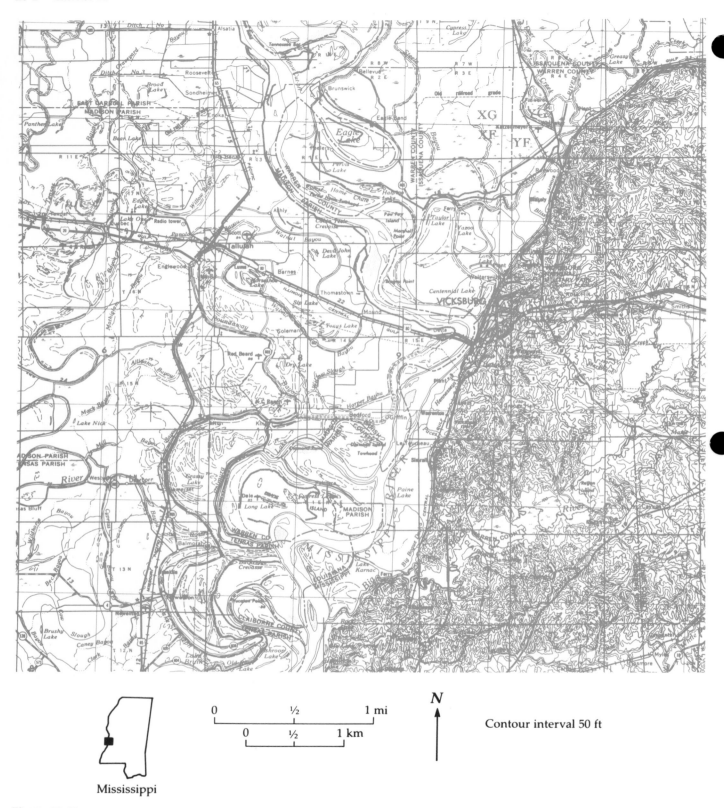

0 ½ 1 mi

0 ½ 1 km

N

Mississippi

Contour interval 50 ft

Figure 12.11
Topographic map, Mississippi River near Vicksburg.

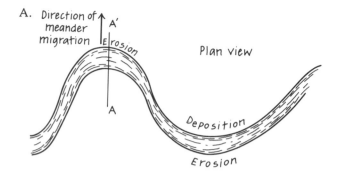

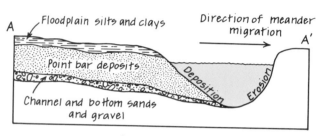

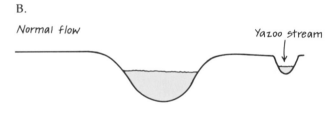

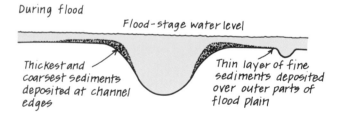

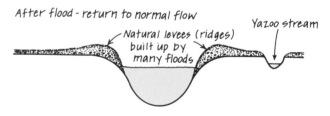

Figure 12.12
A. Deposition and erosion along meander banks. B. Formation of natural levees.

Problem 21

Reexamine Figures 12.11 and 12.13, the Mississippi River at Vicksburg. In what area of the map is stream deposition the dominant process?

Problem 22

Label an example of a point bar.

Problem 23

Label an example of a yazoo stream.

THE RIVER AS AN EQUILIBRIUM SYSTEM: GRADED STREAMS

Rivers and streams, like most of the earth's systems, tend toward an equilibrium with their surroundings so that they reach a steady-state condition. One example of this steady-state equilibrium is the concave-upward shape of longitudinal profiles of rivers and streams, which are remarkably similar for all systems of flowing water. The same shape evolves in small gullies and large rivers, in natural streams and in artificial water courses. This profile seems to be a steady-state equilibrium pattern that evolves from the basic dynamics of water flow. Over time, streambed elevations change by erosion or deposition toward this state of equilibrium, and streams that reach it are said to be *graded streams* (or *at grade*). In graded channels, the discharge is just sufficient to transport the sediment load furnished by the drainage basin. A change in any of the controlling factors—velocity, discharge, sediment supply—will cause a compensating change in the stream system that reestablishes stream equilibrium. In the same sense, the channel form of streams may change to accommodate new stresses on the equilibrium. A sudden increase in coarse sediment load, for example, may cause the channel to change from meandering to braided until the load is dispersed along the stream's course, equilibrium is reestablished, and the stream is again at grade.

Characteristic landforms develop when streams respond to changing surface conditions. *River terraces* are vestiges of former floodplains now located above the present channel due to vertical downcutting by the stream. Because of fluctuations in the sediment supply and changes in the elevation of the channel, there may be several stages of floodplain development

Figure 12.13
Stereopair of aerial photographs near Vicksburg.

0   1 mi   **N**   A.

followed by increased downcutting by the river. A succession of such changes results in multiple river terraces. *Entrenched meanders* are high-sinuosity channels cut deeply into the bedrock. This landform probably records an originally low-gradient meandering channel whose valley has been uplifted slowly enough that stream erosion in the meandering channel was able to keep pace with the uplift.

Alluvial fans and deltas are depositional landforms that result from a sudden drop in water velocity and turbulence. *Alluvial fans* form where a channel that has been confined to a steep narrow valley suddenly opens onto a broad open area. The velocity and turbulence of the water decrease abruptly in the broad area now available to the channel. The main channel divides into smaller *distributary channels*, and sediment is deposited according to size—coarse material

near the head of the fan, fine material farther out in the broad plain.

Sooner or later every major river empties into a large body of standing water, most often the ocean, sometimes a lake. As it does so, the river current slows and ultimately ceases its forward motion. Since fluvial erosion cannot take place without a current, sea level (or lake level) is often referred to as *base level*, the elevation below which fluvial erosion cannot occur. The sediments dropped by the waning river current form the river *delta*, a pile of sediment whose structure resembles, at least superficially, an alluvial fan. If deposition is continuous and erosion by waves, currents, or tides is minimal, the delta will extend outward into the ocean or lake. Deltas forming within lakes eventually enlarge, replacing the lake water with sediment. Figure 12.15 illustrates deltas and alluvial fans.

Figure 12.13
Stereopair of aerial photographs near Vicksburg.

B.

Problem 24

Examine Figures 12.14 and 12.18, the topographic map and stereopair of aerial photographs of Bear Creek near Denver, Colorado. Mark with an E locations where the stream is adjusting to its environment by eroding the channel. Consider both lateral and vertical erosion.

Problem 25

Mark with a D locations where the stream is depositing alluvium in response to external conditions. These deposits may be either overbank or within the channel.

Problem 26

Examine the cross-profile C–C' in Figure 12.17. The profile crosses two terraces and the present floodplains of Bear Creek and Turkey Creek. Locate and label the floodplains on the cross-profile. Color them yellow. Color the terraces green.

Problem 27

What do these terraces and floodplains suggest about the depositional and erosional history of the creeks?

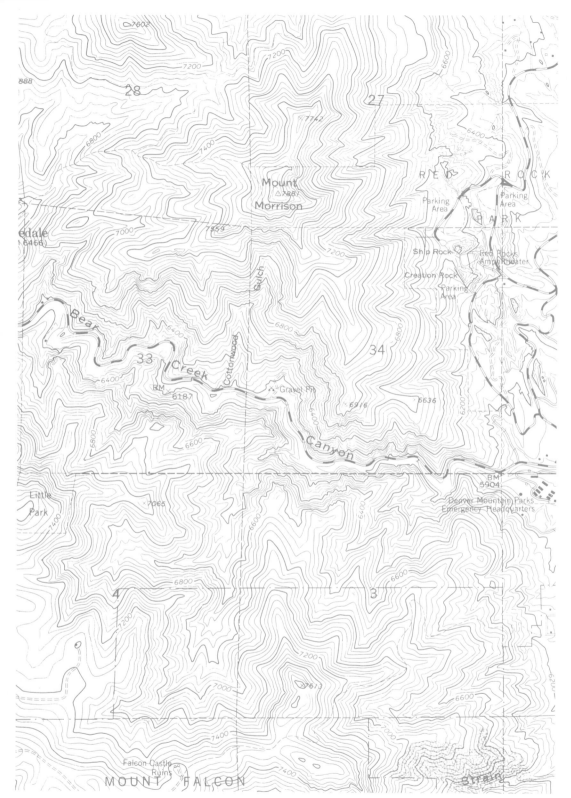

Figure 12.14
Topographic map of Bear Creek, near Denver, Colorado, showing the area in the photographs in Figure 12.18.

A.

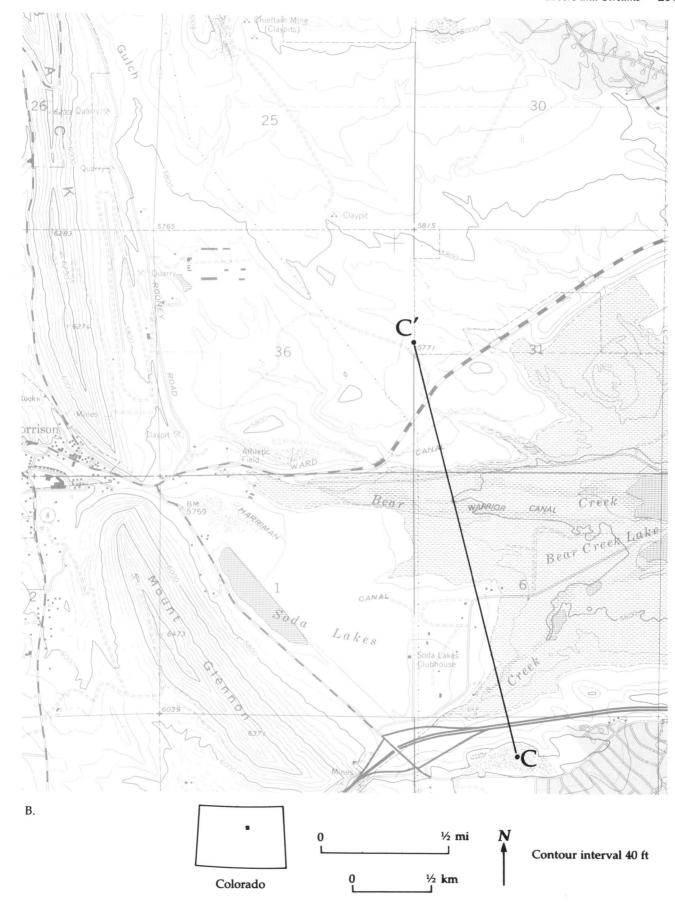

B.

Colorado

0 — ½ mi

0 — ½ km

N

Contour interval 40 ft

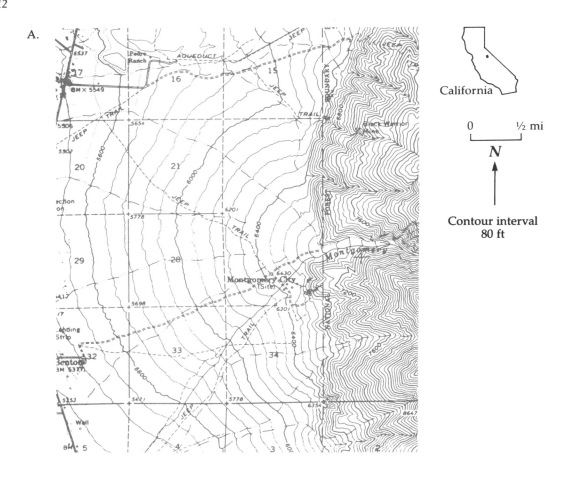

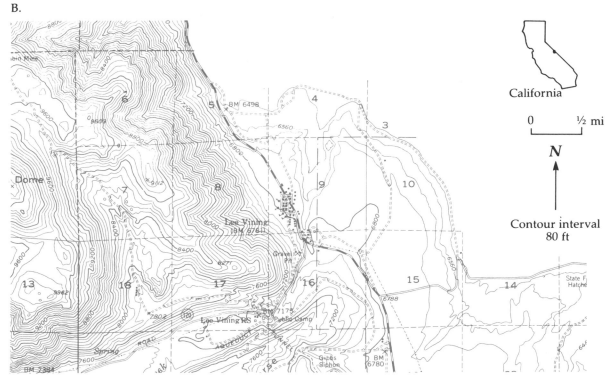

Figure 12.15
Topographic maps showing an alluvial fan (A) and delta (B).

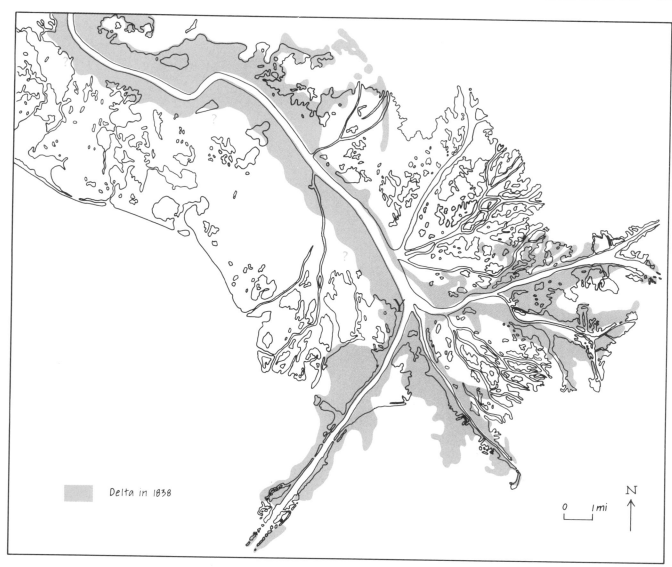

Figure 12.16
Topography near the mouth of the Mississippi River today and in 1838.

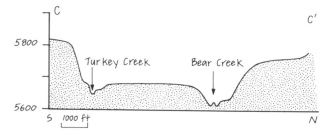

Figure 12.17
Cross-profile of Bear and Turkey
Creeks, Colorado.

Problem 28

The gravel on the terrace between Bear Creek and
Turkey Creek is different in composition and size from
the gravel on the outer terraces. Is the terrace between
the two creeks older or younger than the outer terraces? Explain your answer.

Examine the topography at the mouth of the Mississippi in Figure 12.16.

Figure 12.18
Stereopair of aerial photographs of Bear Creek near Denver, Colorado.
(Rotate these photos 90° clockwise for stereo viewing.)

0 |⎯⎯⎯⎯⎯⎯⎯⎯⎯⎯| 1 mi

N ↑ A.

Problem 29

Where would you expect to find the coarsest sediment on a delta? The finest sediment?

Problem 30

Figure 12.16 also shows the Mississippi delta in 1838. How has the delta changed over the past 150 years?

Figure 12.18
Stereopair of aerial photographs of Bear Creek near Denver, Colorado.

B.

Problem 31

If a hole were drilled at location Y, how would the grain size of sediment change with depth? Illustrate this progression with a stratigraphic column at the side of Figure 12.16.

Problem 32

If the delta continues to enlarge, what size sediment will be deposited over the offshore muds?

It is important to be able to recognize what rocks were formed from stream-deposited sediments in order to interpret the geologic history of a region.

Problem 33

Examine the rocks provided by your instructor and list below the ones formed by lithification of stream deposits.

FIELD STUDY: THE COLORADO RIVER

The drainage basin of the Colorado River encompasses national parks and monuments and is one of North America's most scenic river systems. The main channel is 1440 miles long, extending from the Rocky Mountains to the Gulf of California. The basin area of 243,000 square miles includes parts of Wyoming, Colorado, Utah, New Mexico, Nevada, Arizona, California, and Mexico.

Problem 34

Examine the maps of physical regions and topography (Color Maps 1 and 5). Briefly describe the topography through which the Colorado River flows. What physical regions are included in its drainage basin?

Problem 35

Examine the longitudinal profile of the Colorado River in Figure 12.19. Superimpose in red on Figure 12.19 an idealized equilibrium profile, similar to that shown in Figure 12.5, from the continental divide to sea level. Where and why does the idealized profile differ from the actual profile?

Problem 36

Examine Figure 12.20, the topographic map of the southwest corner of Mesa Verde National Park. What is the shape of the drainage system (dendritic, radial, etc.) in this area?

Problem 37

Draw a cross-profile, A–A', of Soda Canyon in the space below. Describe in words the shape of the profile.

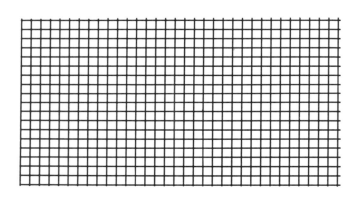

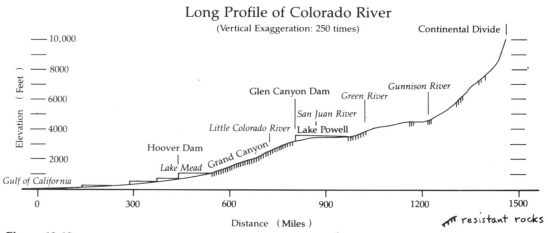

Figure 12.19
Longitudinal profile of the Colorado River.

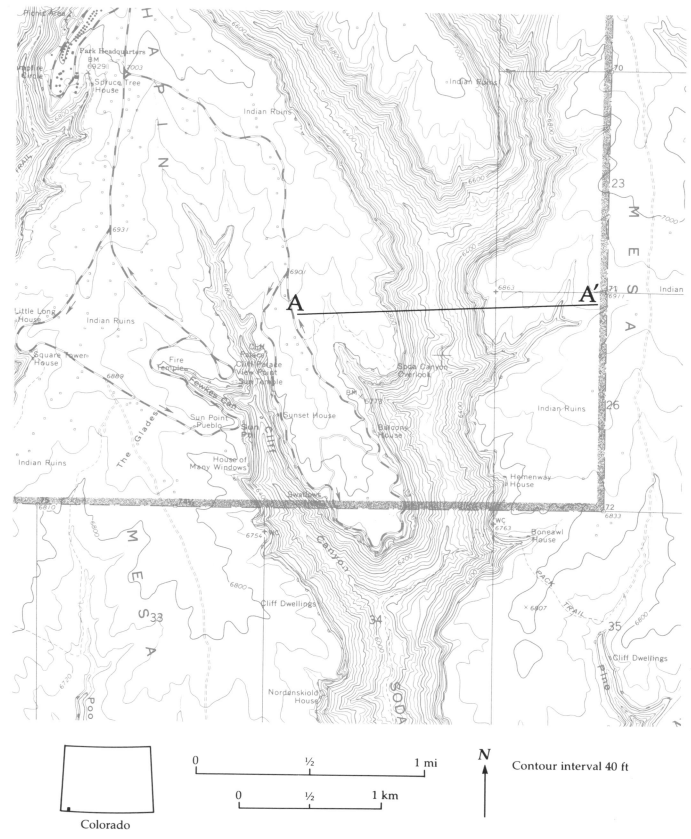

Figure 12.20
Topographic map of part of Mesa Verde National Park, Colorado.

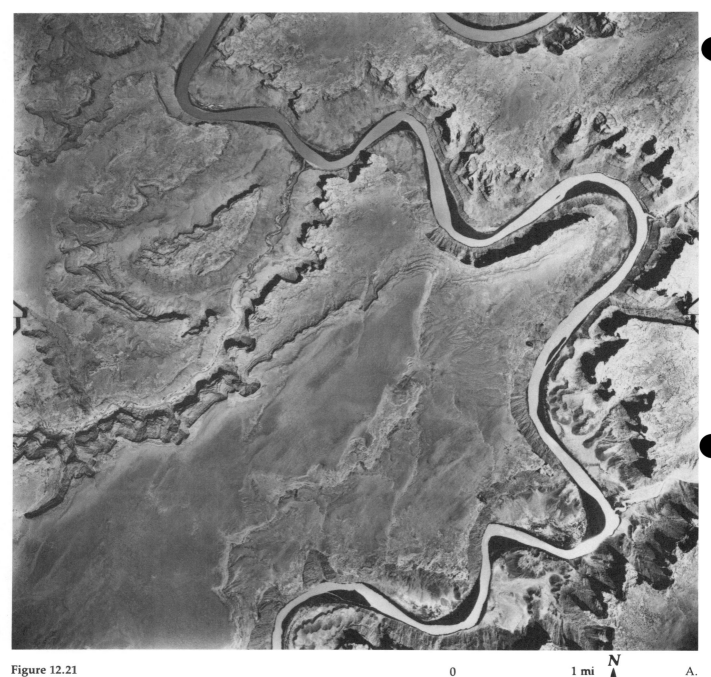

Figure 12.21
Stereopair of aerial photographs, Bowknot Bend quadrangle, Utah.
(Rotate photos 90° clockwise for stereo viewing.)

0   1 mi   N

A.

Problem 38

Is this area undergoing erosion or deposition?

The two photographs shown in Figure 12.22 are unusual—they were taken from the same spot about 100 years apart. The area of the photographs is marked P on the topographic map of Bowknot Bend quadrangle, Utah (Figure 12.23), and on the aerial photographs (Figure 12.21).

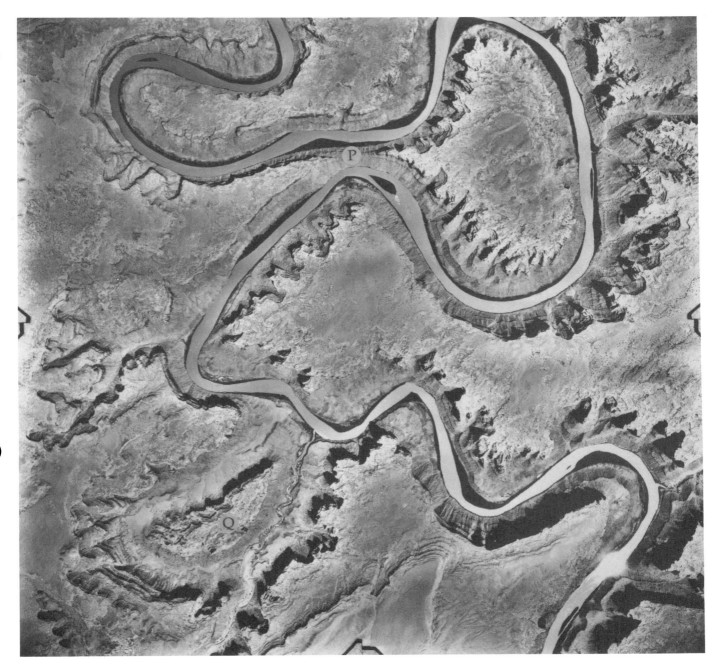

B.

Figure 12.21
Stereopair of aerial photographs, Bowknot Bend quadrangle, Utah.

Problem 39

How has this section of the channel changed in the 100-year period between photographs?

Problem 40

What is the origin of the horseshoe-shaped canyon just southwest of the Green River at point Q?

A.

B.

Figure 12.22
Photographs of Bowknot Bend, Utah, taken 97 years apart. A. 1871. B. 1968.

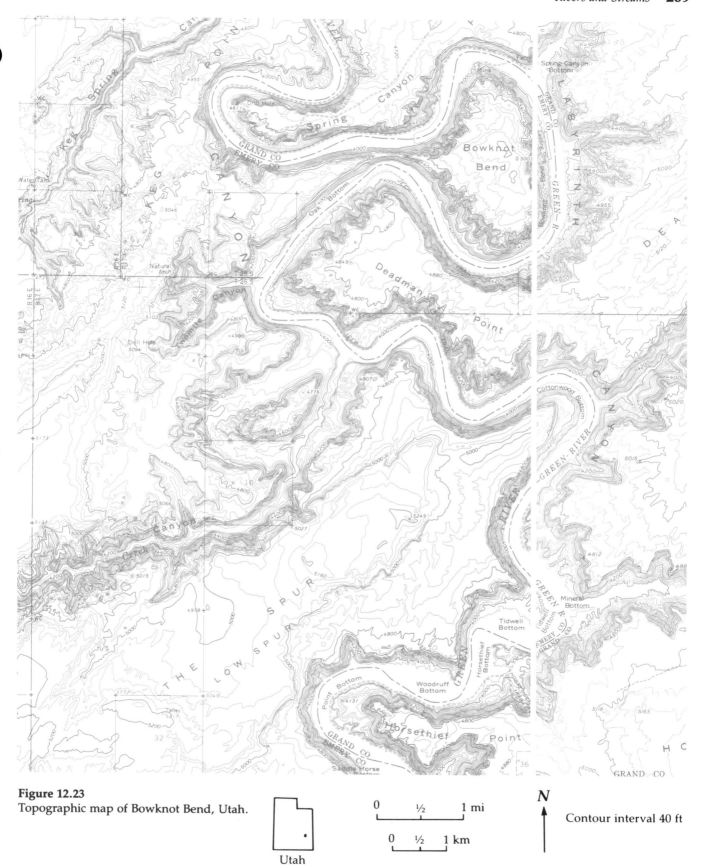

Figure 12.23
Topographic map of Bowknot Bend, Utah.

Utah

0   ½   1 mi

0   ½   1 km

N

Contour interval 40 ft

Problem 41

Trace the former channel of the Green River around Deadman Point with a red pencil. What is the difference in elevation between the floor of the horseshoe canyon and the present river course? If the river is cutting downward at a rate of 1 inch per 100 years, how long ago was the loop abandoned?

Problem 42

Is the present channel path longer or shorter than the historic channel? Which channel therefore has the steepest gradient?

Figure 12.24 shows cross-profiles of the Colorado River near Lees Ferry, Arizona, measured at several times during 1956. The measured values of discharge are given for each date.

Problem 43

Note the fluctuation in discharge of the Colorado River at different periods during the year. Plot these data in the space below.

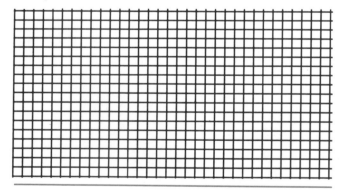

Problem 44

Estimate the width and average depth of the Colorado River on each of the dates given. Measure the width at water level. From the data given for discharge, calculate the flow velocity on those days. The data for June 3 are given as an example. Write the flow velocities in the table in Figure 12.24B.

Problem 45

Notice that periods of high discharge occur in the late spring to early summer. To what factor do you attribute this seasonal cycle of high discharge in the Colorado at Lees Ferry? (Note that these measurements were taken before the construction of the Glen Canyon Dam, just upstream from Lees Ferry.)

Problem 46

How does the bed elevation change as discharge and river velocity increase? Since the elevation of the river bed changes up or down over a period of weeks, the change is not due to erosion of bedrock. What is responsible for the change in bed elevation?

Problem 47

Would you expect sediment deposited in the Colorado River in June to be coarser or finer than sediment deposited after August 15? Explain your reasoning.

Seventy miles downstream from Lees Ferry, the Colorado River flows through Grand Canyon National Park. Reexamine the topographic map of the Bright Angel quadrangle in Figure 8.6.

Problem 48

From the elevation difference between river level and the north rim, how far has the Colorado River eroded into the earth's crust?

Problem 49

Would you say that the load of the Colorado is primarily bed, suspended, or dissolved load?

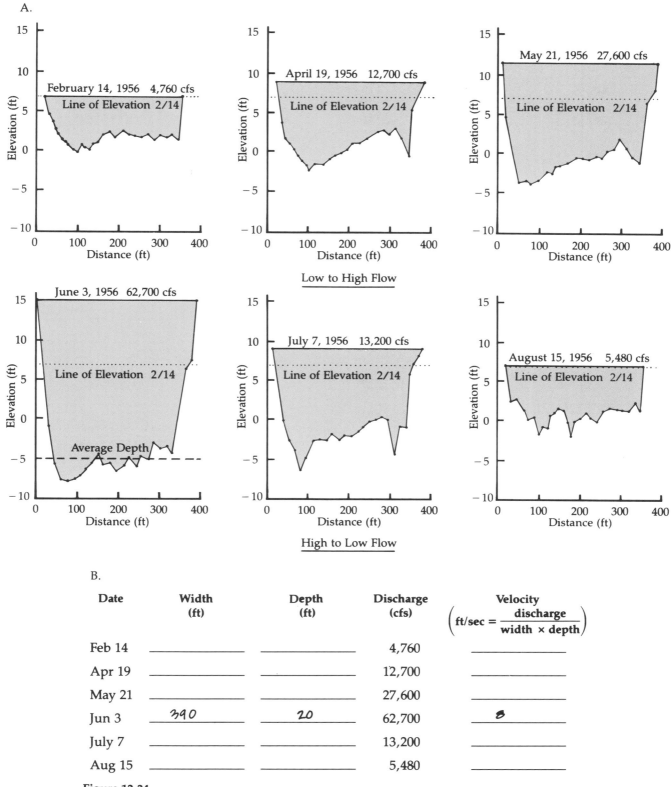

Figure 12.24
Cross-profiles of the Colorado River near Lees Ferry, Arizona.

212

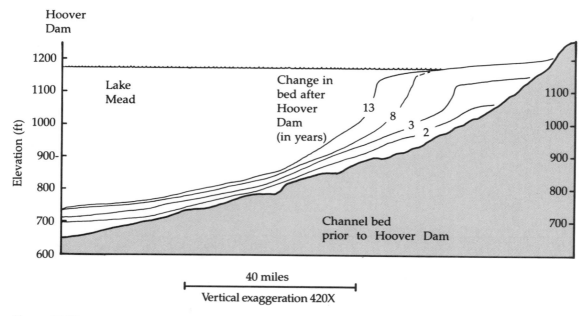

Figure 12.25
Profiles measured across the mouth of the Colorado River at Lake Mead.

Problem 50
Note the location of Roaring Springs near the head of Bright Angel Canyon. From what rock type does it emerge?

Problem 51
Would the water flowing from Roaring Springs be transporting alluvium? What type of load would this stream be carrying?

In 1935, Hoover Dam was constructed on the Colorado River, creating an 80-mile-long reservoir, Lake Mead. The dam increased the elevation of the channel from 600 feet to 1230 feet, forming an artificial base level for most of the drainage basin of the Colorado.

Problem 52
Examine Figure 12.25, which shows profiles measured at different times across the point where the Colorado River enters Lake Mead. What process is occurring here?

Problem 53
What alluvial feature is forming?

Problem 54
How should the size of sediment in this feature change as you go out into Lake Mead?

Problem 55
What is the eventual fate of Lake Mead?

Open Water Bodies and Shorelines

The shattered water made a misty din.
Great waves looked over others coming in,
And thought of doing something to the shore
That water never did to land before. . . .
You could not tell, yet it looked as if
The shore was lucky in being backed by cliff,
The cliff in being backed by continent.

ROBERT FROST

Earth is truly the water planet, as observations from space have so beautifully shown. Earth alone of all the moons and planets in our solar system is blue with open water bodies and white with swirling clouds of water vapor. Open water bodies are the most extensive surface feature of the earth, covering 72 percent of the earth's surface. Our understanding of how the earth works has come about largely through the detailed exploration of the ocean basins. In this exercise we concentrate largely on the interaction between ocean and land at the shoreline.

Despite their essentially horizontal surfaces, open water bodies are dynamic environments. Waves, currents, and tides are the three major agents of erosion, transport, and deposition in open water bodies and along shorelines. In general, the importance of these agents increases with the size of the water body. Ocean waves, for example, generate far more energy than waves in a pond, and although tidal fluctuations can be detected in large lakes, they are extremely small when compared with ocean tides. Waves and tides are active along the shoreline, the boundary of the water with the land. Here the land is sculpted by these shoreline processes through the erosion and deposition of rock and sediment. Figure 13.1 shows landforms that are typical of shoreline processes.

DISTRIBUTION AND DIMENSIONS

The distribution of oceans and continents over the earth's surface is not even. About 81 percent of the Southern Hemisphere is covered by ocean, as compared with 61 percent of the Northern Hemisphere. If the poles of the earth were elsewhere, the distinction between land and water hemispheres would be even more striking.

Problem 1

Use the globe in your laboratory for this question. Imagine that the north pole was in New Zealand. What would be the distribution of continents and oceans in the two hemispheres? Imagine that the pole was over the center of any of the continents. What would be at the other pole—ocean or continent? Generalize your observations with a statement that describes the evenness in the distribution of continents and ocean basins over the globe.

DEPOSITIONAL LANDFORMS

LANDFORM	CHARACTERISTICS	ORIGIN	TOPOGRAPHIC MAP
Barrier island 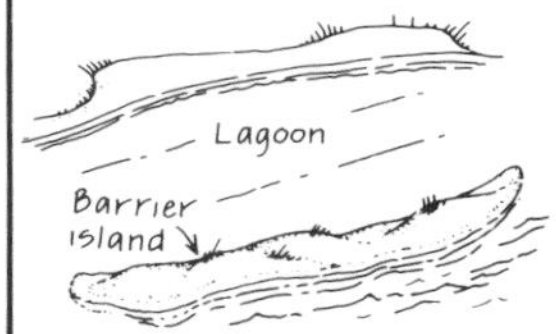	• A long linear island parallel to the shoreline • Composed of sand and gravel	Dependent on local geology	
Lagoon 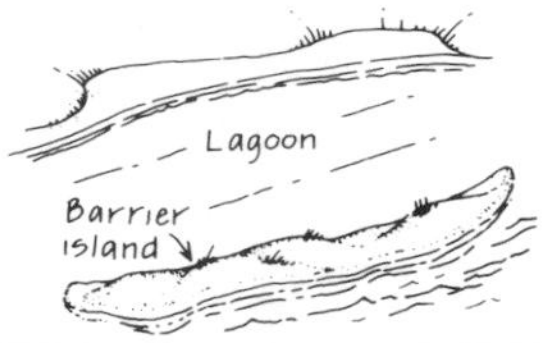	A quiet water area behind a barrier island or carbonate reef	Wave barrier creates still water on its lee side	
Spit 	• A streamlined ridge of sediment extending in the direction of the longshore current • Composed primarily of sand	Deposition by longshore current	
Tombolo 	• A ridge of sediment connecting an island to the shoreline • Composed primarily of sand	Deposition generally caused by wave refraction around island	
Bar 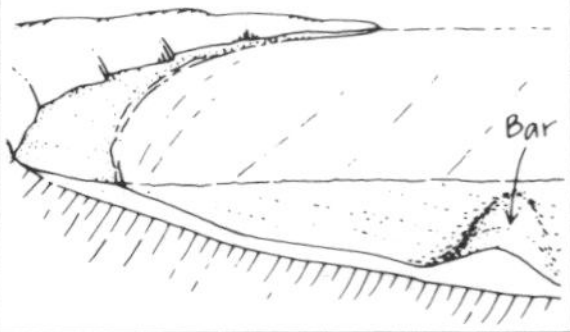	• A submerged depositional mound or ridge • Composed of sand and gravel	Dependent on currents and sediment supply	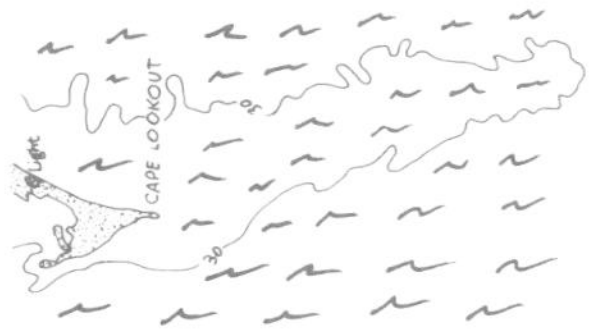
Carbonate reef 	• A submerged mound or ridge formed by organisms, primarily corals • Composed of biogenic calcium carbonate	Organic growth in clear, warm, shallow water	

Figure 13.1
Landforms associated with shoreline processes.

EROSIONAL LANDFORMS

LANDFORM	CHARACTERISTICS	ORIGIN	TOPOGRAPHIC MAP
Sea stack 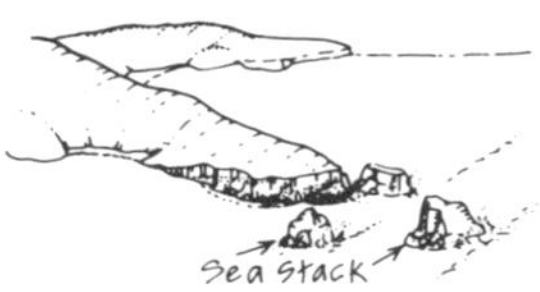	• Isolated, small island just offshore • Composed of local bedrock	Wave erosion	
Wave-cut bench (platform) 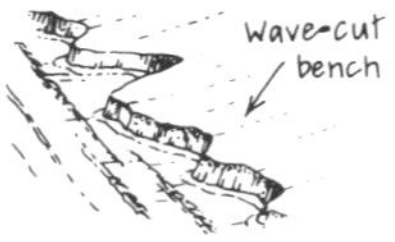	• A flat erosional surface in front of a wave-cut cliff • Carved into local bedrock	Wave erosion and retreat of the shoreline	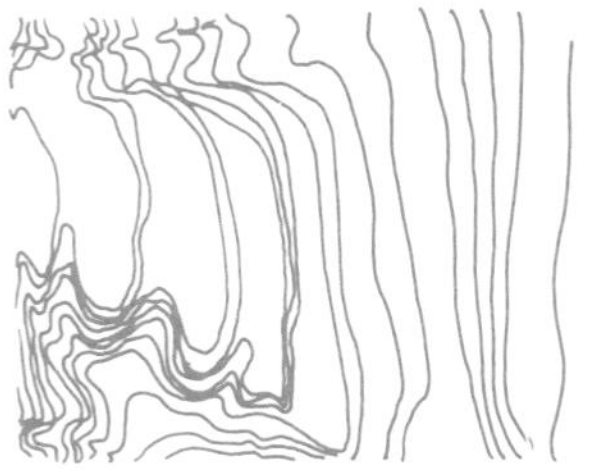
Wave-cut cliff 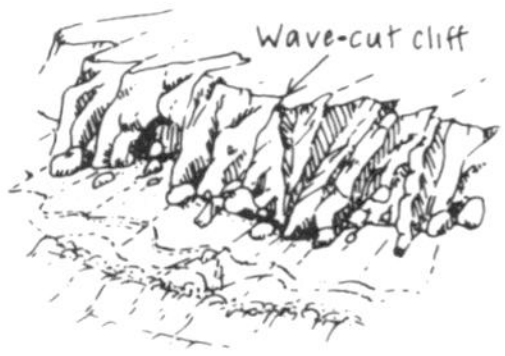	• A steep cliff eroded by waves hitting the base • Composed of local bedrock	Wave erosion directly against the shoreline	
Headland 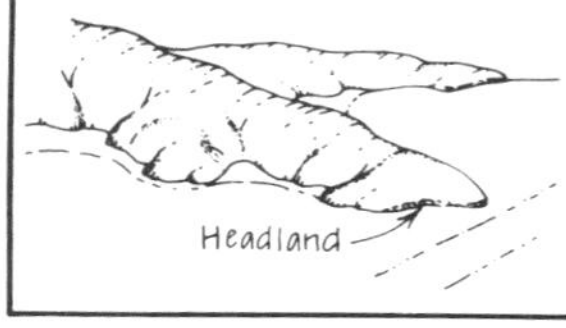	• A point of land extending into an open water body • Composed of local bedrock or sediments	Dependent on local topography	

Figure 13.1
Landforms associated with shoreline processes.

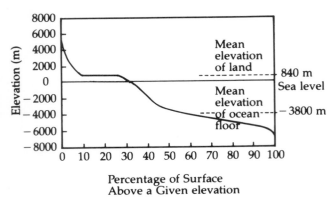

Figure 13.2
Hypsographic curve for the earth.

Elevations of the earth's surface can be plotted against the percentage of the earth's surface at that elevation to give the *hypsographic curve*, shown in Figure 13.2. This graph tells us that the earth's surface is dominated by two major levels: one, the continents, close to sea level (mean elevation 840 m); the other, the ocean basins, with a mean depth of 3800 m. A steep transition zone, the continental slope, occurs between the two.

Problem 2

Inspect the bathymetric map of the earth's oceans in Exercise 18, Figure 18.1. Where does the coastline coincide closely with continental slope? Where does the water line have little or no relationship to the continental slope?

Problem 3

Does water underfill, exactly fill, or overfill the ocean basins?

Problem 4

Return to the map of the earth's major rivers, Figure 12.2. The Amazon and Congo Rivers account for a total of one-quarter of the earth's river discharge. From this information and from the number of rivers that discharge into the Atlantic and Pacific Oceans, which of the two ocean basins should have the greatest thickness of terrigenous (land-derived) sediment?

Problem 5

Inspect the topography of the Atlantic and Pacific basins (Figure 18.1). Does your answer to Problem 4 explain the fact that peaks and hills are much more abundant on the floor of the Pacific than the Atlantic?

WAVES AND TIDES

Waves

Waves form from the pressure of the wind against the water surface; the faster, longer, and more extensive the wind, the larger the waves. Waves form in rows oriented at 90° to the wind direction, and these rows move in the same direction as the wind. *Swells* are low broad waves formed during storms at sea. The swells move across the water surface, sometimes hundreds or thousands of miles, until they reach the shoreline and become *breakers*.

Wave Dimensions.   The dimensions of length, height, and time describe waves. The horizontal distance between adjacent wave crests is the *wavelength*. The *wave height* is the vertical distance from crest to trough, and the time for adjacent waves to pass a stationary point is the *wave period*. Wave motion is oscillatory; that is, an object floating in the open water far from shore does not travel with the wave but describes a nearly circular path, returning to the same point after each wave passes. The circular motion decreases with depth until, at a depth equal to half the wavelength, the orbital movement has practically ceased. The dimensions of a wave are illustrated in Figure 13.3.

Waves change shape as they move into shallow water near the shoreline. Approaching the land, the base of the wave slows because of friction against the bottom, while the wave crest continues at its original velocity. The wave becomes steeper with shallower depths until the crest collapses. The collapsing wave, called a *breaker*, continues as foaming water toward the shoreline. The area of wave collapse is a high-energy environment called the *surf zone*.

The wave dimensions and the water depth control the transition of a wave from an open-water swell to a shoreline breaker. The slowing of the wave base begins where the water depth is approximately half the length of the wave. Breakers form when the wave height is about three-quarters of the water depth.

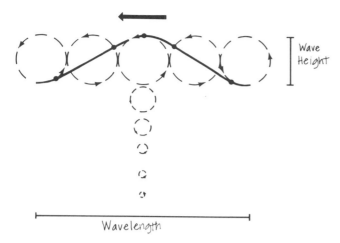

Figure 13.3
Wave dimensions and motions. The circles show the path of an individual water molecule or floating object as a wave passes. The circle of oscillation becomes smaller and smaller as water depth increases.

Problem 6

The chart below lists measurements of waves at different wind speeds. Calculate the wave velocity for each wind speed, and enter the values in the chart (wave velocity = wavelength/wave period).

Wind Speed (mi/hr)	Wave Height (ft)	Wave-length (ft)	Wave Period (s)	Velocity (ft/s)	(mi/hr)
20	2	128	5		
30	6	184	6		
40	16	250	7		
60	23	415	9		

Is the wave velocity as high as the wind speed?

Problem 7

Examine the shoreline topography shown in Figure 13.4, and contour the water depth at 5-foot intervals. Using the wave characteristics in the table above, determine the location of the surf zone at different wind speeds. Mark these positions in red on Figure 13.4, and label them with the appropriate wind speed. Do higher velocity waves break close to shore or far out?

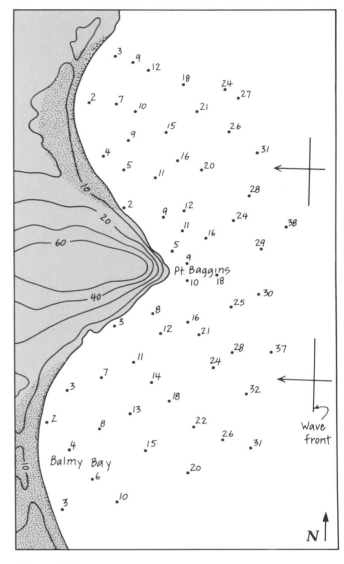

Figure 13.4
Shoreline topography.

Wave Refraction. In addition to shortening and steepening near the shoreline, the wave may bend laterally. This bending, called *wave refraction*, occurs along shorelines of irregular depth or when waves approach a shoreline at an angle. As the wave approaches the shoreline, the velocity decreases in shallow areas but continues unchanged in deeper areas. The net effect is to cause the wave front to bend toward the shallow water, where its velocity is less. This process is illustrated in Figure 13.5. Wave refraction is an important shoreline process because it concentrates the energy of the advancing wave front on land areas that protrude into the ocean and dissipates that energy in bays.

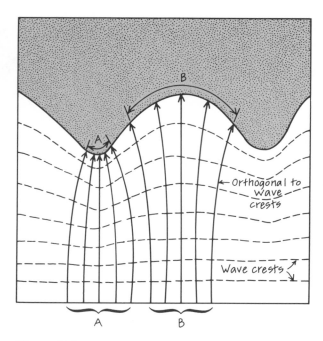

Figure 13.5
Wave refraction where depth contours are parallel to the shoreline.

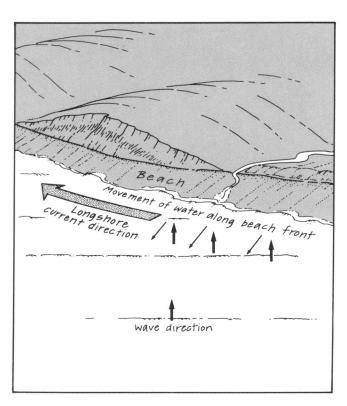

Figure 13.6
Origin of the longshore current.

Problem 8

Wave crests approach the shoreline in Figure 13.4 from the east, as shown by the dark line and arrow. In red, draw a line to show the direction of the wave crests approaching the shoreline from the arrow.

Problem 10

Is longshore drift active along this shoreline? How can you tell?

Problem 11

In what direction do the waves strike the shoreline?

Longshore Drift. A current running parallel to the shoreline develops when waves approach the shoreline at an angle. This current, the *longshore current*, forms because water runs up onto the beach in the direction of the approaching wave but slides back to sea in a direction perpendicular to the shoreline in response to gravity, as illustrated in Figure 13.6. *Longshore drift* is the movement of material, primarily sand sized, down a beach by a longshore current.

Problem 12

Return to Figure 13.4 and notice the focus of wave energy represented by the lines you drew in Problem 5. Is Point Baggins undergoing deposition or erosion? Balmy Bay? Place arrows along the coast to indicate the direction(s) of longshore drift.

Problem 9

Examine the aerial photograph in Figure 13.7. Where along this shoreline is the water turbid with sediment?

Figure 13.7
Aerial photograph of the shoreline near Key Largo, Florida.

$\uparrow N$

Problem 13

The processes of wave refraction and longshore drift are changing the shape of this coastline. In blue, sketch the ultimate shape of the coastline.

Many beaches, like graded streams, are systems in equilibrium. The construction of breakwaters or groins, much like the construction of a dam across a stream, may throw the system out of equilibrium. Sand transported parallel to the beach by a longshore current piles up behind the groins, and downstream from the groin the existing supply of sand is eroded by the current. Examine Figure 13.8.

A.

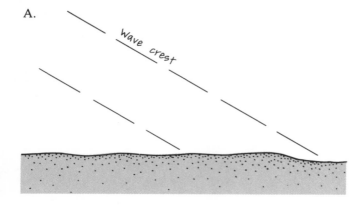

B.

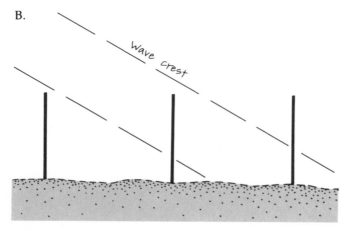

Figure 13.8
Beach modification by groins.

Problem 14

In which direction does the longshore current flow along this coast? Place an arrow on Figure 13.8A to indicate the direction of sediment transport.

Problem 15

Now imagine that a series of groins is constructed, as shown in Figure 13.8B. Where will the sand pile up along the beach? Where will the sand be depleted? Draw on Figure 13.8B the new shoreline of the beach after the groins have been in place long enough to establish the new equilibrium.

Tides

The combined relative motions of earth, sun, and moon cause sea level to fluctuate. A bulge in sea level, the *tidal bulge*, forms from the gravitational attraction of the moon and sun on the ocean water. As the earth rotates, the tidal bulge travels around the shorelines of open water bodies, causing sea-level oscillations. These rhythmic fluctuations are the *tides*.

The high and low tides vary in height and number during the day in different areas of the globe. The variations in height and number define several types of tides:

- *Diurnal*   One daily high and low tide.

- *Semidiurnal*   Two daily high and low tides of equal height.

- *Mixed*   Two daily high and low tides of unequal height.

The tidal range is the difference in water elevation between high and low tide. The minimum tidal range is a *neap tide*; the maximum tidal range is a *spring tide*. Neap tides occur at the first and third lunar quarters when the sun, earth, and moon form a right angle. Spring tides develop at new and full moons when the sun, moon, and earth are aligned. These configurations are illustrated in Figure 13.9.

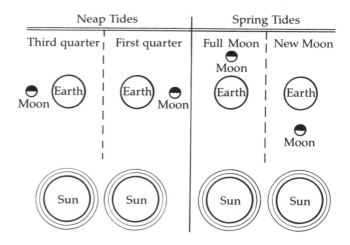

Figure 13.9
Configuration of sun, moon, and earth at various tides.

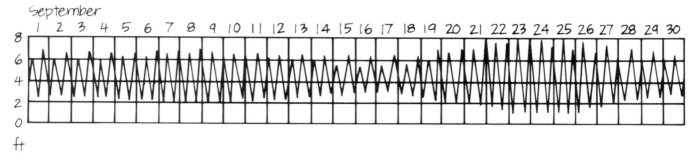

Figure 13.10
Tidal fluctuations at New York, September, 1966.

Problem 16
Study the oscillations in sea-level elevation during September, 1966, at New York, shown in Figure 13.10. Record the following information in the table provided.

1. The average number of oscillations per day (column A).

2. The date of the maximum change in sea level (column B).

3. The maximum sea-level elevation change for one day (column C).

4. The date of the minimum change in sea level (column D).

5. The minimum sea-level elevation change for one day (column E).

	A	B	C	D	E
New York					

What type of tide does New York have—diurnal, semi-diurnal, or mixed?

Problem 17
Label neap and spring tides during September above the appropriate places in the tidal cycle in Figure 13.10.

Tides are also agents of erosion and deposition along a coast, especially if the water moving in tidal flow must pass through a narrow channel. Erosion will occur in the constricted area as the tidal current is forced to move faster; deposition occurs as the current enters the open water on either side of the constricted channel. Figure 13.11 is a photograph of Long Key, Florida, showing an inlet through a chain of islands. The Atlantic Ocean is at the bottom of the photographs.

Problem 18
What is the origin of the lighter tones in the water?

Problem 19
In what direction was the current moving at the time of the photograph?

Problem 20
Where would a boat sailing through the inlet encounter the deepest water? The shallowest?

Figure 13.11
Aerial photograph of the Florida coast at Long Key.

↑N

SHORELINE PROCESSES

Wave erosion and deposition create distinctive land-forms along a shoreline. Erosional shorelines that lack abundant sediment expose bedrock directly to the breaking waves; this forms a rugged steep topography, commonly with small rocky offshore islands. Conversely, depositional shorelines accumulate sediment that covers and protects the bedrock. Smooth, flat, straight or gently curved shorelines characterize wave deposition.

Erosion

Waves erode horizontally, with little vertical down-cutting. As a result, wave erosion causes the land to retreat parallel to the shoreline. The erosion and subsequent retreat of the shoreline create a smooth, gently sloping subsea surface, the *wave-cut platform*, which is sometimes dotted with knobs of resistant rock (*sea stacks*). The shoreline usually displays *wave-cut cliffs*, sometimes with fronting beaches. These features are all shown in Figure 13.12. Along shorelines where

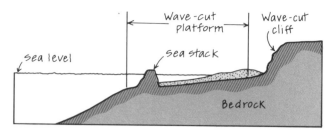

Figure 13.12
Formation of wave-cut bench.

wave refraction occurs, erosion attacks the areas that protrude into the ocean, the *headlands*. Sometimes these protruding points of land become isolated from the mainland to form stacks.

Figures 13.13 and 13.16 show part of the Santa Cruz quadrangle, California. To see the distinctive topography of this area, draw a set of generalized topographic contours on Figure 13.16. Generalized contours are smoothly curving lines that follow the main trend of certain contour lines, ignoring local features such as gullies. The generalized contour line for the 100-foot contour elevation, for example, is given approximately by the trace of State Route 1.

Problem 21

Using a colored pencil, draw generalized contours at elevations of 40, 100, 200, and 300 feet on Figure 13.16.

Problem 22

Sketch a north–south cross-profile of this area that shows clearly the slope of the land between the four generalized contours you have drawn.

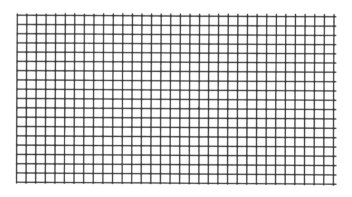

Problem 23

Of the flat areas you noted in your cross-profile, which one has been least eroded by streams? Most heavily eroded?

Problem 24

Which flat area is therefore youngest? Oldest?

Problem 25

What are these flat areas called? Why might they be above sea level?

Problem 26

In a few sentences, summarize the geologic history of the area shown in Figure 13.13.

Problem 27

You have a friend who is a soils scientist. How might your knowledge of the geologic history help your friend to understand the soils formed on the flat areas? How would you expect soils to differ from the lowest platform to the highest?

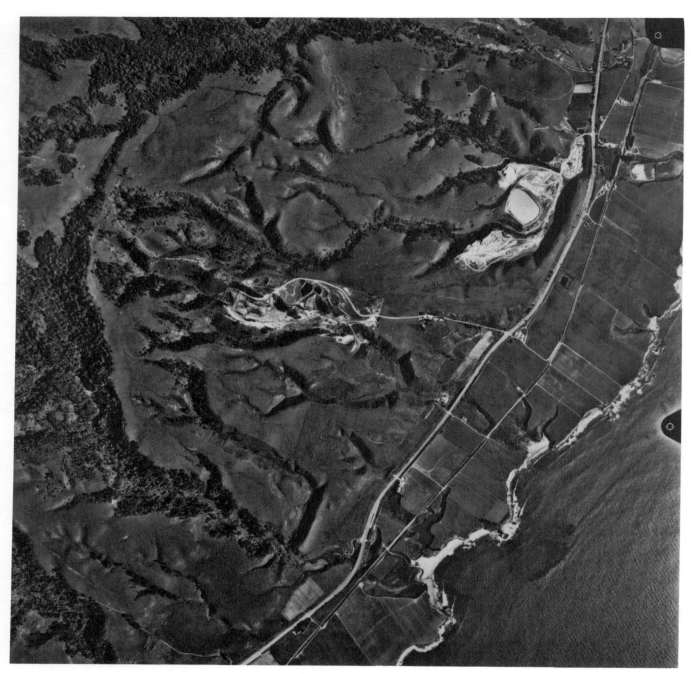

Figure 13.13
Aerial photographs of part of the Santa Cruz
quadrangle, California.

N

0   1 mi

A.

Deposition

Deposition along a shoreline occurs in two primary
forms, clastic and carbonate. Clastic deposition is the
accumulation of rock and mineral fragments. Carbonate deposition is the accumulation of biogenic calcium
carbonate.

Clastic.   Deposition of sediment directly at the land–
water boundary forms a *beach*. Beaches are generally
smooth sloping surfaces above sea level composed of
sand or gravel. Offshore deposits are called *bars. Barrier islands* are long narrow islands parallel to the
shoreline. Because barrier islands break the incoming

Figure 13.13
Aerial photographs of part of the Santa Cruz quadrangle, California.

B.

wave energy, a quiet water environment called a *lagoon* lies between the shoreline and the barrier island. Where a longshore current is active, ridges of sediment sometimes extend from the land in the direction of the current; these are *spits*. A *tombolo* is a wave deposit that connects an island to the mainland.

Problem 28

Figure 13.14 shows the development of a sand spit at the south end of Assateague Island, near Cape Hatteras, North Carolina. What is the rate at which the shoreline is moving, in feet per year?

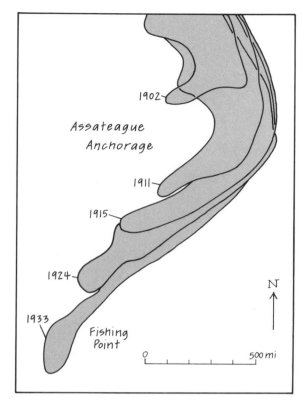

Figure 13.14
Fish Hook spit at the south end of Assateague
Island, North Carolina.

Problem 29

Notice the streamlined shape of the spit. From which
direction is the sediment coming?

Carbonates. Carbonate deposition occurs in warm
shallow marine waters that receive little or no clastic
sedimentation. These conditions promote the growth
of marine plants and animals that utilize calcium carbonate in their tissues, skeletons, and shells. The principal shoreline landforms built by organic activity are
carbonate reefs. These rigid mounds are built by various animals and plants living together in large thick
colonies. Because many of the organisms need light,
the reef framework begins only in shallow water and
grows upward toward the surface. Waves breaking on
the reef carry oxygen and nutrients to the colony, which
promotes the growth of the reef. As the colony grows,
older organisms die, and newer ones grow on the old
skeletons, thereby enlarging the reef over time.

Biogenic calcium carbonate is preserved in the rock
record in many forms. Broken fragments of reef material are typically found on the ocean side of a reef in

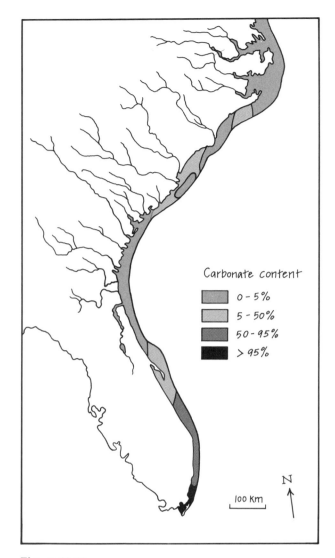

Figure 13.15
Distribution of carbonate sediment along the
southeastern United States.

---

a high-energy environment and may form a limestone
breccia. Carbonate-rich sand and mud are often
deposited in the quiet waters on the sheltered side of
the reef. These sediments may be preserved as finer-
grained limestones. In some cases the whole reef
structure may be preserved intact, creating a thick
massive layer of limestone that extends for many miles.

Problem 30

Examine Figure 13.15, which shows the distribution
of carbonate sediment along the southeastern shoreline of North America. Where is carbonate sediment
most abundant?

California

Contour interval 20 ft

0 ½ 1 mi

0 ½ 1 km

← N

Figure 13.16
Topographic map of part of the Santa Cruz quadrangle, California.

A.

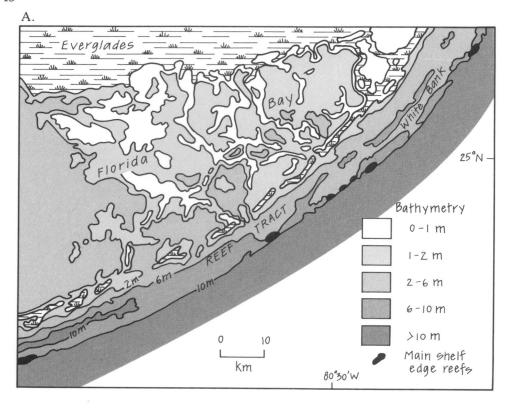

B.

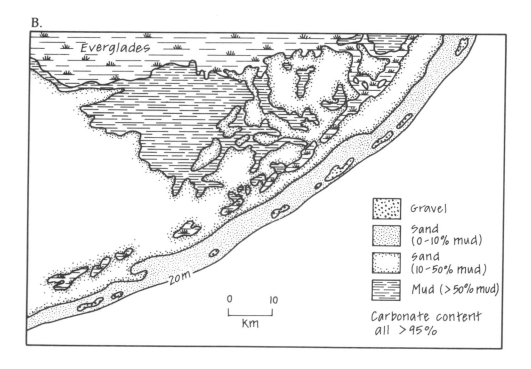

Figure 13.17
Maps showing (A) water depth along Florida Keys and (B) size distribution
of carbonate sediment along Florida Keys.

Problem 31

Do carbonates occur near the mouths of rivers? Why or why not?

Problem 32

Figure 13.17A shows the depth of the water along the Florida Keys. Do carbonate reefs occur in shallow or deep water? What is the reason for this?

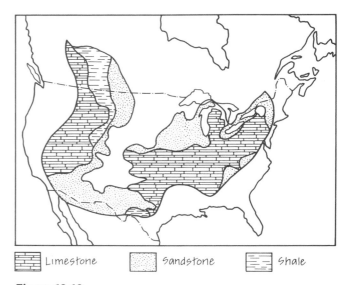

Limestone   Sandstone   Shale

Figure 13.18
Cambrian limestones in North America.

Problem 33

Examine the size distribution of carbonate sediment along the Florida Keys illustrated in Figure 13.17B. How does the sediment size change from the land seaward across the reef? How might you explain this distribution?

Problem 35

Abundant hard-shelled organisms first appeared early in the Cambrian Period. Examine Figure 13.18, which shows the distribution of Cambrian limestone in North America. What does this distribution suggest about the sedimentary environment 550 million years ago?

Problem 34

Summarize characteristics of modern carbonate sedimentation along the southeastern shoreline of North America.

TYPES OF SHORELINES

Both land and standing water bodies change in elevation through time, causing shorelines to move laterally and vertically. Erosion and deposition, fluctuations in water volume, and crustal movements are processes that affect the shoreline location. Shorelines migrate at various rates and over different distances, depending on the processes that cause the change. Over a human time scale, the shoreline may move hundreds of feet; over geologic time, the shoreline may move thousands of miles.

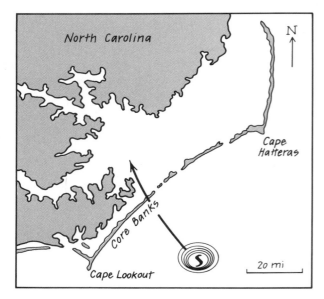

Figure 13.19
Cape Hatteras, Virginia and North Carolina, and
path of Hurricane Ginger in 1971.

Problem 36

Examine the cross-profiles of Core Banks near Cape
Hatteras, North Carolina (Figures 13.19 and 13.20),
on different dates. Hurricane Ginger struck the Cape
Hatteras area in September, 1971. The storm generated winds above 100 km/hr and waves 4 to 5 meters
high. Describe the shoreline changes caused by Hurricane Ginger.

Shoreline movements occurring over thousands to
millions of years develop distinct landforms. Shorelines where the water level has fallen, or the land has
risen, expose elevated wave-cut platforms and beach
deposits. These shorelines are *emergent*. In contrast,
submergent shorelines are those that have experienced a
rise in water level or subsidence of the land. Flooded
river valleys *(drowned valleys)* characterize submergent
shorelines.

Large-scale crustal movements cause ocean shorelines to migrate hundreds or thousands of miles across
the continents. *Transgressive shorelines* advance over the
continents, and *regressive shorelines* retreat from the
continents. Since the Cambrian Period roughly 600
million years ago, North America has had six major
transgressions and regressions of the ocean. The last
major transgression and subsequent regression
occurred during the Cretaceous Period, which ended
about 65 million years ago.

Problem 37

Return to Figure 13.16, the Santa Cruz quadrangle.
Study the character of the valley floors within half a
mile of the ocean. Instead of typical V-shaped stream
valleys, these valleys have broad flat floors similar to
river floodplains, yet they are not floodplains because
the sediments in them are beach sands. Notice that a
tidal marsh or lagoon lies at the mouth of most of these
valleys near where the streams enter the ocean. Do
these features suggest a rise in sea level or a lowering
of sea level? Why?

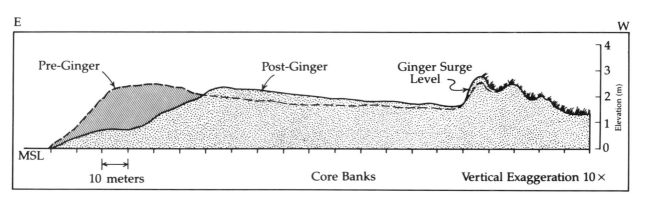

Figure 13.20
Cross profile of Core Banks before and after Hurricane Ginger. Vertical
exaggeration is 10×; MSL is mean sea level.

Problem 38

Do the marine terraces you identified earlier suggest a rise in sea level or a lowering of sea level? Why?

Problem 39

Modify your geologic history of the Santa Cruz area to take into account this new evidence.

Problem 40

Return to Figure 13.19, the shoreline map of Virginia and North Carolina. Most of the bays along this shore show a pattern of small branching arms. Is this shape typical of shorelines formed by wave erosion?

Problem 41

Is this shoreline a submergent or emergent shoreline?

Problem 42

Examine the topographic maps provided by your instructor. Record the shoreline landforms present in the areas in the chart below. Put the name or general location of the feature in the proper column. If the landform is not present, record NP. What processes are occurring along each shoreline? Support your answer with evidence. Are these shorelines emergent or submergent?

Shoreline Landform	Map 1	Map 2	Map 3
• Beach			
• Spit			
• Tombolo			
• Barrier island			
• Lagoon			
• Headland			
• Stack			
• Wave-cut cliff			
• Wave-cut platform			
• Processes that are occurring			
• Submergent or emergent			

In this exercise you have learned what features characterize sediments deposited by waves and tides in an open body of water and carbonate deposits of offshore origin.

Problem 43

Examine the rocks provided by your instructor, and list the ones formed by lithification of beach or lagoon or carbonate deposits of offshore origin.

FIELD STUDY: CAPE COD
NATIONAL SEASHORE

Even to the casual observer, Cape Cod shows an intriguing outline on a map—that of a bent arm, as if a man were showing off his biceps. The cape extends about 25 miles east of the mainland of Massachusetts, then about 30 miles north and northwest to Race Point. It is nowhere more than 10 miles wide or over 200 feet in elevation. Although it is a peninsula of New England, Cape Cod is geologically very unlike the mainland to which it is attached. Most of New England's bedrock is very old, but Cape Cod is very young, a product of the ice sheets of the Pleistocene Epoch. As the continental ice sheet pushed its way southward, it pushed and carried with it fragments of rock and sediment. After reaching its line of farthest advance, the ice margin began to recede, but it left behind a ridge of boulders, gravel, sand, and clay in an arc-shaped ridge. This ridge is the southern part of Cape Cod, the biceps of the arm. Along the north-trending arm of the cape, the material is water-sorted and layered gravel and sand, deposited by meltwater from the glacier. The rest of the peninsula has been shaped by shoreline processes.

Figure 13.21
Hypothetical shoreline of Cape Cod as it might have been 3500 years ago.

Problem 44

Figure 13.21 shows Cape Cod as it might have been after the glacial retreat. The dotted line shows the present shoreline. Is the irregular shoreline typical of wave-shaped shores? Why or why not?

Problem 45

Mark on Figure 13.21 the area that seems to have undergone the most conspicuous erosion in postglacial times. What predominant wave direction does this area of erosion imply? Draw an arrow on the maps (both Figures 13.21 and 13.22) to indicate this direction.

Problem 46

Shoreline studies done in the 1800s indicate that the average yearly westward movement in the north-trending shoreline is about 3 feet each year. At this very rapid rate, how long will it take for Cape Cod to be eroded through at its narrowest point, near South Wellfleet (see Figure 13.22)?

Problem 47

Reexamine Figure 13.22 and notice the very obvious streamlined forms of many parts of the shoreline. The geologic map of Cape Cod indicates that all these areas are composed of sand. Careful consideration of the curvature of these beaches along with the knowledge of the predominant wave direction will enable you to determine the direction of longshore drift along the coastline of Cape Cod. Place arrows along the entire shoreline of Cape Cod in Figure 13.22 to indicate the direction of local longshore drift.

Problem 48

Explain, in terms of wave refraction around obstacles, why longshore drift flows in so many directions around Cape Cod.

Problem 49

In Wellfleet Harbor are several islands of glacial material. Actually the "islands" are connected to the mainland by narrow necks of sand. What are these connecting links of sand called?

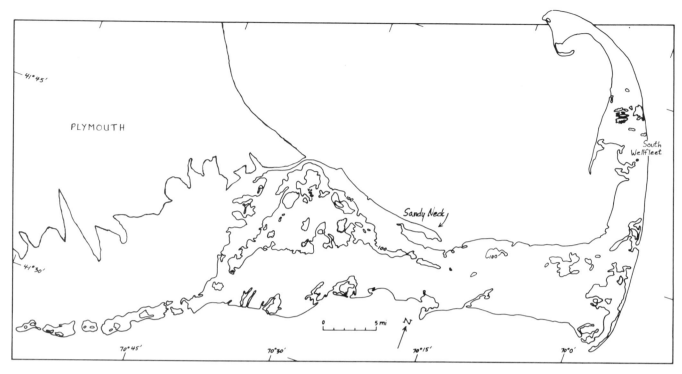

Figure 13.22
Topographic map of Cape Cod.

Problem 50

Tides have had a major role in the shaping of the shoreline of Cape Cod. Tidal currents drag sand and pebbles along the bottom of channels, scouring out deep areas. They are also swift enough to keep in suspension smaller silt and clay particles. Imagine, then, a tidal current sweeping through an inlet such as the one at the east end of Sandy Neck on the north shore of the cape. The water filling the bay from an incoming tide is turbid with suspended sediments. At high tide the current ceases, and the water is nearly still for several hours. What happens to the silt and mud in still water?

Problem 51

With the answer to the preceding question in mind, explain how the mud flats and salt marshes inland from Sandy Neck developed.

Problem 52

Suppose the National Park Service wants to build a fancy new Visitors Center with a lovely ocean view. With erosional and depositional processes in mind, decide on a safe site for the new building. Mark your proposed site with an X on Figure 13.22.

Glaciers

The glaciers creep
Like snakes that watch their prey
From their far mountains, slow rolling on . . .

Percy B. Shelley

Glaciers are masses of ice that persist on the land throughout the year. At present, "permanent ice" covers nearly 6 million square miles, or approximately 10 percent of the earth's surface area. Except for Australia, glaciers exist on every continent. Antarctica contains most of the ice, accounting for 86 percent of the world's total.

Low temperature and snowfall are the primary climatic conditions that promote the formation of glaciers. These conditions prevail in two general areas of the earth's surface, mountains and polar regions. Glaciers of mountainous regions, called *alpine glaciers*, accumulate in high-altitude hollows and valleys and flow downslope through preexisting stream valleys. Alpine glaciers are long and narrow in shape, a few miles wide, and tens of miles long. In contrast, *continental glaciers* are extensive sheets of ice that form at high latitudes. Continental ice sheets can be over a mile thick and extend over millions of square miles. Like many other descriptive categories, the difference between alpine and continental glaciers is not always distinct. The continental glaciers covering Greenland at present, for example, divide into narrower tongues of ice at the periphery of the ice sheet and flow down valleys to the sea, much like alpine glaciers. In addition, present-day alpine glaciers could become local centers of continental glaciation if global temperatures were to drop significantly. Figure 14.3 shows landforms associated with alpine and continental glaciation.

DISTRIBUTION AND DIMENSIONS

Although glaciers are relatively rare surface features today in North America, stratigraphic evidence from surface deposits suggests that at least four advances of glacial ice have occurred across the continent in the last 2 to 3 million years. As a result, the topography in much of the northern and eastern parts of the continent reflects erosion and deposition by glaciers. Mapping the orientation and distribution of glacial landforms has given geologists evidence to reconstruct the most recent glacial environment in North America. However, the reasons for past ice ages are not well understood, and prediction of future glacial advances remains only speculative.

Distribution of Glaciers

Examine the distribution of glaciers in the world today, as illustrated in Figure 14.1.

Problem 1

Note the elevations given for some of the world's alpine glaciers. Plot glacial elevation with respect to latitude on the graph in Figure 14.2. We know that temperatures are colder near the poles and warmer near the equator. We also know that temperature is colder at higher elevations. Explain the relationship shown on the graph.

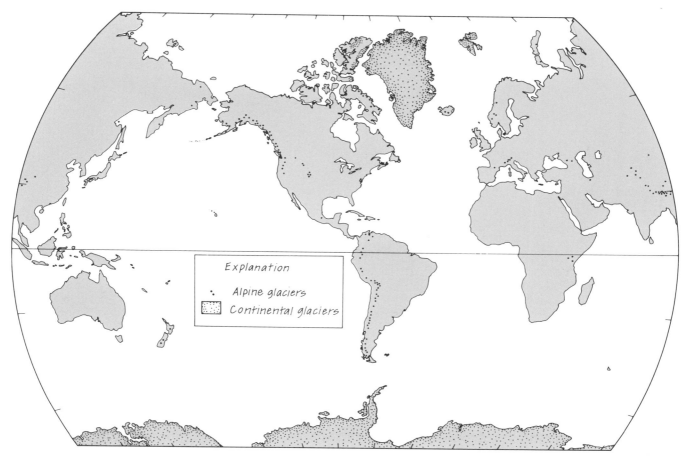

Figure 14.1
Worldwide distribution of glaciers.

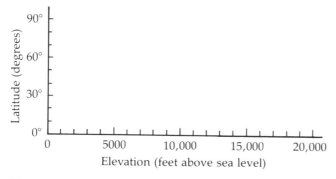

Figure 14.2
Graph showing the elevation of
glacial ice with respect to
latitude.

Problem 2

The average annual sea-level temperature is recorded in Exercise 9, in Figure 9.1. What is the relationship between average annual sea-level temperature and continental glacial distribution?

Problem 3

In what physical regions of the United States are there glaciers? What is responsible for this distribution?

DEPOSITIONAL LANDFORMS

LANDFORM	CHARACTERISTICS	ORIGIN	TOPOGRAPHIC MAP
Moraine 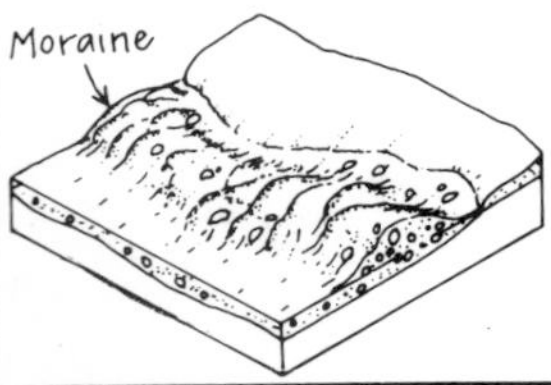	• Composed of till • Size and shape variable, depending on specific origin	• Rock rubble deposited directly from glacial ice • Common to all glaciers	
Drumlin 	• An elongate oval hill • Principally composed of till, sometimes with bedrock core • Often in clusters	• Unique to continental glaciers • Origin uncertain	
Esker 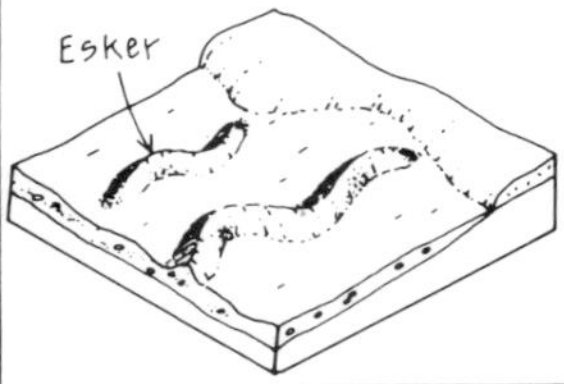	• A meandering narrow ridge • Composed of stratified drift	• A subglacial stream channel deposit • Common in continental glaciers.	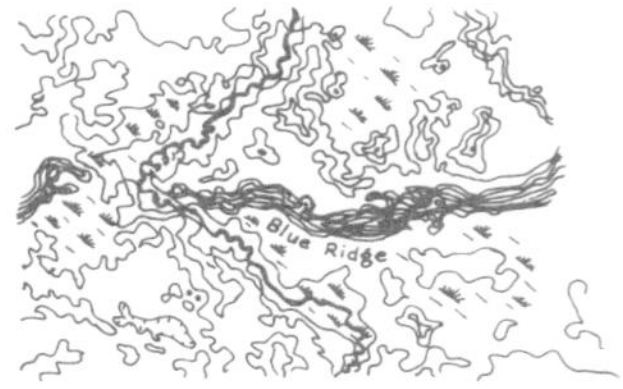
Kame 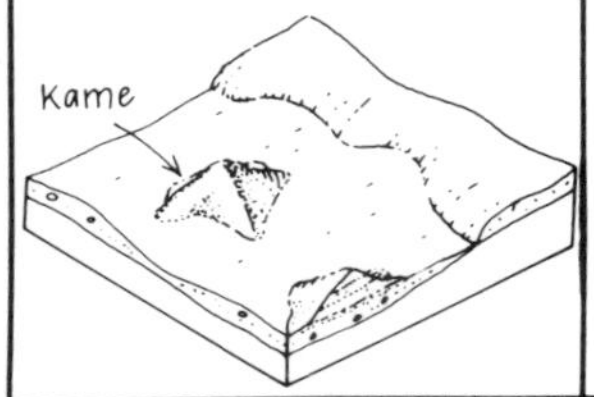	• An irregularly shaped topographic high • Composed of stratified drift	• A subglacial delta or fan deposit • Common in continental glaciers	
Outwash plain 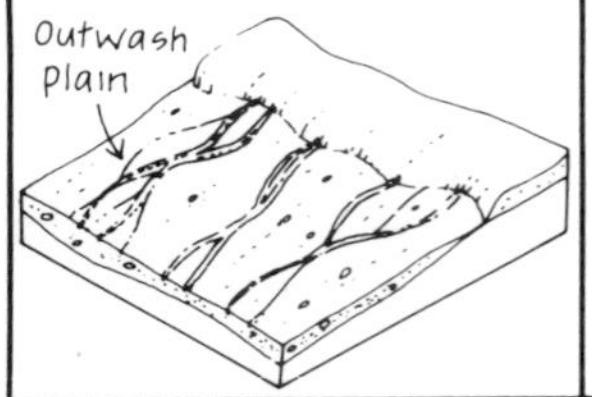	• An alluvial apron in front of the glacial terminus • Composed of stratified drift	• Drift deposited by melt-water streams	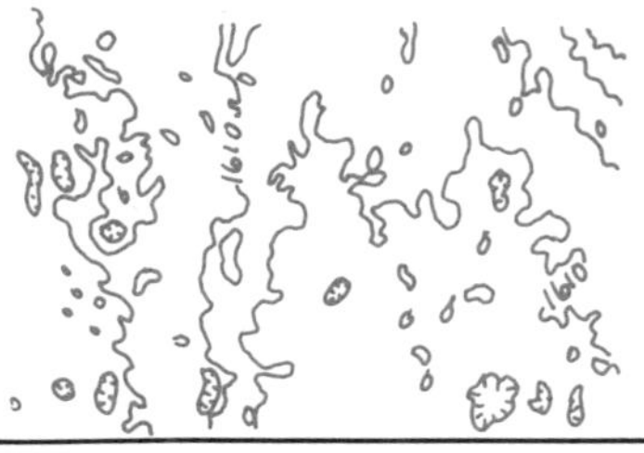
Kettle 	• A closed depression in the outwash plain • Often circular in shape • Commonly filled by water	• The melting of remnant ice in the outwash plain	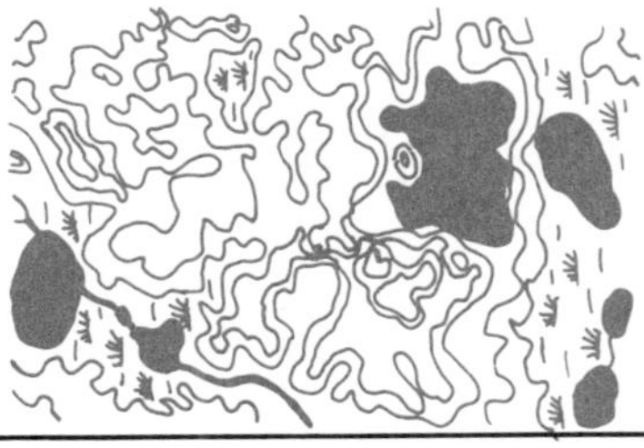

Figure 14.3
Landforms of continental and alpine glaciation.

EROSIONAL LANDFORMS

LANDFORM	CHARACTERISTICS	ORIGIN	TOPOGRAPHIC MAP
Cirque 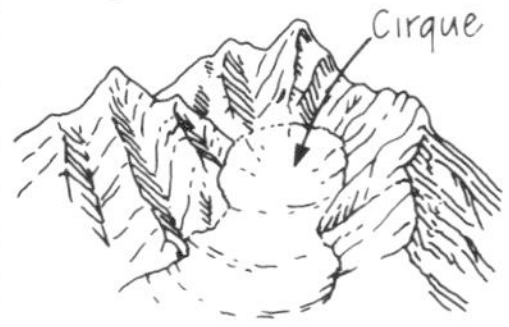	• A bowl-shaped depression near the crest line of ridges and peaks	Erosion at the head of an alpine glacier	
Tarn 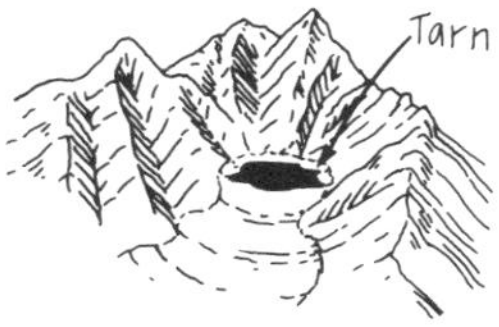	• A small lake or pond within a cirque	Water-filled cirque depression	
Horn 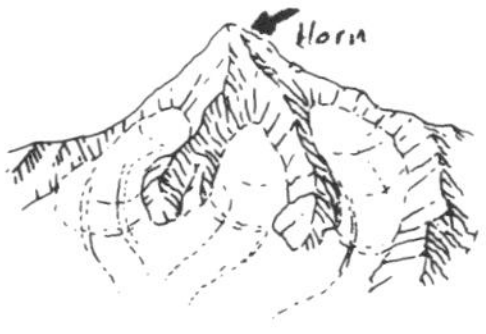	• An isolated rock spire separating several cirques	Headward erosion of several glaciers around a mountain peak	
Arête 	• A jagged, narrow ridge separating two or more cirques	Headward erosion of two or more glaciers from opposite sides of a ridge	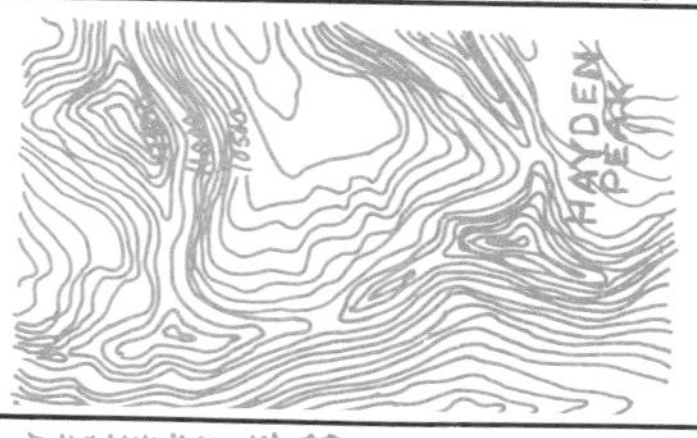
Col 	• A gap in an arete connecting two cirques	Headward erosion of glaciers	
Glacial valley 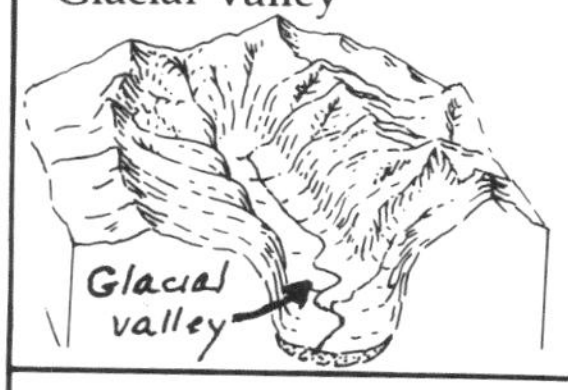	• A deep, steep-sided valley with a flat bottom and U-shaped cross-profile	Erosion of a preexisting stream valley by a glacier	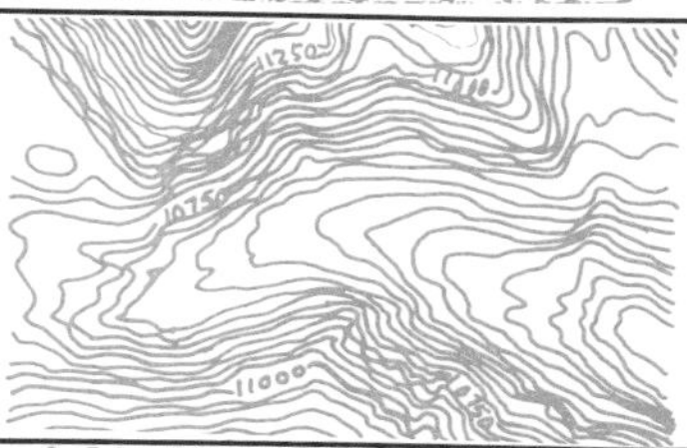
Hanging valley 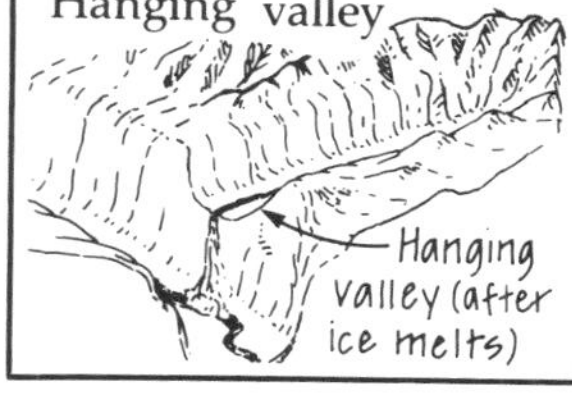	• A tributary valley that intersects the main valley at a higher elevation	The intersection of a tributary glacier with the main trunk glacier	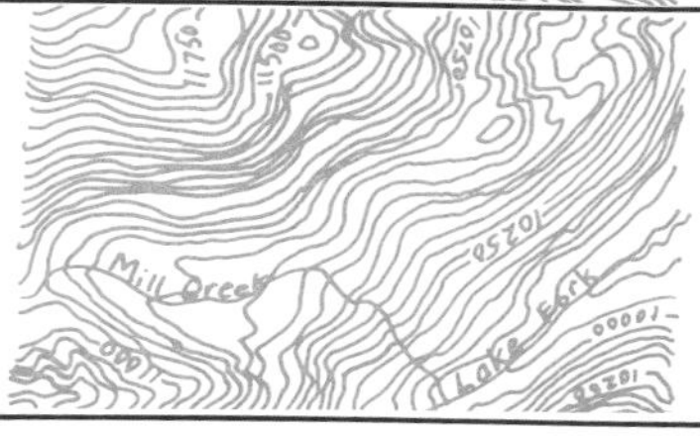

Figure 14.3
Landforms of continental and alpine glaciation.

Dimensions

Examine the topographic map of part of Mount Rainier (Figure 14.5).

Problem 4

Describe the shape and size of the glaciers in this area. How long are they? How wide? Are they wider at one end than the other?

Problem 5

What is the elevation range of the glaciers on Mount Rainier?

Problem 6

What type of glaciers are these?

Problem 7

Is the topography of the ice surface parallel to that of the surrounding bedrock surface?

Examine the topographic maps of Antarctica in Figure 14.4; one illustrates the ice surface, and the other shows the bedrock surface beneath the ice.

Problem 8

What type of glacier is present on Antarctica?

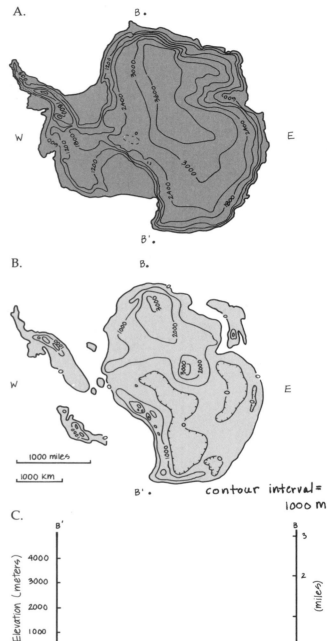

Figure 14.4
Topographic maps of ice and bedrock surfaces, Antarctica.

Problem 9

On the coordinates provided in Figure 14.4, draw a topographic profile across the Antarctic ice surface between points B and B′.

Problem 10

Using the same coordinates, draw a topographic profile across the rock surface of Antarctica between the points B and B′. Label each profile.

0   5 mi

0   5 km

N

Contour interval 100 ft

Figure 14.5
Topographic map of Mount Rainier, Washington.

Problem 11

Are the ice and bedrock surfaces similar in shape? Compare this result to the case of Mount Rainier.

GLACIAL ICE

Glacial ice forms wherever the winter's snowfall does not completely melt in the summer. Over many years, the accumulated snowfall is gradually transformed to glacial ice. First, pressure from the weight of overlying snow, plus some melting and refreezing, transforms snow to granules of ice called *firn*. Firn then compacts into a solid mass of ice under the increasing pressure from the mounting mass of overlying snow. When the ice mass is about 100 feet thick, it begins to flow plastically outward and downhill under the force of gravity, and the surface area of the glacier expands.

Problem 12

Review the three criteria of a mineral. Do both snow and ice satisfy these criteria? Explain.

Figure 14.6 illustrates samples of snow and ice from Antarctica. Sample A is snow collected at the surface. Samples B and C were collected from cores at different depths within the ice, at 71 and 300 meters, respectively.

Problem 13

Describe what happens to snow as it is buried. Mention changes in crystal shape and size and the overall density of the sample.

Problem 14

What rock process is analogous to the changes observed in the ice with increasing depth?

A. Snow

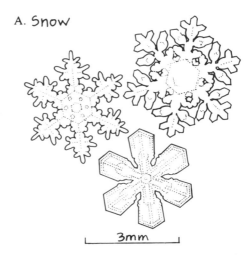

B. Ice

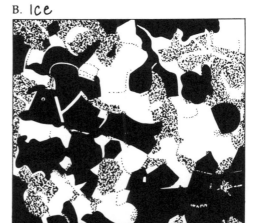

C. Ice

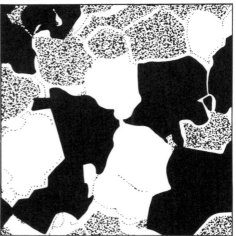

Figure 14.6
Samples of snow and ice from Antarctica.

Problem 15

Radiocarbon dating of material in the ice at the 300-m depth suggests that this ice was deposited as snow approximately 1600 years ago. Using some reasonable value for the average thickness of the Antarctic ice sheet, estimate the age of the ice sheet. Show all your calculations.

GLACIAL PROCESSES

Glacial ice moves slowly from the highest elevation of the glacier, the *head*, to the lowest elevation, the *terminus*. Typical velocities range from less than an inch to a few feet per day. During glacial *surges* (unexplained periods of faster glacial movement), however, the ice may flow over 50 feet in a day. As the glacier flows, rock and other debris that have fallen onto the ice surface also move, creating the distinctive landforms of glaciated regions. Rock material transported by glacial ice, either under the ice sheet or on top of it, is called *drift*.

The ice movement extends beyond the area of net yearly snow accumulation into the area of net yearly ice loss. These two areas are, respectively, the *zone of accumulation* and the *zone of wastage* (or *ablation*). The boundary between the two zones is called the *equilibrium line*.

The position of the terminus of the glacier depends on the balance between the generation of ice and the rate of wastage. If the volume of ice moving down from the zone of accumulation exceeds the wastage, the terminus advances downslope; the terminus melts back or retreats when wastage predominates. The terminus remains stationary when the inflow of ice from the zone of accumulation equals the rate of wastage at the terminus. These three conditions are illustrated in Figure 14.7.

Figure 14.9 is a series of drawings made from aerial surveys and satellite images of Lowell Glacier (a long alpine glacier that descends from Mount Kennedy in Kluane National Park Reserve in Alberta, Canada) in 1954 and 1973. The dark areas in the drawings represent drift present on the surface of the ice. Points A, A', B, and B' are fixed points on the bedrock surface adjacent to Lowell Glacier. Points C and D are distinctive patterns in the drift on top of the ice.

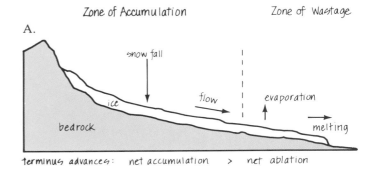

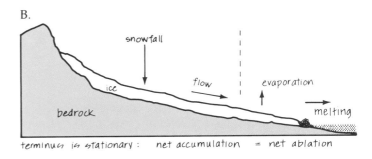

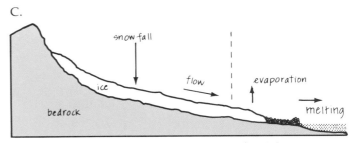

Figure 14.7
Three glacial conditions: A. Advancing. B. Stationary. C. Retreating.

Problem 16

On Figure 14.9 draw line segments across the glacier from A to A' and B to B'. Note the location of points C and D on the 1954 map of Lowell Glacier, and then find and label their positions on the 1973 map.

Problem 17

Measure the displacement between 1954 and 1973 of points C and D, using the line segments A–A' and B–B' as reference locations.

• C

• D

Problem 18

What is the average yearly rate of ice movement based on the displacement of point C? Of point D? Show your calculations.

Problem 19

How do you account for the difference in the two rates?

Problem 20

What happens to the shape of the trains of rock rubble near the terminus? What geologic structure do they resemble?

Problem 21

Is the terminus of the glacier undergoing compression or extension?

Problem 22

Within a 14-month period between the spring of 1983 and the summer of 1984, the terminus of Lowell Glacier advanced 1.45 miles. Calculate the yearly rate of ice motion during this period, and compare this value with the one obtained for ice movement from 1954 to 1973. What glacial event occurred during 1983 and 1984?

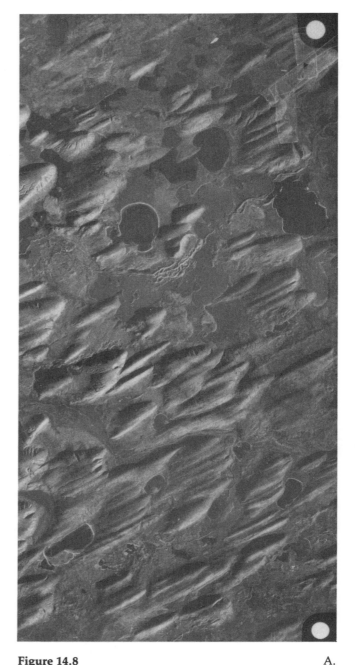

Figure 14.8   A.
Stereo aerial photographs of the Canadian shield.

GLACIAL LANDSCAPES

Like streams, glaciers generally erode the upper parts of their courses and deposit debris near the lower ends. Alpine and continental glaciers form different characteristic landforms, some the result of erosion, others the result of deposition. Some of the most spectacular mountain scenery is the result of erosion by alpine glaciers because they flow as isolated streams through preexisting valleys and leave higher areas untouched, thus exaggerating the relief of the landscape through which they flow. Continental glaciers, in contrast, cover entire areas with great thicknesses of ice, grinding off high points and leveling the topography over which

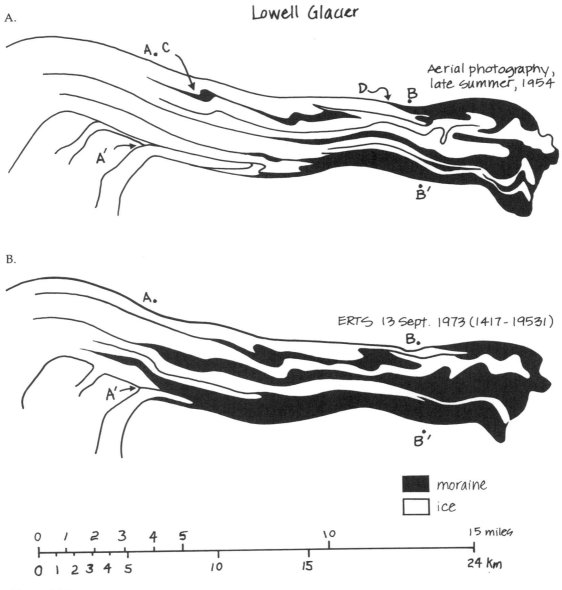

Figure 14.9
Ice surface maps, Lowell Glacier, Alberta, Canada.

they flow. These basic differences in topography between regions of alpine and continental glaciation are readily discernible on topographic maps.

Alpine glaciers erode distinctive mountain landforms. Cirques, arêtes, and horns are formed near the head of a glacier. A *cirque* is a bowl-shaped basin. It has a shape similar to a human heelprint in sand and is often occupied by a small lake *(tarn)* when the ice melts from the area. An *arête* is a sawtoothed divide between two cirques, where two glaciers moved in

opposite directions on opposite sides of a high ridge. A *horn* is an isolated peak formed by the intersection of three or more cirques. Where headward erosion by glaciers breaks through the arête, a low pass called a *col* forms in the ridge.

Below the head, alpine glaciers erode the base and sides of the valleys through which they flow (usually preexisting stream valleys), flattening and widening the valley floor and steepening the sides. The cross section of the resulting valley has a characteristic U-

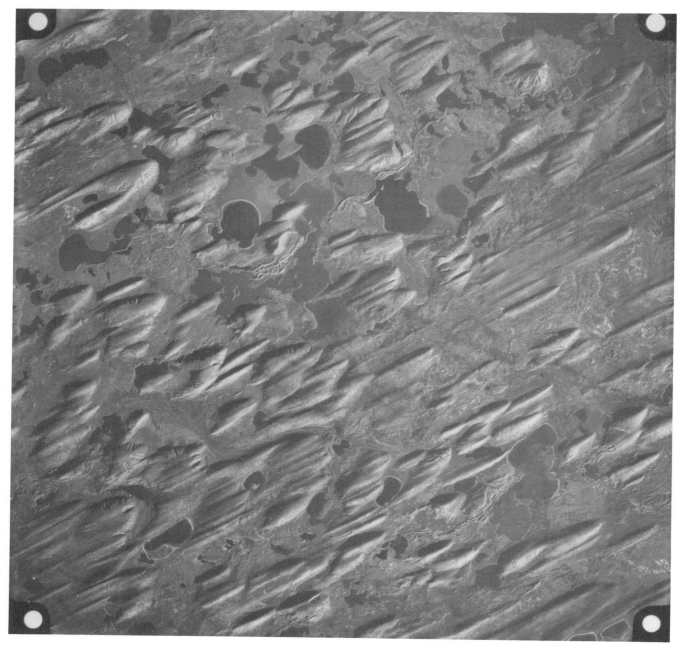

Figure 14.8
Stereo aerial photographs of the Canadian shield.

B.

N ↑

shape. One particularly scenic landform occurs at the intersection of a tributary glacier and the main glacier. Unlike tributary streams, the bottoms of tributary glaciers do not extend down to the bottom level of the main glacier. When the ice melts, the mouth of the tributary valley is exposed high above the main valley. Glacial valleys of this type are termed *hanging valleys*.

Waterfalls commonly occur at the mouths of hanging valleys.

In contrast to alpine glaciers, continental glaciers flatten the topography by scraping and scouring the underlying bedrock. This action leaves large and small depressions in the rock surface, which often fill with water or sediment upon the retreat of the ice. A sur-

Color Plate 66 shows glacial scratches; Color Plate 67 shows granodiorite from the Sierra Nevada ground flat and polished to a high sheen by glacial action.

Problem 23

Color Plate 66 shows four distinct directions of glacial movement. If we arbitrarily place north at the top of the photograph, these directions are:

A. S40°E

B. S15°E

C. S65°W

D. S15°W

Applying the principle of crosscutting relationships, number these directions in order of decreasing age, 1 being the oldest.

Glaciers also create depositional landforms because, like streams, they deposit the sediment they carry. Rock material deposited directly and indirectly from a glacier is *drift*. Drift may be stratified (layered) or unstratified. Unstratified drift is called *till* and is typically very poorly sorted, with particle sizes ranging from clay to boulders. Till is deposited directly by the glacier. Reworking of glacial material by wind and water forms stratified drift.

Moraine is a general term used to describe most landforms composed of till. The nature and location of the till define several types of moraines. The till accumulation at the terminus of a glacier, the *end* or *terminal moraine*, marks the farthest limit of the glacier's advance. Behind the terminal moraine a retreating glacier may deposit smaller moraines of much the same shape called *recessional moraines*, indicating that the glacier paused in its retreat periodically. If the glacier's retreat was continuous, a sheet of till, the *ground moraine*, is deposited. Other forms of moraines are *lateral* and *medial moraines*, ridges of till present along the sides and in the center of alpine glaciers. *Interlobate moraines* are similar to medial moraines but occur between continental ice lobes.

One special type of glacial landform is a *drumlin*, a streamlined oval-shaped hill composed entirely or partially of till. Characteristically, these hills occur in clusters and are known as *drumlin fields*. Drumlins are unique to continental glaciers and are oriented with their long axes parallel to the direction of ice flow and the steepest slope in the up-ice direction.

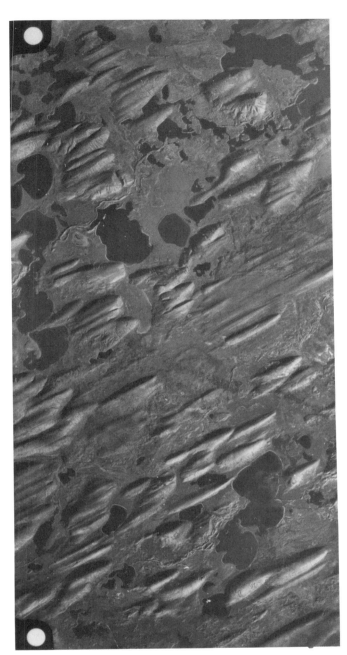

Figure 14.8
Stereo aerial photographs of the Canadian shield. C.

face of lakes, ponds, or marshes is typical of regions that have undergone continental glaciation.

Small-scale erosional features typical of both alpine and continental glaciers are grooves, scratches, and polished surfaces in the bedrock formed by coarse to very fine rock material dragged over the bedrock by the ice.

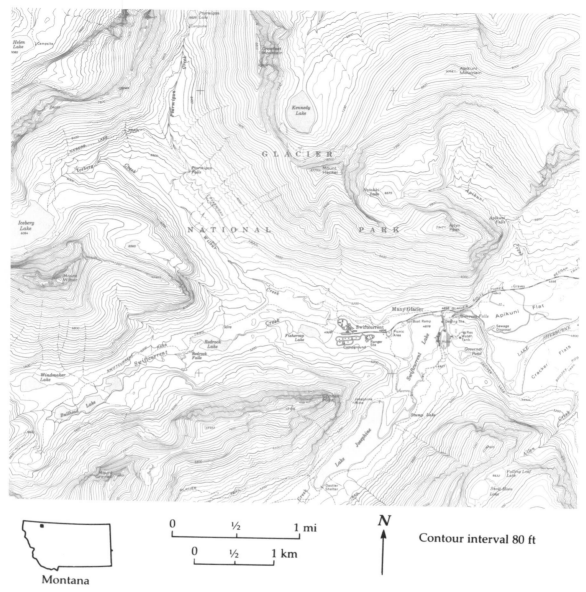

0   ½   1 mi

0   ½   1 km

Montana

N

Contour interval 80 ft

Figure 14.10
Topographic map of part of Glacier National Park, Montana.

Eskers, kames, and outwash plains are depositional landforms composed of stratified drift deposited by meltwater from the glacier. *Eskers* are narrow, winding ridges of gravel deposited by streams flowing under or within the glacial ice. *Kames* are irregular hills or mounds of gravel deposited as deltalike forms when glacial streams deposit their load of gravel. The drift deposited in front of the glacial terminus by meltwater streams is called outwash, and the entire sheet of sediment is the *outwash plain*. Depressions in the outwash plain are *kettles*. Kettles originate by the melting of isolated blocks of ice left by a retreating glacier. Some kettles subsequently fill with water, creating lakes and ponds in the outwash plain.

Examine the stereo photographs of the Canadian shield (Figure 14.8, pages 242, 244, and 245).

Problem 24

Is this topography the result of alpine or continental glaciation?

Problem 25

On one of the stereoscopic photographs, indicate with a number the location of the following features:

1. Drumlin

2. Kettle

3. Esker

4. Kame

Indicate the direction of ice flow with an arrow.

Examine Figures 14.10 (a topographic map of part of Glacier National Park, Montana) and 14.13 (stereo photographs), pages 248, 249, and 251.

Problem 26

To orient yourself as to the location of ridges and valleys on the map, mark the major ridge crests on the map with a red pencil. Are the ridge crests rounded or angular in profile?

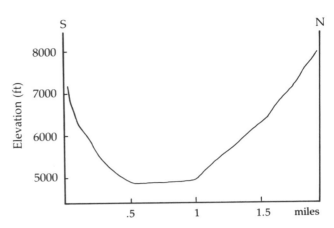

Figure 14.12
Cross-profile of Swiftcurrent Valley, Montana.

Problem 27

There are many fine examples of alpine glacial erosion on this map. Using place names or map coordinates, give an example of one of each of the following features, and describe briefly the pattern of topographic contours by which you recognize it.

• Cirque

• Arête

• Horn

• Tarn

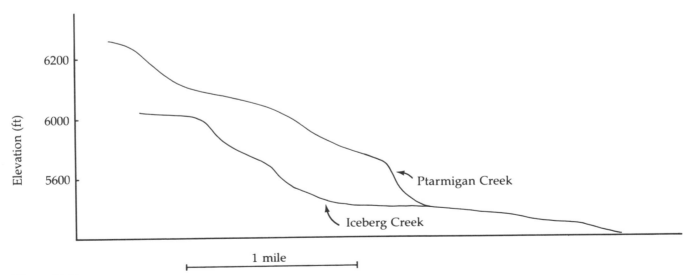

Figure 14.11
Longitudinal profiles of Ptarmigan and Iceberg Creeks, Montana.

Figure 14.13
Stereopairs of aerial photographs, Glacier National Park, Montana.

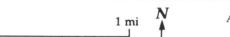

0   1 mi   **N**   A.

Problem 28

Figure 14.12 is a cross-profile of Swiftcurrent Valley. Is the shape of the profile typical of stream erosion or of glacial erosion?

Problem 29

Figure 14.11 shows longitudinal profiles of Ptarmigan and Iceberg Creeks. What is unusual about the profile of Ptarmigan Creek where it joins Iceberg Creek? What type of valley is Ptarmigan Creek?

Figure 14.13
Stereopairs of aerial photographs, Glacier National Park, Montana.

B.

Problem 30

Mark large blue arrows in the center of each valley in Figure 14.10 to show the direction of ice flow.

Problem 31

By locating the *trim line*, the elevation below which the valley sides have been ground smooth by glacial ice, we can determine that the glaciated valleys in this area were about half full of ice at the fullest extent of glaciation. How deep would you estimate the ice was over Swiftcurrent campground?

Figure 14.14
Glacial deposits in the Great Lakes region.

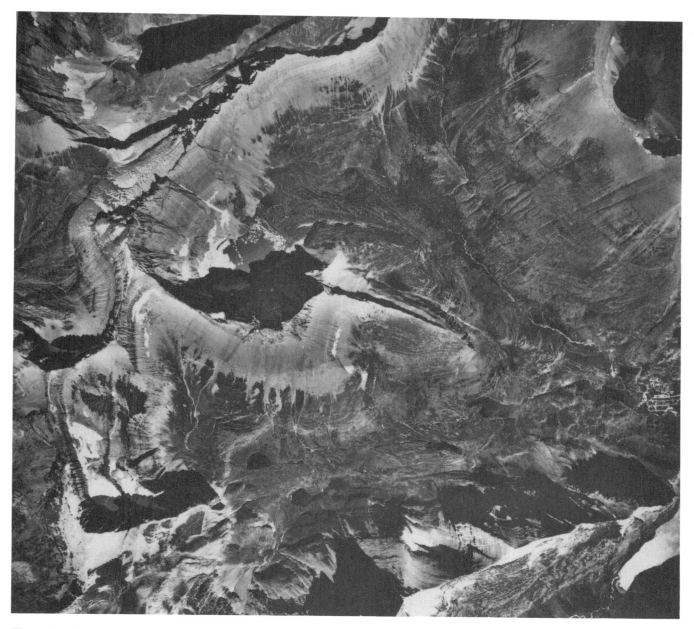

Figure 14.13
Stereopairs of aerial photographs, Glacier National Park, Montana.

C.

Problem 32

If you were to hike from Swiftcurrent campground to Grinnell Point, how could you tell from looking at the rock when you had passed the upper limit of the former glacier?

Problem 33

Examine the rocks provided by your instructor. Have any of them formed from glacial deposits? If so, which ones?

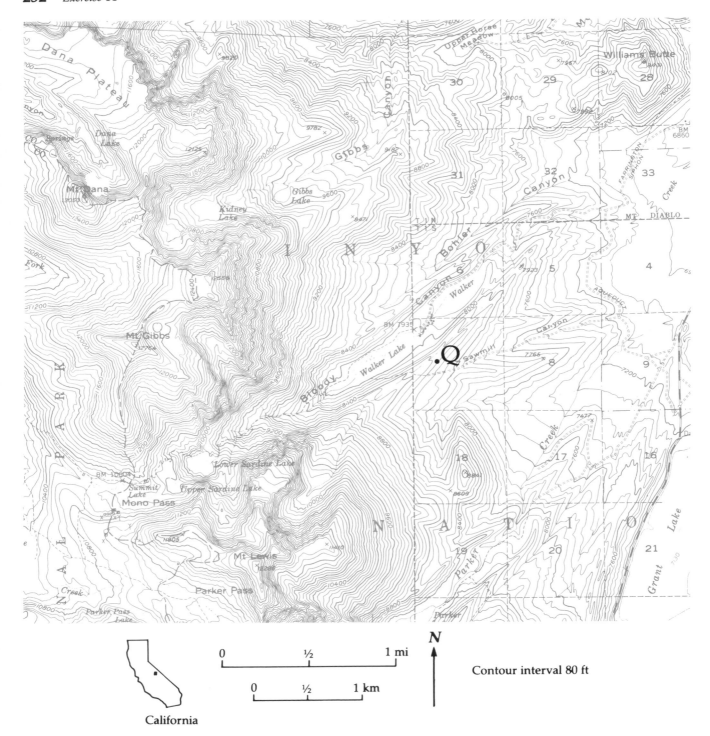

California

Contour interval 80 ft

Figure 14.15
Topographic map, eastern slope of the Sierra Nevada, California, near Yosemite
National Park.

FIELD STUDY: CONTINENTAL GLACIATION IN THE GREAT LAKES AREA

Examine the map of glacial deposits from Pleistocene glaciation in the Great Lakes region (Figure 14.14). The numbers on the map refer to carbon-14 dates from organic material trapped in the glacial deposits. From the glacial deposits and radiometric dates, you will try to reconstruct the environment in the Great Lakes region during the last ice age.

Problem 34

Draw generalized contours (trending approximately east–west) for the radiometric dates on Figure 14.14. Use a contour interval of 1,000 years. Was the ice advancing or was it retreating across the continent approximately 10,000 years ago?

Problem 35

Outline the terminal moraines in red. What does this red line represent?

Problem 36

Color the recessional moraines orange. What do these areas record in the movement of the glacier?

Problem 37

What factors would cause a variation in the width of the different moraines?

Problem 38

Color the outwash deposits green. What do these areas represent in the glacial environment? Where was the glacial front when these deposits were forming?

Problem 39

Color the glacial lake deposits blue. Where do the glacial lake deposits occur with respect to the Great Lakes? What does this suggest about the origin and subsequent history of the Great Lakes?

FIELD STUDY: ALPINE GLACIATION IN THE SIERRA NEVADA

Figures 14.15 and 14.16 are the topographic map and stereoscopic aerial photographs of the eastern slope of the Sierra Nevada near Yosemite National Park.

Problem 40

To orient yourself on the aerial photographs, find similar features on the photographs and topographic map. Place a north arrow on the aerial photographs.

Problem 41

Trace the ridge line between Mount Gibbs and Mount Lewis in red.

Problem 42

Imagine that you are standing at Mono Pass. The bedrock at your feet is granodiorite. Looking toward Mount Lewis, you see a sharp ridge leading up to the peak. What is this glacial feature?

Problem 43

Mono Pass is a low point in this ridge. What is the geological name of this low point, and how did it form?

Problem 44

Looking east down Bloody Canyon, you can see two small lakes, Upper and Lower Sardine Lakes. What is the geological name for the basin in which these lakes occur? What is the origin of these lakes?

Continue walking down the trail to Walker Lake. The high ridges adjacent to the lake are not bedrock but consist instead of a mixture of boulders, cobbles, sand, and silt.

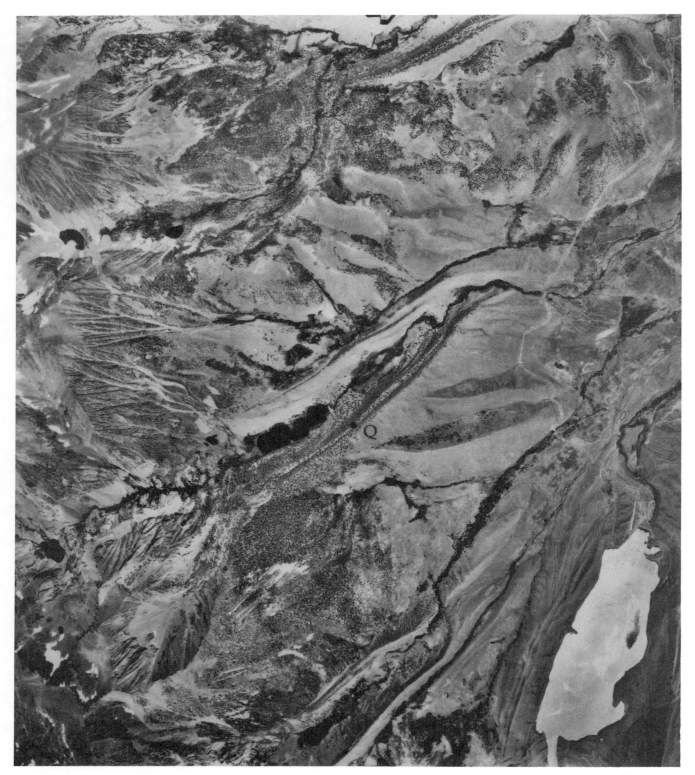

Figure 14.16
Stereopair of photographs, eastern slope of the Sierra Nevada, California, near
Yosemite National Park.

0   1 mi

N

A.

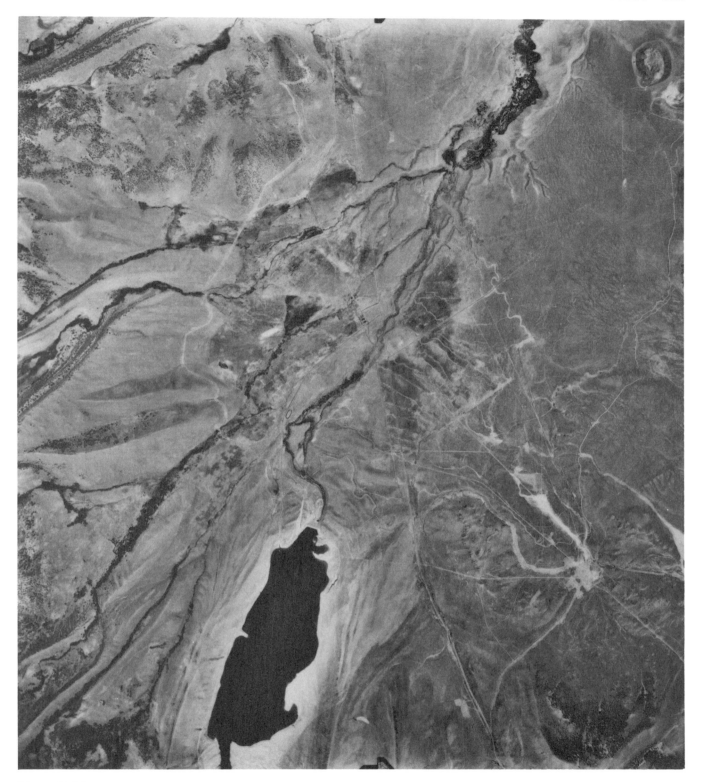

Figure 14.16
Stereopair of photographs, eastern slope of the Sierra Nevada, California, near
Yosemite National Park.

B.

Problem 45

What is probably the main rock type of the boulders? Would you expect this material to be layered? Why or why not?

Problem 46

What is the geologic term for these ridges?

Problem 47

Using the aerial photographs, examine the horseshoe-shaped ridge at the northeast end of Walker Lake. What type of feature is this ridge?

Problem 48

Is the origin of Walker Lake the same as the origin of Sardine Lakes? If not, how did Walker Lake form?

You now walk off the trail to point Q marked on the topographic map and the aerial photographs. Looking east, you see two smaller parallel ridges trending almost due east.

Problem 49

What are these ridges? How did they form?

Problem 50

Which of the ridge systems is older, the northeast-trending ridge you are standing on or the eastward-trending ridges? How do you know?

Problem 51

On the topographic map, make a geologic map of glacial features you see in Gibbs Canyon and Bloody Canyon and along Parker Creek.

1. Trace in green pencil the crests of all moraines—lateral, terminal, and recessional.

2. Use blue pencil to shade in the area covered by ice from the head of the glacier to its farthest extent.

3. Indicate the location of erosional features, such as cirques, aretes, and horns.

4. Draw a line that separates areas of glacial erosion from areas of glacial deposition.

Problem 52

What is the dominant land surface form in the Great Lakes area? In the Sierra Nevada?

Wind

No wind serves him who addresses his voyage to no certain port.

CHARLES COTTON

Wind is the movement of air more or less parallel to the earth's surface. Geologic processes associated with wind are called *eolian* processes (from Aeolus, Greek god of the wind). Global wind patterns are relatively stable because they are caused by interactions between the earth's rotation and the large influx of solar heat at the equator. Global patterns of air circulation are responsible for the distribution of the earth's major deserts. The movement of air on a more local scale is due to differences in air pressure, which in turn develop from variations in the density of the air in different places. An average column of air 1 square inch in area extending from sea level to the top of the atmosphere weighs 14.7 pounds. That is, it exerts a pressure of 14.7 pounds per square inch (1 atmosphere, or roughly 1 bar) on the surface. However, if the column of air were heated, it would expand outside the column, its density would decrease, and it would weigh less, therefore exerting a lower pressure. Conversely, a column of cold, denser air would weigh more than average and therefore exert a higher pressure. In the atmosphere, areas of cold and warm air result in areas of high and low pressure. Winds develop when air moves from areas of high pressure to areas of low pressure.

The difference in pressure between the high and low areas determines the wind speed. If there is little difference in pressure between the areas, light winds develop. Conversely, where there is a large difference, strong, gusty winds blow. Figure 15.1 shows landforms associated with wind action.

When we discuss the geologic work of the wind, we often discuss arid regions because wind is effective as an agent of erosion and transport only when loose sediments are dry. Even a thin film of water between grains will prevent wind erosion because of the very high surface tension of water that holds particles together. The "Dust Bowl" of the 1930s was an area of prolonged drought in the southwestern states.

Absence of water (and vegetation) led to erosion by the wind of many millions of tons of valuable topsoil. We should stress at the outset that, even in deserts, where the work of the wind is best observed, running water is the most important agent in forming the landscape.

WIND AND ATMOSPHERIC PRESSURE

Contour lines of atmospheric pressure, called *isobars*, represent the distribution of high and low pressure over the earth's surface. The configuration of the isobars indicates the general direction and speed of the wind. Wind flows across the isobars and toward the direction of lower pressure. The spacing between isobars is indicative of the wind speed; closely spaced lines suggest fast winds, and widely spaced lines suggest slow winds. High-pressure areas are associated with clear dry atmospheric conditions, and low-pressure areas are associated with precipitation.

Examine Figure 15.3, the average atmospheric pressure over the earth during January and July.

Problem 1

At what latitudes do high-pressure areas persist in both winter and summer?

Problem 2

Would you expect these latitudes to be areas of high precipitation or low precipitation? Check your prediction on Figure 9.4, the map of global precipitation.

DEPOSITIONAL LANDFORMS

LANDFORM	CHARACTERISTICS	ORIGIN	TOPOGRAPHIC MAP
Dunes	• An accumulation of sand and silt as a mound on the ground surface	Wind deposition	
Barchan dune 	• A crescent-shaped dune, concave in the direction of wind flow • Migration common	• Wind deposition • Limited sand supply • No vegetation • Constant wind direction	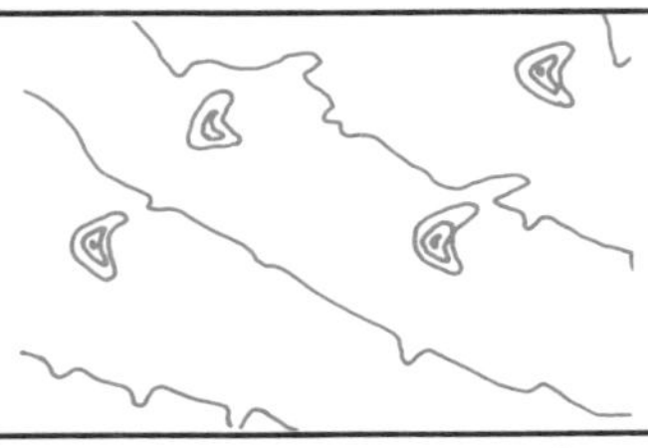
Parabolic dune 	• A crescent-shaped dune, convex in the direction of wind flow • Migration occasional	• Wind deposition • Some vegetation	
Longitudinal dune 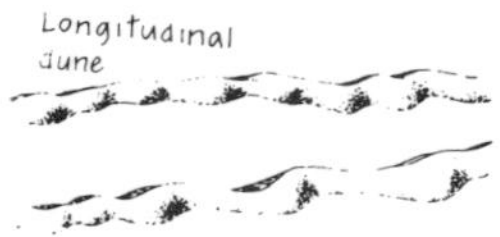	• Ridges of sand parallel to the direction of wind flow	• Wind deposition • Moderate sand supply • Wind variable but from some quadrant	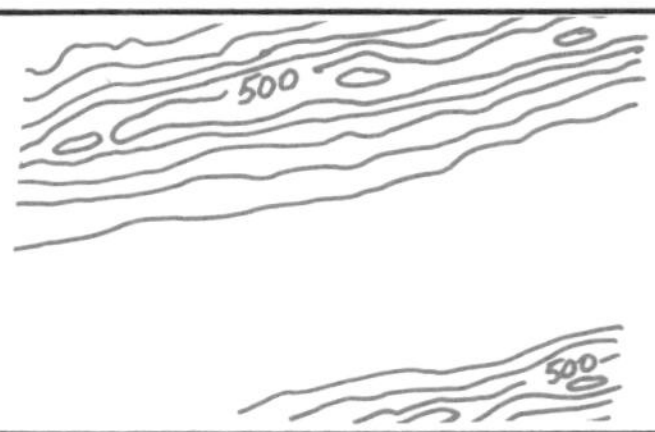
Transverse dune 	• A ridge of sand perpendicular to the direction of wind flow	• Wind deposition • Abundant sand • No vegetation • Constant wind direction	
Irregular dune 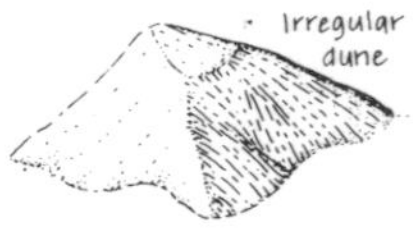	• An accumulation of sand without pattern	• Wind deposition • Abundant sand • Variable wind direction	

EROSIONAL LANDFORMS

Desert pavement 	• A flat gravel-covered surface • Present in arid areas	A residual gravel sheet formed from the erosion of finer particles	

Figure 15.1
Landforms associated with the wind.

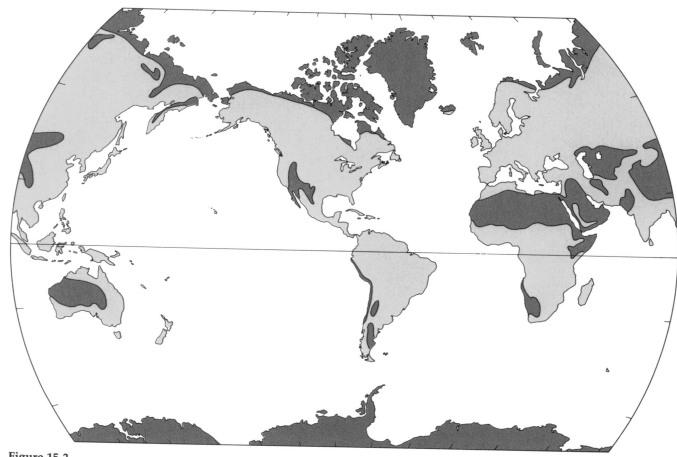

Figure 15.2
Global distribution of deserts.

Problem 3

Notice that low-pressure areas are concentrated south of the equator in January and north of the equator in July. What is the astronomical reason for this pattern?

DESERTS: DIMENSIONS AND DISTRIBUTION

Figure 15.2 shows the global distribution of deserts. Notice that many of the world's major deserts lie near latitudes 25°N and 25°S. This configuration is due to global patterns of atmospheric circulation. The atmosphere is strongly heated near the equator by the sun. This hot air rises, is deflected north and south at high altitudes, cools by expansion, and sinks back to the ground at about latitude 25° north and south of the equator. These areas, near the Tropics of Cancer and Capricorn, are under virtually stationary areas of high pressure due to the continually subsiding air.

Problem 4

From the discussion in the introduction to this exercise and the questions above, explain why most equatorial regions are moist, whereas regions near the Tropics of Cancer and Capricorn are dry.

Problem 5

From the extent of polar deserts shown in Figure 15.2, is the air over the poles rising or sinking?

Naturally, not all deserts occur at the two tropics. The deserts of the southwestern United States, for example, are rain-shadow deserts. They lie east of high mountain ranges that intercept the moist westerly winds off the Pacific Ocean. As the moist air is deflected upward over the mountains, it cools and its moisture condenses as rain on the west slopes of the mountains, leaving warm dry air to continue eastward.

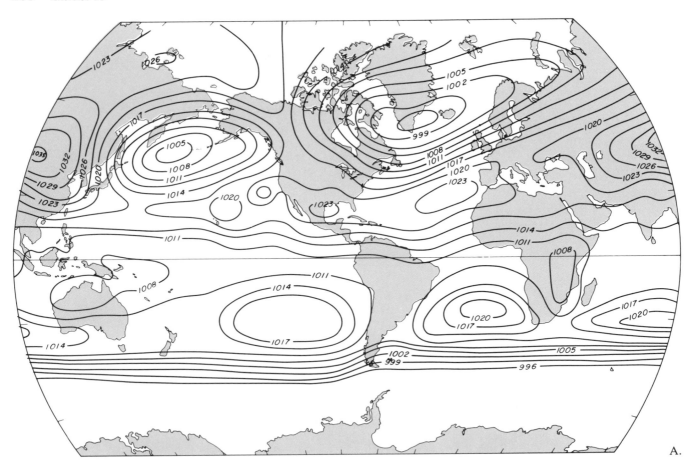

Figure 15.3
Average atmospheric pressures over the earth during (A) January and (B) July.

Problem 6

In what physical regions of the United States are there deserts? Do you see any factors in the geology, structure, or topography (Color Maps 2, 3, and 5) to account for the distribution of deserts? Is there any correlation of desert to land surface form (Color Map 6)?

Problem 7

Where in South America might you expect to find rainshadow deserts analogous to those of the United States?

EOLIAN PROCESSES

Wind is an important sedimentary agent in arid regions, where the surface lacks vegetation, and loose particles lie exposed to the wind. Because of the great differ-

ence in density between air and rock material, wind primarily affects sediment of small grain size. Even at velocities of 50 miles an hour, wind can erode and transport particles no larger than sand; particles of larger size are virtually unaffected by wind. Within this limited size range, however, wind is an effective sedimentary agent. Single wind storms may erode and transport over 100 million tons of sediment. Vegetation can be effective in preventing erosion by the wind because it slows the moving air at the ground surface and protects the underlying loose material from erosion. In addition, vegetated areas are usually moist, and thin films of water plus organic residue from plant decomposition bind particles together to prevent erosion by wind. Wind may be an active erosional agent locally in humid areas where abundant fine-grained sediment is exposed at the surface (shorelines, for example). Erosion of fine-grained material from the surface by wind is called *deflation*. Where deflation is extensive, a coarse pebbly surface called *desert pavement* often develops as the wind leaves behind pebble-

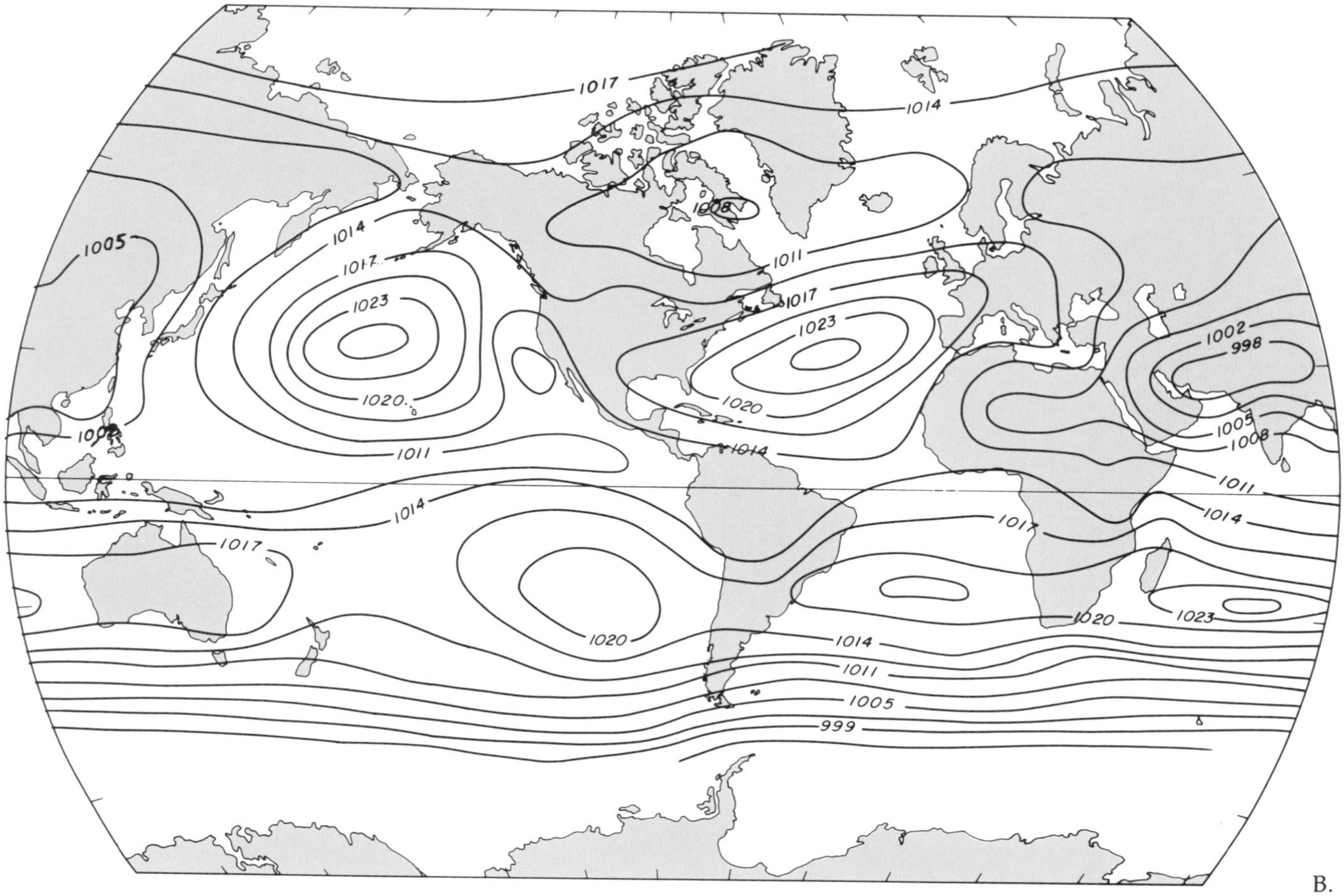

Figure 15.3
Average atmospheric pressures over the earth during (A) January and (B) July.

sized and coarser fragments, and a sheet of gravel eventually covers the surface. Once formed, desert pavement protects the underlying finer particles from the wind, and erosion ceases. Ventifacts (see Color Plate 68) are larger cobble-sized fragments whose surfaces have been smoothed and polished by wind carrying smaller sediments.

Problem 8

How would a strong wind affect the following surfaces?

- Bare sand

- Grass-covered sand

- Bare surface of pebbles

- Bare surface of pebbles, sand, and silt

Problem 9

A desert pavement is fragile, and any disruption of the closely packed pebble surface may expose more fine-grained material to the erosive power of the wind. What would be the result of off-road vehicles traversing desert pavement?

Eolian sediment is rock material that is eroded, transported, and deposited by the wind. Well-sorted fine-grained deposits are characteristic of eolian sediment. Although clay, silt, and sand particles may be eroded together, they are separated during transport and deposited by size. Wind sweeps clay and silt high into the atmosphere, carrying it as suspended load and separating it from sand, which travels near the surface as bed load. This winnowing process results in an extremely well-sorted deposit. In addition, eolian sands are typically well rounded. During transport, especially over long distances, softer mineral fragments disintegrate, and the harder mineral fragments become rounded. Well-rounded quartz sands are characteristic deposits in an eolian environment.

EOLIAN LANDFORMS

Sand *dunes* are characteristic landforms developed by wind deposition. Dunes occur downwind from a source of loose sand. The sand accumulates, generally near obstructions such as bushes and large rocks, where the wind velocity decreases. The deposited sand adds to the wind interference, and with time more sand accumulates. The pile of sand continues to enlarge by this process, and a dune forms.

Dunes are classified by their shape and their orientation with respect to wind direction. Figure 15.4 shows the major types of dunes. When viewed in aerial photographs, the type of dune can suggest characteristics of the area, such as the prevailing wind direction, sand supply, and vegetative cover.

1. *Barchan dunes* are crescent-shaped dunes with the horns pointing downwind. Barchans occur as solitary dunes or in groups, and they suggest an area of scant sand supply, no vegetation, and a relatively uniform wind direction.

2. *Transverse dunes:* Where sand becomes abundant, barchans may coalesce into transverse dunes, long wavy dunes oriented at right angles to the relatively constant prevailing wind.

3. *Parabolic dunes* are crescent-shaped but are often stretched out into hairpinlike forms. The points of a parabolic dune point upwind (unlike barchans) because the wind blows out the center of the dune, leaving behind the sides, which are anchored by vegetation.

4. *Longitudinal dunes* form in regions of moderate sand supply and winds of varying direction (but always from the same quadrant of the compass). These dunes can be very large, as high as 250 meters (800 feet) and extending for tens of kilometers, as in the Arabian desert.

Dunes migrate across the land surface by continuous erosion, transport, and deposition of individual grains. As illustrated in Figure 15.5, dunes have a gentle windward face and a steep leeward (downwind) face. The distinction between the steep and gentle faces defines the wind direction. Wind blowing over the

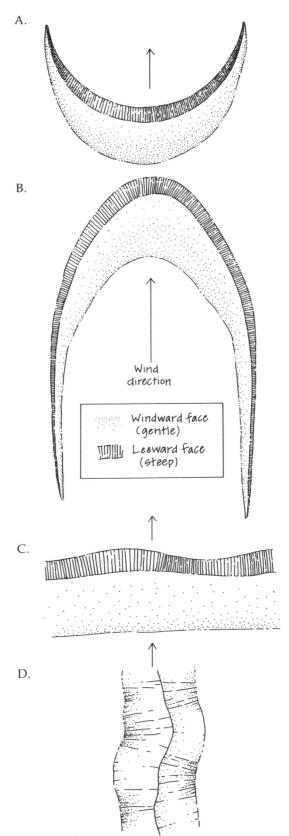

Figure 15.4
Different types of sand dunes: A. Barchan. B. Parabolic. C. Transverse. D. Longitudinal.

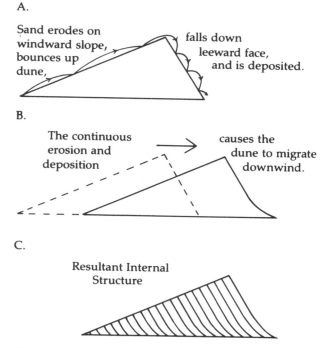

A. Sand erodes on windward slope, bounces up dune,

falls down leeward face, and is deposited.

B. The continuous erosion and deposition

causes the dune to migrate downwind.

C. Resultant Internal Structure

Figure 15.5
Dynamics of sand dunes: A. Movement of sand in a dune. B. Migration. C. Crossbedding produced by sand movement.

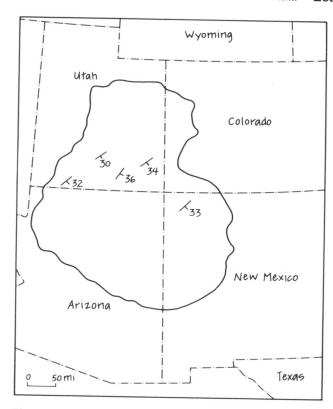

Figure 15.6
Orientation of crossbedding in the Navajo Sandstone.

dune erodes grains from the windward side, transports them up the gentle slope, and drops them on the steep leeward face, and the dune migrates in a downwind direction. This process develops an internal stratification parallel to the leeward face that may be several feet high. The characteristic layering is called *crossbedding* because changes in wind direction cause different orientations of the nonhorizontal layers. When preserved in rocks, crossbedding can indicate the ancient wind direction that deposited the sand. Crossbedding is also common in sediments deposited by water, but water-laid crossbeds are usually much smaller than those associated with wind. See Color Plate 45.

Problem 10

The Navajo Sandstone is a prominent rock unit within the Colorado Plateau. This thick, crossbedded, quartz-rich sandstone was deposited during the Early Jurassic period about 163 million years ago. Figure 15.6 is a map illustrating the strike and dip of the crossbedding preserved in the sandstone at several locations. What was the prevailing wind direction during the deposition of the sand?

Problem 11

The Navajo Sandstone is commonly hundreds of feet thick and has an areal extent of hundreds of square miles. What type of environment does this suggest? What modern geographic location might this Jurassic environment resemble?

Wind is also an important process in volcanic eruptions, especially violent, ash-rich explosions. During an eruption, the volcanic ash blown into the atmosphere is carried by the prevailing winds away from the volcanic vent. As the cloud of ash travels, particles fall to earth, forming a layer of volcanic ash. Typically, these airfall deposits increase in area, decrease in particle size, and become thinner in the downwind direction. If preserved widely in the geologic record, a volcanic ashfall deposit can record the wind direction at the time of the eruption.

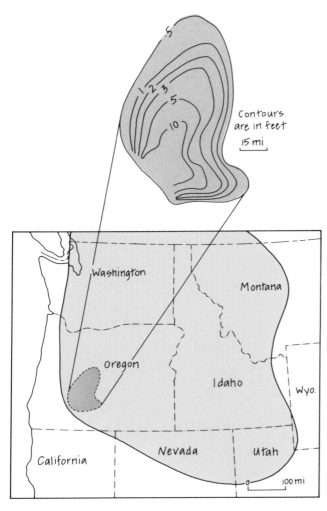

Figure 15.7
Areal distribution of 700,000-year-old volcanic ash.

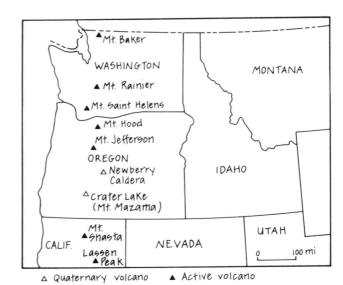

Figure 15.8
Active and extinct volcanoes in the northwestern
United States.

Problem 12

Examine Figure 15.7, a map of the distribution of a
700,000-year-old volcanic ash found throughout the
northwestern United States. Mark a star where you
think the volcano was located. Give two reasons for
your decision.

Problem 13

How would the size of the ash particles vary over the
deposit?

Problem 14

Examine the map of active and extinct volcanoes in
this area shown in Figure 15.8. What volcano was the
source of the ash? What further evidence would you
seek to confirm your conclusion?

Rocks formed from windblown deposits often have
a very distinctive appearance. Quartz is the most
prevalent mineral. The grains are usually very well
rounded, and their surface is often frosted.

Problem 15

Examine the rock samples provided by your instructor. Which of the samples are eolian in origin? Explain
your reasoning.

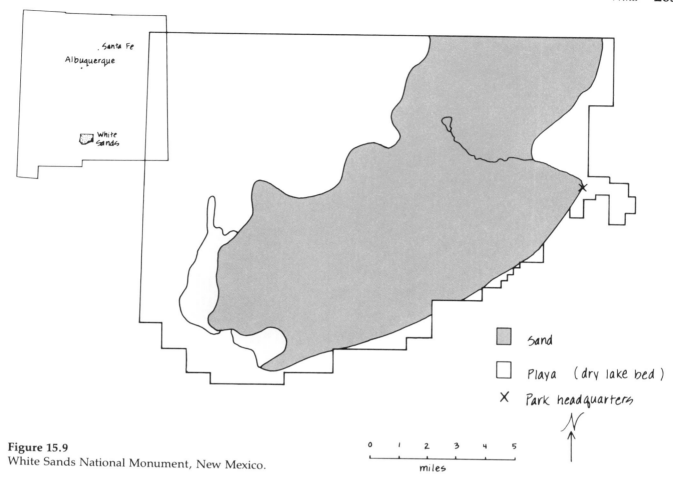

Figure 15.9
White Sands National Monument, New Mexico.

FIELD STUDY: WHITE SANDS NATIONAL MONUMENT, NEW MEXICO

White Sands National Monument in southern New Mexico derives its name from the unusually white sand in the area. The sand is also unusual in its composition: It is gypsum rather than quartz.

Problem 16

What is the sedimentary origin of rock gypsum: clastic, biogenic, or chemical?

Problem 17

Figure 15.9 is a map of White Sands National Monument and the surrounding area. What is the most likely source of the gypsum of which the sand is composed?

Figure 15.10 is a stereopair of aerial photographs of the northeast section of White Sands National Monument. North is approximately at the top of the photograph.

Problem 18

What type of sand dune is in the southwestern corner of the dune field?

Problem 19

What does this type of dune suggest about the vegetation and the sand supply in the area?

Problem 20

What wind direction is indicated by the orientation of the dunes and the position of the leeward faces?

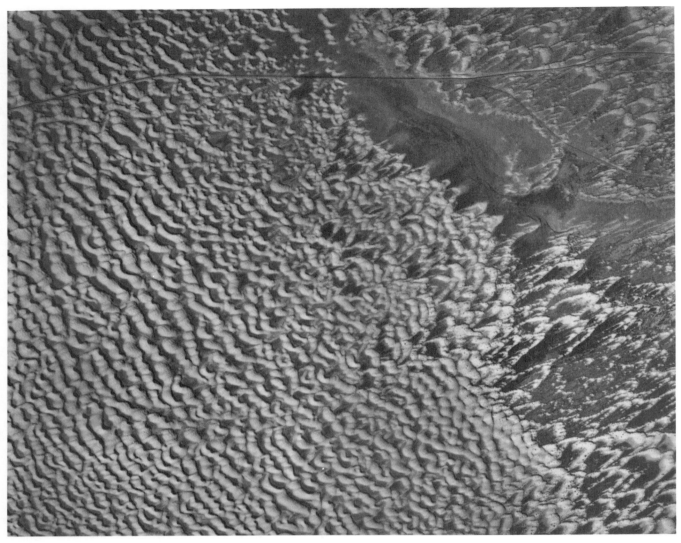

Figure 15.10
Stereopair of aerial photographs of the northeastern section of White Sands
National Monument. (Turn 90° for stereo viewing.)

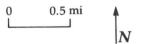

0   0.5 mi

N

A.

Problem 21

In the northwest corner of photograph B, the dunes
are more isolated. What type of dune are they?

Problem 22

What does this type of dune suggest about vegetation
and sand supply?

Problem 23

What type of dune occupies the northeast quadrant
of the map?

Problem 24

What does this type of dune suggest about vegetation
and sand supply?

B.

Figure 15.10
Stereopair of aerial photographs of the northeastern section of White Sands National Monument.

Problem 25

Look closely at the photographs. The small dark dots in the northeast quadrant are bushes. In other areas where there are bushes (southeast corner), do you see the same type of dunes?

Problem 26

From the distribution and orientation of the dunes, do they appear to be stationary or are they moving?

Problem 27

Speculate on how the National Park Service headquarters might be affected by the dunes in the future.

Figure 15.11 shows a parabolic dune located in White Sands National Monument. The dune was mapped at different times from December, 1962, to February, 1965.

A. December 1962

B. January 1964

C. February 1965

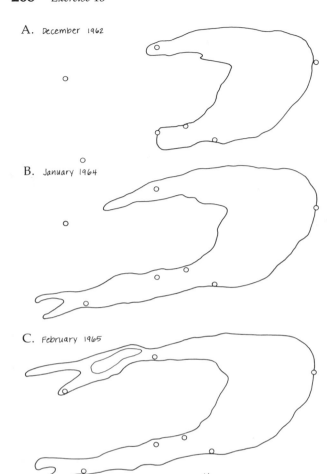

0   100   200   N↑   ○ Shrub
feet

Figure 15.11
A parabolic dune at White Sands National
Monument.

Problem 28

What is the wind direction indicated by the shape of
the dune? Explain your answer.

Problem 29

From your knowledge of the slope angles on the
windward and lee sides of dunes, estimate the
approximate strike and dip direction of the internal
crossbedding in this dune. Explain your reasoning.

Problem 30

How far did the dune move over the period of the
study?

Problem 31

Reexamine your speculations on the fate of the park
headquarters. Is there likely to be a problem in the
near future?

REFERENCES FOR PART C

Books

Bagnold, R.A., *The Physics of Blown Sand and Desert Dunes.* Methuen, London, 1941.

Bascom, W., *Waves and Beaches.* Doubleday, New York, 1980.

Bloom, A.L., *Geomorphology: A Systematic Analysis of Late Cenozoic Landforms.* Prentice-Hall, Englewood Cliffs, N.J., 1978.

Carroll, D., *Rock Weathering.* Plenum Press, New York, 1970.

Dunn, T., and L.B. Leopold, *Water in Environmental Planning.* W.H. Freeman, New York, 1978.

Flint, R.F., *Glacial and Quaternary Geology.* John Wiley & Sons, New York, 1971.

Hunt, C.B., *The Geology of Soils.* W.H. Freeman, New York, 1972.

————. *Natural Regions of the United States and Canada.* W.H. Freeman, New York, 1974.

Jennings, J.N., *Karst.* MIT Press, Cambridge, Mass., 1971.

Kennett, James, *Marine Geology.* Prentice-Hall, Englewood Cliffs, N.J., 1982.

Leopold, L.B., *Water, a Primer.* W.H. Freeman, New York, 1974.

Leopold, L.B., M.G. Wolman, and J.P. Miller, *Fluvial Processes in Geomorphology.* W.H. Freeman, New York, 1974.

McGinnies, W.G., B.J. Goldman, and P. Paylore, eds., *Deserts of the World: An Appraisal of Research into their Physical and Biological Environments.* University of Arizona Press, 1968.

McKee, E.D., ed., *A Study of Global Sand Seas.* U.S. Geological Survey Professional Paper 1052, 1979.

Patterson, W.J.B., *The Physics of Glaciers,* 2nd ed. Pergamon, London, 1981.

Ritter, D.J., *Process Geomorphology.* Wm. C. Brown, Dubuque, Iowa, 1986.

Sweeting, M.M., *Karst Landforms.* Columbia University Press, New York, 1973.

Thoreau, H.D., *Cape Cod.* Crowell, New York, n.d.

Time–Life Books, Planet Earth Series: *Arid Lands; Atmosphere; Edge of the Sea; Flood; Forest; Glacier; Grasslands and Tundra; Ice Ages; Restless Oceans; Rivers and Lakes; Storm; Underground Worlds.* Arlington, Va., 1982.

U.S. Geological Survey, *National Water Summary 1983– Hydrologic Events and Issues.* U.S. Geological Survey Water Supply Paper 2250, 1984.

Waltham, T. *Caves.* Crown, New York, 1975.

Weisberg, Joseph S., *Meteorology: The Earth and Its Weather.* Houghton Mifflin, Boston, 1976.

Young, A., *Slopes.* Oliver and Boyd, Edinburgh, 1972.

U.S. Government Publications

U.S. Geological Survey Information Pamphlets, available free from

Distribution Branch
U.S. Geological Survey
Box 25286, Federal Center
Denver, CO 80225

The Amazon: Measuring a Mighty River, 1976.

The Channeled Scablands of Eastern Washington, 1982.

Geologic History of Cape Cod, 1981.

Geology of Caves, 1984.

Glaciers, A Water Resource, 1980.

Glaciers: Clues to Future Climate? 1986.

The Great Ice Age, 1978.

Ground Water, 1982.

Ground Water Contamination—No "Quick Fix" in Sight, 1980.

The Hydrologic Cycle, 1981.

John Wesley Powell's Exploration of the Colorado River, 1981.

Marine Geology: Research Beneath the Sea, 1984.

Permafrost, 1983.

River Basins of the United States: A Series (Colorado, Columbia, Delaware, Hudson, Potomac, Suwannee, and Wabash), 1981.

State Hydrologic Unit Maps, 1983.

Submarine Landslides, 1981.

Water of the World, 1984.

What Is Water? 1982.

Why Is the Ocean Salty? 1984.

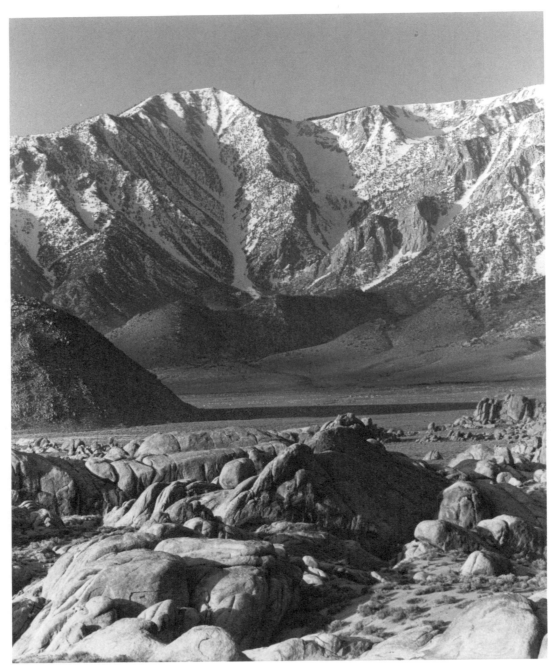

The eastern face of the Sierra Nevada from the Alabama hills, California. Still active normal faulting along the eastern front of the Sierra Nevada has separated the valley floor in the foreground from the peaks in the background.

Internal Processes

In Part C we described the surface processes of the earth, which are driven by energy from an external source, the sun. The processes investigated in Part D are driven by the internal energy of the earth, which is derived from the earth's molten interior. In general, surface processes wear down the earth's crust, and the internal processes build it up. Internal processes include volcanism and plutonism (which add material to the continental crust), earthquakes (which often thrust rocks upward, increasing the elevation of the continents), and plate tectonics. We also discuss here the range of physical properties of the earth as a whole that are due to the nature of the earth's interior. It will become apparent that the earth is by no means a cold, lifeless body but is very active.

Earthquakes are the subject of Exercise 16. One well-recorded earthquake is used as an example of determinations of epicenter, origin time, magnitude, and sense of motion. Records of other earthquakes show how the interior structure of the earth can be studied with seismic waves. Exercise 17 covers the physical properties of the earth—its size and surface relief, heat flow, magnetism, and gravity.

One of the most exciting and important ideas to emerge in scientific thought in recent decades is the theory of plate tectonics. Although the idea that the continents have rifted apart has been the subject of conjecture for centuries, strong scientific evidence in support of a theory of continental motion has been a long time coming. Only since World War II has the necessary technology been available to explore the ocean basins, and it was there that the conclusive evidence for plate tectonics was found. Exercise 18 begins with the evidence from the ocean basins and constructs the various facets of plate-tectonic theory from that evidence. It also recalls previous exercises to add bits of data from various specialties, much as the geologists of the 1950s and 1960s did. Exercise 18 concludes with a working model of the East Pacific region that you can construct. Exercise 22 contains a plate-tectonic "movie" of continental motion since early in the Mesozoic Era, 200 million years ago.

Earthquakes

In nature things move violently to their place, and calmly in their place.

Francis Bacon

Earthquakes are events within the earth that generate elastic waves. They are usually sudden displacements of rock masses along *faults,* zones of weakness in the crust. The waves from the disturbance propagate outward, transmitting the energy of the earthquake to other parts of the earth. The particles of matter through which the waves move simply vibrate back and forth about their normal positions. This process is analogous to transmission of the energy of a storm at sea to a distant shore by means of ocean waves. The water itself simply oscillates; only the energy of the wave is transmitted.

Seismic waves (from the Greek *seismos,* shaking) may vibrate back and forth in the direction in which the wave front travels, like sound waves. These are compressional waves. They are called primary or *P*-waves because they are the first to arrive from a distant earthquake. Secondary or *S*-waves travel more slowly than primary waves. In *S*-waves, the motion of particles is from side to side, perpendicular to the direction in which the wave is traveling. These waves are analogous in their motion to transverse light waves. Both types are illustrated in Figure 16.1.

S- and *P*-waves are called body waves because they travel through the interior of the earth. Their velocities depend on the compressibility, rigidity, and density of the rocks through which they travel, and much has been learned about the interior structure of the earth through the study of the complex routes and travel times of these waves. In a sense the body waves of earthquakes can be used to x-ray the earth, although the images obtained are far from simple.

The first part of this laboratory exercise explores some practical aspects of seismology—how to determine the *epicenter* (the point on the earth's surface directly above the origin of the earthquake), the origin time, and the magnitude of an earthquake, and how

to determine the sense of motion of the rock masses that have moved.

To make these measurements requires a permanent record of the ground movements produced by the earthquake. Seismic waves are recorded by *seismographs,* instruments containing a heavy mass suspended in such a way that its inertia keeps it relatively immobile while the ground and the rest of the instrument vibrate. There are different ways to record the vibrations, but in a fairly simple system ground motions are amplified and recorded as a line on a revolving drum. A continuous line that is drawn on the seismograph paper as the drum revolves will appear as a series of parallel lines after the paper has been removed from the drum. Figure 16.1B shows part of one line of such a recording—a *seismogram.*

DESCRIBING AN EARTHQUAKE

Epicenter

On April 28, 1979, a moderate earthquake occurred in northern California. Its seismic waves were recorded on an extensive array of instruments, three of which are identified in Figure 16.2. BRK represents the seismographic station on the University of California, Berkeley campus; MHC is the station at Mount Hamilton (Lick Observatory); and PCC is the one at Pilarcitos Creek, south of San Francisco. Figure 16.3 shows seismograms obtained from this event at these three stations. Notice that the arrival of the *P*-wave is followed several seconds later by the arrival of the slower moving *S*-wave. The farther the recording station is from the epicenter of the earthquake, the greater is the difference between *P*- and *S*-wave arrival times. This feature may be used to calculate epicentral dis-

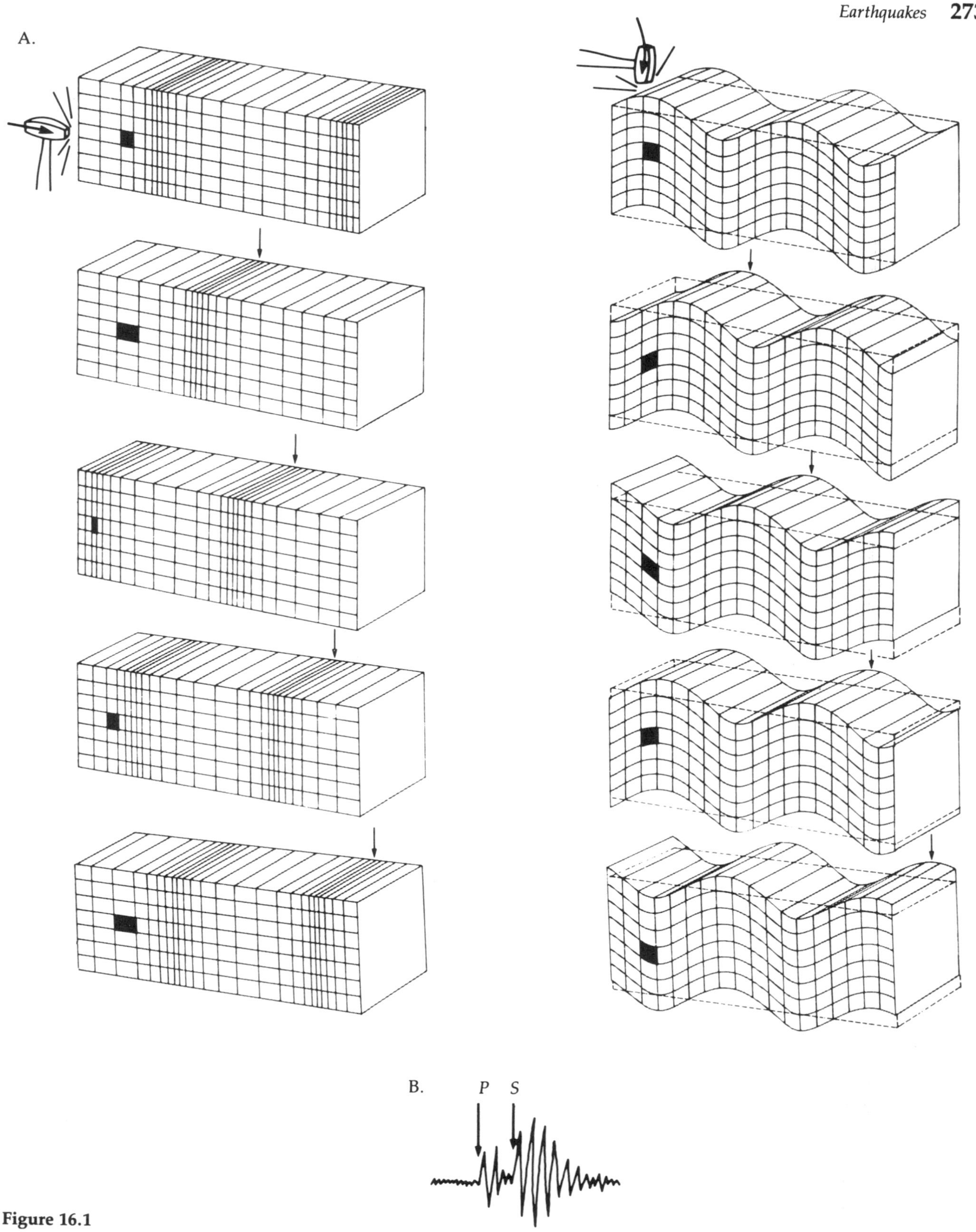

A.

B.

Figure 16.1
A. Diagrammatic representations of *P*-waves (left) and *S*-waves (right). B. Seismogram with *P*- and *S*-waves. (From *Earth*, 4th ed., by Frank Press and Raymond Siever. Copyright © 1974, 1978, 1982, 1986 W.H. Freeman and Company. Reprinted with permission.

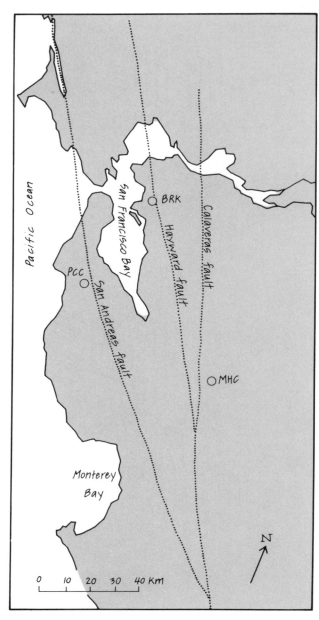

Figure 16.2
Three seismographic stations and major faults of the
San Francisco Bay Area.

tances. Using data from many earthquakes, seismologists have constructed *travel-time curves*, graphs in
which the distances from the epicenters of earthquakes to the recording station are plotted against the
times taken by the seismic waves to reach the same
station. It has been found that these average travel-
time curves are valid for events within the region
studied. Travel-time curves for *P*- and *S*-waves and
the *S–P* intervals for northern California are shown
in Figure 16.4.

For the following problems, refer to the seismograms in Figure 16.3.

Problem 1
At what time did the *P*-wave arrive at MHC?

Problem 2
What was the corresponding *S*-wave arrival time?

Problem 3
What was the difference (*S* − *P*) in the time of arrival between the secondary wave (*S*) and the primary
wave (*P*)?

Problem 4
Using the *S–P* interval and the travel-time curves in
Figure 16.4, estimate as accurately as you can the distance in kilometers from the epicenter to each of the
three stations.

• Epicenter to BRK

• Epicenter to MHC

• Epicenter to PCC

Problem 5
Using the map of the San Francisco Bay Area (Figure
16.2), draw arcs around the three stations using the
appropriate epicentral distances as radii.

Problem 6
Where was the epicenter of the earthquake?

Problem 7
Was it on or near a known fault?

A. Berkeley (BRK)

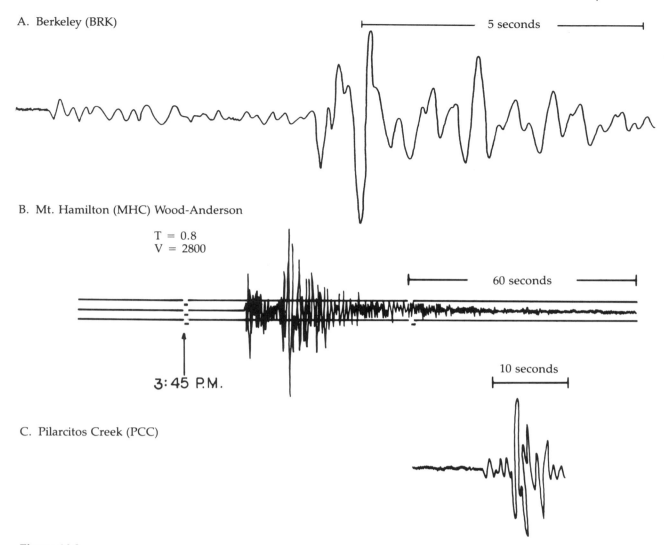

B. Mt. Hamilton (MHC) Wood-Anderson

T = 0.8
V = 2800

3:45 P.M.

C. Pilarcitos Creek (PCC)

Figure 16.3
Seismograms from stations BRK, MHC, and PCC, April 28, 1979.

Origin Time

The origin time of the earthquake may be calculated from the MHC record since the exact arrival time of the *P*-wave is known for this station. The small, regular displacements in the seismogram lines are timing marks placed on the record by internal clocks. The beginning of the first timing displacement on the second line in the MHC record occurs at 3:45:00 P.M. Given the time scale to the right of the MHC record, you can determine the exact time at which the *P*-wave arrived at that station. You have already calculated the distance from MHC to the epicenter, so you may read the travel time for the *P*-wave directly off the travel-time curve for *P*-waves in Figure 16.4. The origin time of the earthquake is its arrival time minus its travel time.

Problem 8

What was the origin time of this earthquake?

Magnitude

The well-known *Richter scale* was developed by the late Charles F. Richter of the California Institute of Technology as a means of comparing earthquakes worldwide on a quantitative scale that is independent of the damage they cause. Richter defined the *magnitude* of a local earthquake as "the logarithm to the base 10 of the maximum seismic wave amplitude (in mm) recorded

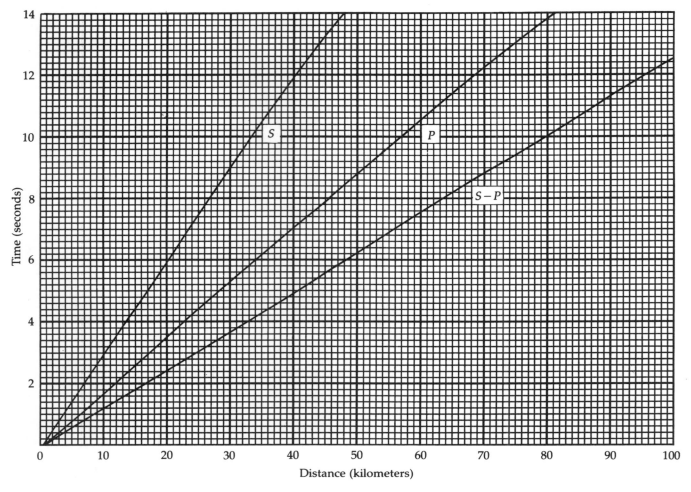

Figure 16.4
Travel-time curves for northern California.

on a standard seismograph at a distance of 100 kilometers from the earthquake epicenter." As the magnitude goes up 1 unit, therefore, the amplitude of the seismic waves increases by a factor of 10.

Just as the size of the ripples made by dropping a stone into water decreases as they spread, so the amplitude of seismic waves diminishes with increasing distance from the epicenter. Since seismographic stations are necessarily at random distances from earthquake epicenters, Richter devised a method of compensating for differences in epicentral distance when calculating the Richter magnitude. Figure 16.5 is a nomograph for obtaining the magnitude of a local earthquake once the epicentral distance and the maximum amplitude have been determined. To use Figure 16.5, place a straightedge between the appropriate points on the distance (left) and amplitude (right) scales, and read the magnitude directly from the center scale.

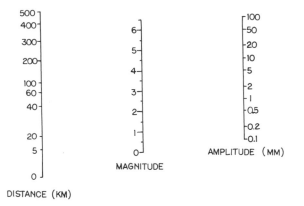

Figure 16.5
Nomograph for Richter magnitude calculation.

Problem 9

The seismogram from Mount Hamilton (MHC) in Figure 16.3 was recorded on a standard Wood–Anderson horizontal seismograph. What is the distance from MHC to the epicenter?

Problem 10

The amplitude of the maximum deviation of the seismogram trace is most accurately determined by measuring the peak-to-peak distance and dividing by 2. What is the maximum deviation of the trace on the seismogram?

Problem 11

What is the local Richter magnitude of the earthquake?

It is a popular belief that a reasonable number of moderate earthquakes will relieve the stress along an earthquake fault sufficiently to prevent a great earthquake (one of magnitude greater than 8). Seismologists estimate that the energy released in an earthquake changes by a factor of 32 for each integral step in the Richter scale. For example, an earthquake of magnitude 2 releases 32 times the energy of an earthquake of magnitude 1, and an earthquake of magnitude 3 releases 1024 (32^2) times the energy released by an earthquake of magnitude 1.

Problem 12

The average recurrence interval for a great earthquake on the San Andreas fault is approximately 100 years. How many moderate earthquakes (magnitude 5) every 100 years would be needed to dissipate the energy associated with a great earthquake (magnitude 8) along the San Andreas fault? How many earthquakes of magnitude 5 per year does this correspond to? Show all your calculations below.

DETERMINATION OF THE SENSE OF MOTION

Seismograms show the direction in which the ground at the instrument moved, and geophysicists use this information to interpret the type of faulting that caused the earthquake. The first deviation of the seismogram trace (produced by the *P*-wave) shows the *first motion* of the ground. *P*-waves are pushes and pulls—alternating compressions and dilatations—of the rocks. If, for example, the ground at the seismograph first moves away from the epicenter and toward the seismograph, then the first *P*-wave shows a compressional phase, and the first deviation of the seismograph trace is up. If the ground moves toward the epicenter and away from the instrument, then the first movement shows a dilatational phase, and the first deviation of the trace is down. Examination of the first *P*-wave motions over a wide area surrounding the epicenter reveals a rather simple pattern of pushes and pulls on the earth's surface, centered at or near the epicenter.

Faults are divided into categories that are based on the sense of motion along the fault plane. The *fault plane* is the surface along which the motion of an earthquake takes place. It has a certain orientation in space, described by its strike and dip, just as contacts and bedding planes do (see Exercise 7). Faults are divided into two broad categories, depending on whether the predominant motion of the sides of the fault plane is parallel to the dip or parallel to the strike of the fault plane.

Dip-Slip Faulting

In *dip-slip* faulting, the two sides of the fault slip in a direction parallel to the dip of the fault plane. On Figure 16.6, the fault plane is the shaded surface, and X is the point on the fault plane at which the earthquake originated—the *focus*, or *hypocenter*.

In a *normal fault*, the movement of block B, the upper block, is down relative to block A. We cannot determine the absolute motion. We can only determine the relative motion of B with respect to A because we have no fixed frame of reference on the earth's surface.

In a *reverse fault* the movement of block B is up relative to A.

Imagine the movement that seismographs would detect in these two cases, normal and reverse faulting. During a normal faulting event (earthquake) with the focus at X on the fault plane, block B slips downward relative to A, away from the seismographs located on the surface of B. The crust of the earth on B is stretched vertically, causing a dilatational first motion. Block A, conversely, moves upward (toward the instrument) relative to B; the crust is compressed vertically, and

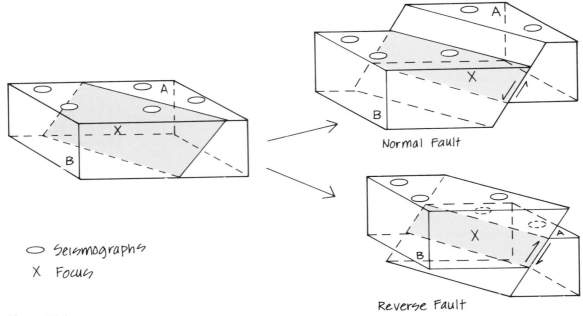

○ Seismographs

X Focus

Figure 16.6
Dip-slip faults. Point X is the focus or hypocenter, and the circles on the top surface represent seismographs.

instruments on A record a compressional first motion. These motions are represented in Figure 16.7A.

In a reverse faulting event, block B moves upward relative to A, toward the seismographs, which register a compressional first motion. Stations on block A register a dilatational first motion. Figure 16.7B shows the pattern recorded for reverse faulting.

Problem 13

X' is the epicenter of the earthquake, the point on the earth's surface vertically above the focus where the movement along the fault plane began. Why is X' not exactly on the surface trace of the fault?

Problem 14

Inspect Figure 16.6 closely. In one case the part of the earth's crust shown in the diagram is being pulled apart horizontally, or extended; in the other case, the crust is being compressed horizontally. What type of faulting, normal or reverse, is associated with extension of the crust?

Problem 15

What type of faulting is associated with crustal compression?

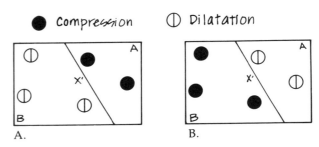

Figure 16.7
Map of dip-slip fault zone. A. Normal fault. B. Reverse fault. X' is the epicenter, and the symbols represent motions recorded at seismographic stations.

The matter of horizontal crustal compression and extension is very important in deciding what processes occur in various regions of the earth's crust. We shall refer to crustal compression and extension again in Exercise 18 on plate tectonics.

Strike-Slip Faulting

In the case of *strike-slip* faulting, movement of the two sides of the fault is in a horizontal direction, that is, along the strike of the fault plane (Figure 16.8). Strike-slip faults are designated as *left-lateral*, or *sinistral*, if the block of crust on the opposite side of the fault appears to move to the left to an observer facing the fault, and *right-lateral*, or *dextral*, if the opposite block appears to move to the right. Notice that it does not

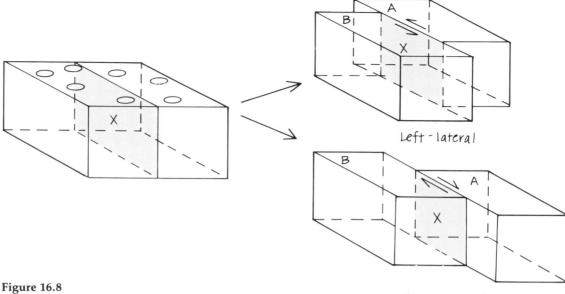

Figure 16.8
Strike-slip faults. Same symbols as in Figure 16.6.

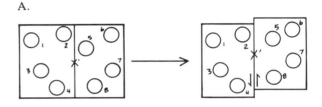

Left - lateral

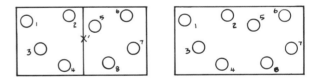

Right - lateral

matter which side of the fault the observer is on as long as the observer faces the fault.

Use the same kind of reasoning that you used for the pattern of compressional and dilatational first motions in dip-slip faulting to predict the patterns that would be observed in strike-slip faulting. Consider the motion of the ground in a strike-slip earthquake originating at point X, and imagine the first motions recorded by seismographs at the locations indicated in Figure 16.8. Figure 16.9 is a map view of the same area shown in Figure 16.8; X' is the epicenter of the earthquake. Consider stations 1 and 2. In left-lateral motion along the fault, shown by arrows along the fault line, the ground moves first toward the epicenter and away from the seismograph. Instruments at stations 1 and 2 show dilatational first motions. At stations 3 and 4, the first movement of the crust is away from the epicenter and toward the instruments, which register compressional first motions.

A.

B.

Figure 16.9
A. Map of strike-slip fault zone before and after left-lateral movement along the fault. B. Schematic map of seismographic stations.

Problem 16

What is the sense of first motion at the other stations, dilatational or compressional?

5.

6.

7.

8.

Problem 17

The left map of Figure 16.9B represents the case of right-lateral motion along the fault. Place the appropriate symbols for compression and dilatation at the eight stations.

Problem 18

If you did not know the surface trace of the fault on which the earthquake occurred, a second possible interpretation of the data exists. To see this alternate interpretation more clearly, replot the data of the first map of Figure 16.9B onto the other map, which does not have the fault line drawn in. What other orientation for the fault would be possible for the data shown? Draw it on the map, and place arrows to indicate the sense of motion.

Problem 19

What is the strike of the second possible fault trace?

Problem 20

Is the sense of motion along the second fault right- or left-lateral?

The two possible fault planes are called *nodal planes*, and it is not possible to distinguish between them on the basis of a first-motion study alone. The choice of the actual fault trace must be made on the basis of field evidence for faulting or by a study of *aftershocks* (small earthquakes following a major event) whose epicenters lie in the same fault plane.

A first-motion study will distinguish between dip-slip and strike-slip faulting using the local patterns of compressional and dilatational first motions plotted on a map. In the case of dip-slip faulting, one line separates the region of compressional first motions from the region of dilatational first motion on a local map. (On a global map this is not the case.) In strike-slip faulting, two lines that intersect near the epicenter separate areas of compressional from dilatational first motion. Look back at the examples given previously to satisfy yourself that this is the case.

In a closely monitored seismically active region, such as the San Francisco Bay Area, there are many seismographic stations. Some of these stations are shown in Figure 16.10. The major faults are shown as dotted lines.

Problem 21

Figure 16.11 shows seismograms from a wide area around the epicenter of the April 1979 earthquake. The small peaks starting near the left margin of the records and recurring about every inch are timing marks placed automatically on the records. Notice that the records are synchronous, so that the farther the station is from the epicenter, the later (farther to the right) the *P*-wave arrives. Plot the first movement at each seismograph as either compressional or dilatational on the map of Figure 16.10. (Remember that an "up" first motion is compressional and a "down" first motion is dilatational.)

Problem 22

Is the faulting dip-slip or strike-slip?

Problem 23

Draw in all possible fault traces on Figure 16.10, and indicate with arrows the relative motion of the ground for each case.

Problem 24

Does the pattern of first motions determine unequivocally the trace of the fault plane? What other evidence might be useful for this task?

WORLDWIDE OCCURRENCE OF EARTHQUAKES

The *focus* or *hypocenter* of an earthquake is its point of origin, from which the earthquake rupture propagates along the fault plane. Focal depths range widely from very shallow to about 700 km beneath the earth's surface. For simplicity's sake, focal depths have been categorized as shallow (less than 70 km), intermediate (between 70 and 300 km), and deep (greater than 300 km).

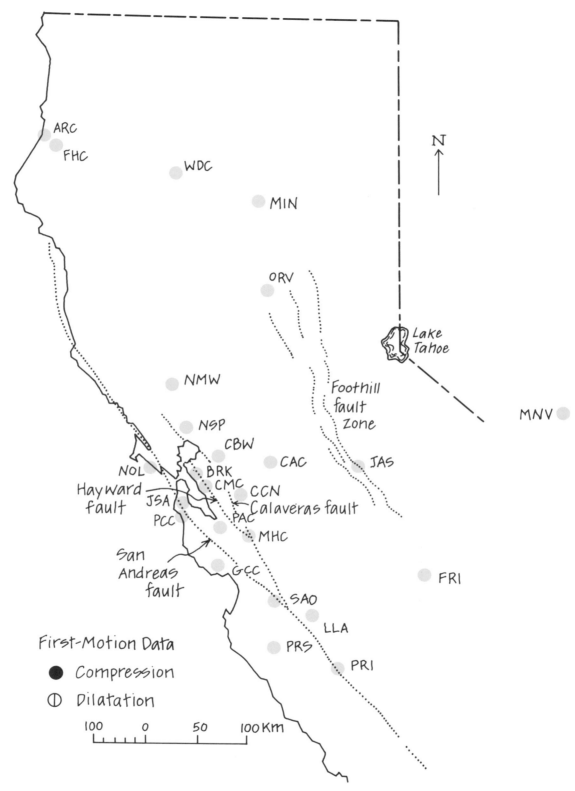

Figure 16.10
Seismographic stations in northern California.

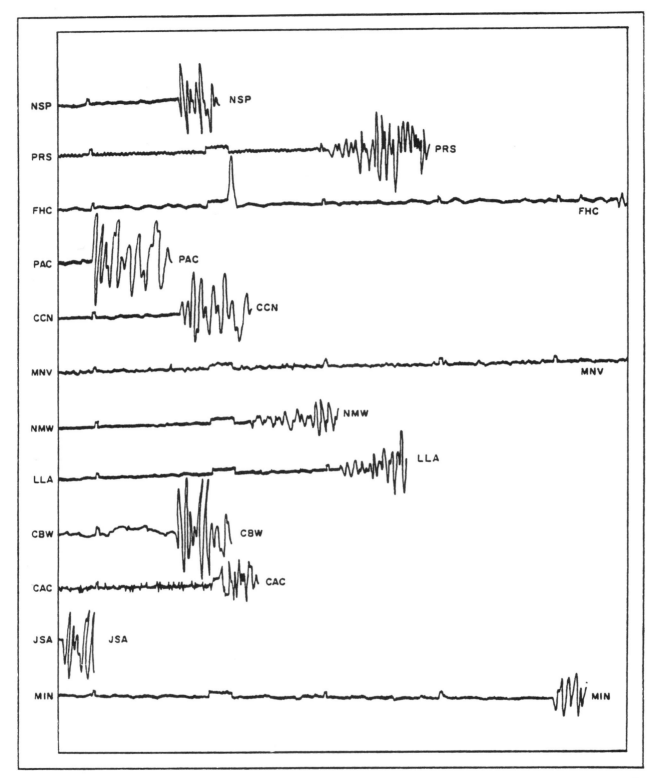

Figure 16.11
Seismograms of April 28, 1979, earthquake recorded at
northern California stations.

A.

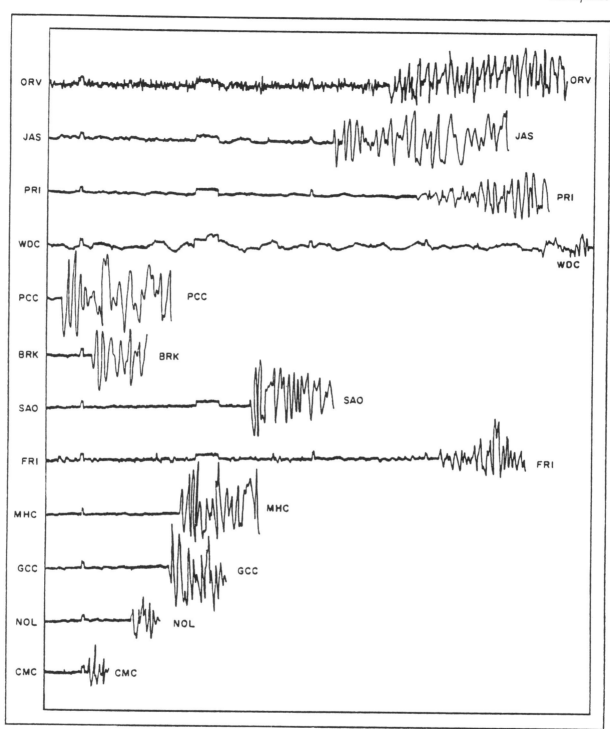

Figure 16.11
Seismograms of April 28, 1979, earthquake recorded at
northern California stations.

B.

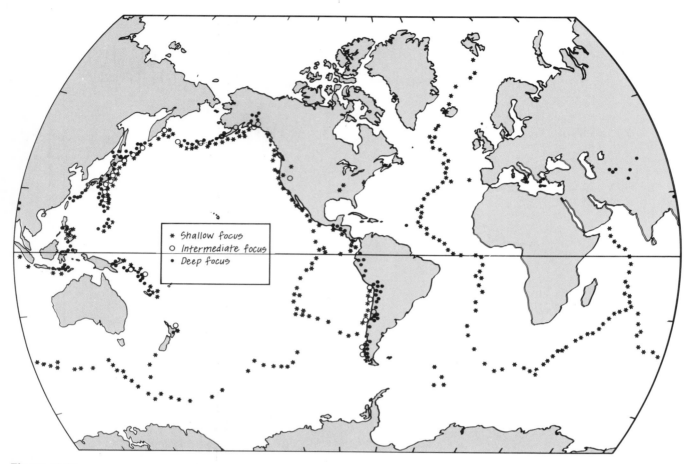

Figure 16.12
Map of global seismicity.

Problem 25

Figure 16.12 is a map of global seismicity on which shallow, intermediate, and deep earthquakes have been distinguished. Are intermediate- and deep-focus earthquakes associated with continental margins or with mid-ocean regions?

Problem 26

Do shallow-focus earthquakes occur primarily in ocean basins or on continents?

THE INTERIOR STRUCTURE OF THE EARTH

Much of our information about the internal structure of the earth comes from the study of earthquake waves, particularly those of distant earthquakes whose waves have traveled through most of the earth's interior. By plotting travel-time curves for the *P*- and *S*-waves of many earthquakes and noting anomalous features, the British seismologist R.D. Oldham proposed in 1906 that the interior of the earth is not homogeneous but has a central core with a density greater than that of the crust. In 1909 Andrija Mohorovičić of what is now Yugoslavia published additional evidence that the earth's interior has a structure. When he plotted *P*-wave travel times for earthquakes in the Balkans, Mohorovičić found a sharp bend in the curves at an epicentral distance of about 200 km. He explained this abrupt rise in *P*-wave velocity by proposing that there is an abrupt change in the properties of the earth's interior at a depth he calculated to be about 40 km.

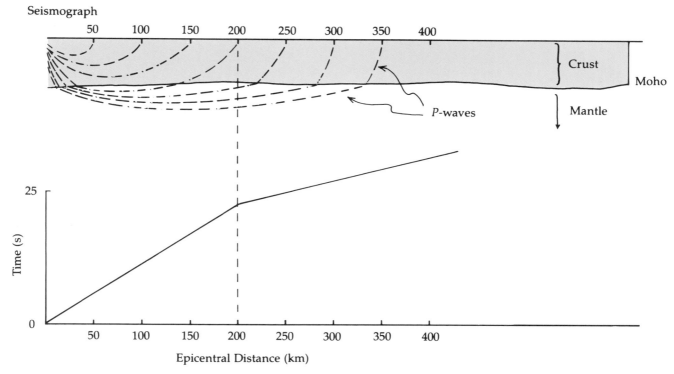

Figure 16.13
Seismic *P*-wave paths and travel-time curve for the crust-mantle boundary, the
Mohorovičić discontinuity (Moho).

This change in *P*-wave velocity marks the boundary (the Mohorovičić discontinuity, or Moho) between the outermost layer of the earth (the crust) and the middle layer (the mantle). See Figure 16.13.

Since these earliest discoveries, the analysis of seismograms has revealed the presence of many seismic waves that are not direct *P*- or *S*-waves. Many of these "extra" waves are reflections of ordinary *P*- and *S*-waves from the mantle–crust boundary; others are reflections of *P*- and *S*-waves from the underside of the earth's surface. Such reflections occur in the same way a light beam is partly reflected during its passage through a pane of glass. One important discovery was the presence near the top of the mantle of the *asthenosphere*, a zone of partially molten rock. It is also called the low-velocity zone, reflecting the fact that seismic waves travel more slowly there than in the rest of the upper mantle. Figure 16.14 shows the paths traveled by some of the common seismic waves identified on seismograms.

Figure 16.15 is a seismogram from a distant earthquake showing typical complexity. Over a dozen different seismic waves arrived during the first 30 minutes of the record. The regularly spaced marks are timing marks at one-minute intervals. Once in each line appears a wider timing mark, an hour mark. The earthquake record you will analyze begins shortly after 11:28:00.

Notice the very large amplitude waves beginning at 11:56. These waves are not body waves (simple *P*- or *S*-waves or variations of them) but are *surface waves*. Surface waves travel only around the surface of the earth, not through its interior. They are of two types, designated by L_Q and L_R.

Problem 27

Although the earthquake itself lasted for only a very few minutes, the disturbance on the seismogram is spread out for well over an hour. Why is this so?

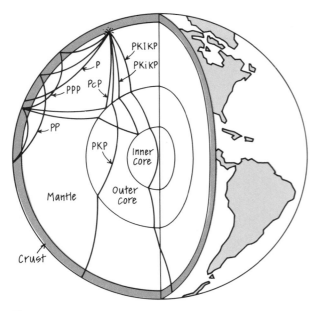

Figure 16.14
Cross section of the earth including major structural boundaries and the paths of some common compressional (*P*) waves.

P: The direct *P*-wave travels through crust and/or mantle from focus to detector.

PP: This wave is a reflection from the underside of the earth's surface.

PPP: Here there are two reflections from the earth's surface.

PcP: This reflected *P*-wave was bounced off the mantle-outer core boundary.

PKP: These waves have traversed the mantle and part of the outer core, having been bent (refracted) at the interface between mantle and outer core.

PKiKP: This wave has traversed the mantle and the outer core and has been refracted by the boundary between outer core (liquid) and inner core (solid).

PKIKP: Here the *P*-wave has passed through all the major zones of the earth, crust, mantle, outer core, and inner core.

Phases analogous to many of these *P*-wave derivatives have also been observed for *S*-waves.

Figure 16.16 shows a series of travel-time curves similar to those of Figure 16.3, which you used to determine the epicentral distance of a local earthquake. There are many more phases in Figure 16.16, and the epicentral distances are much larger. Note that the distance to the epicenter is given in degrees rather than in kilometers. The earth's full circumference is considered as 360°, a more convenient notation for distant earthquakes. To identify the various seismic waves in the seismogram of Figure 16.15, you will need to know accurately the relative time at which each wave arrived at the seismograph. On the straight edge of a paper held against the seismogram, plot and label the arrival times of the eight most distinct pulses

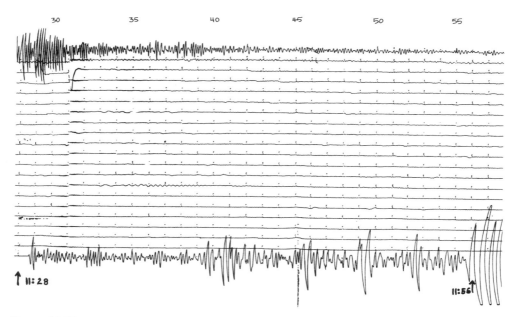

Figure 16.15
Seismogram from a distant earthquake on July 28, 1977.

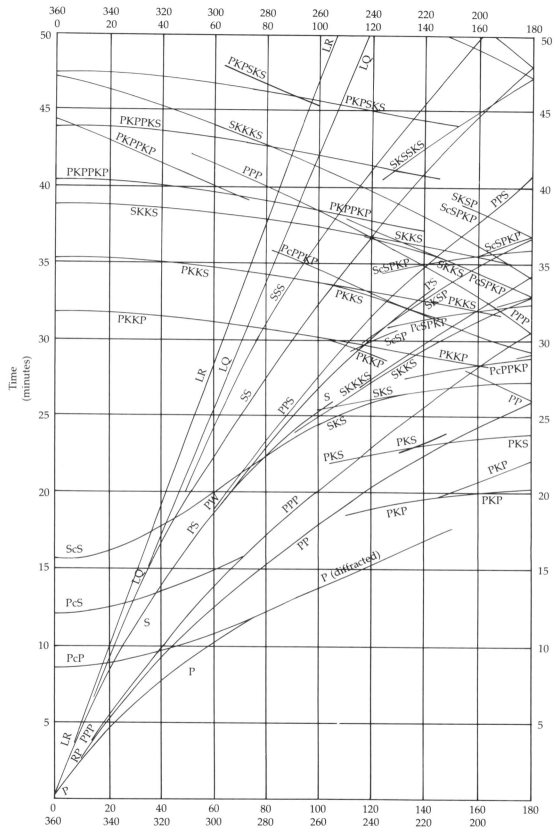

Figure 16.16
Travel-time curves for distant earthquakes.

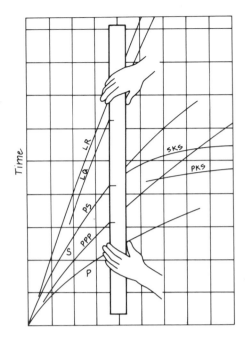

Figure 16.17
Method of identifying phases present in a seismogram.

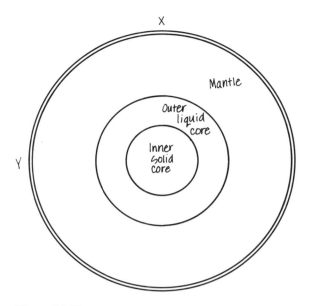

Figure 16.18
The earth's interior structure.

of energy. Place the strip of paper vertically on the travel-time curves with the two time scales *parallel* (This is very important!) and move the strip, keeping the two time scales parallel, until the time marks that you have plotted on the strip match a set of travel times on the chart. Figure 16.17 shows this procedure.

Problem 28

Identify eight of the waves present in the seismogram by placing the appropriate symbols (*P, S, PP, PcP*, etc.) on Figure 16.15 above each phase.

Problem 29

What was the distance to the epicenter in degrees? In kilometers?

Problem 30

On the diagram of the earth's interior in Figure 16.18, sketch the paths from the focus, X, to the recording station, Y, of the phases you identified. Label each path.

Physical Properties of the Earth

Geologists have turned in despair from the [constitution of the earth] leaving its center as a playground for mathematicians.

R. D. OLDHAM

We know more about the interior of the atom than we do about the interior of the earth. Thus, it is surprising, in view of the size and complexity of the earth, that some fundamental questions about the physical properties of the earth may be answered by relatively simple means. The manipulation of a few observations, many of which you could make yourself, can tell much about the size, mass, and density of the earth. Information and calculations concerning the internal heat, gravity, and magnetism of the earth are a bit more complicated but still are within the understanding of the beginning student of geology.

In our study of the various physical properties of the earth—topography, heat, gravity, and geomagnetism—this exercise will emphasize the study of the ocean basins. The continents have been relatively well explored for centuries, but it has been only since the 1950s, with the advent of modern geophysical measuring techniques, that we have been able to study the seven-tenths of the earth's surface that is obscured by water. It is no coincidence that the development of a successful and comprehensive theory of the nature of the earth has coincided with observations of the ocean floor; most of the important evidence for basic plate-tectonic theory lies concealed there. Since direct observation of the ocean floor is difficult and expensive, most of our information has been derived from remote measurements of its topography, heat flow, gravity, and magnetism.

PHYSICAL PARAMETERS OF THE EARTH

Topography and Size

The size of the earth was estimated in the third century B.C. by the Greek astronomer Eratosthenes. He accepted the minority opinion among people of his time that the earth is a sphere. This assumption, together with a few measurements and a knowledge of simple geometry, is all one needs to calculate the circumference of the earth.

Suppose you are in Mazatlan, Mexico, on the longest day of the year, June 21. You notice that, when the sun reaches its highest point in the sky, your body casts virtually no shadow; the sun is directly overhead. This is because Mazatlan lies very close to the Tropic of Cancer (latitude 23° 27' north). Because you are a quick-witted scientist, you immediately make a phone call to Taos, New Mexico, which is almost due north of Mazatlan, and ask a friend there to run outside at once and measure the length of the shadow cast by a vertical rod 2 meters long. Your friend soon informs you that the length of the shadow is 45 centimeters. By consulting a map (Figure 17.1) you can measure the distance from Taos to Mazatlan. You now have all the data you need to calculate the circumference of the earth.

A.

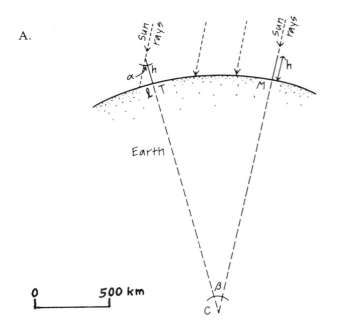

500 km

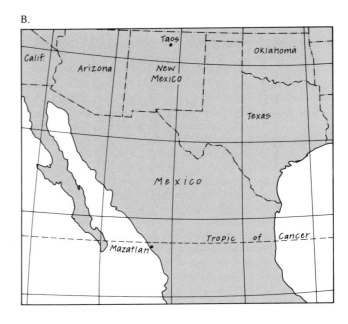

Figure 17.1
A. Incidence of the sun's rays at zenith, Mazatlan, Mexico (M) and Taos, New Mexico (T). B. Map of Taos and Mazatlan areas.

Problem 1

Examine Figure 17.1A. Since you have the length of the pole, h, and the length of the shadow, l, you may compute the angle α (alpha) by trigonometry. When measuring the sides and internal angles of right triangles (triangles containing a 90° angle), certain useful quantities are defined. The tangent of a given angle is defined as the length of the side opposite the angle divided by the length of the side between that angle and the 90° angle. Calculate the tangent of angle α, and then find the value of angle α from the Partial Table of Tangents in Appendix 8. Show your calculations.

Problem 2

Assuming that the sun's rays are parallel, angles α and β (beta) must be equal according to the laws of geometry. Measure the map distance between Taos and Mazatlan from Figure 17.1. Knowing that this distance, TM, is the part of the earth's circumference subtended by the angle β, calculate the distance corresponding to an angle of 360°; that is, calculate the circumference of the earth. Show your calculations.

Problem 3

Calculate the radius of the earth, knowing that the circumference equals πd, where d is the diameter.

Problem 4

It is important in scientific work to have some idea of how reliable your results are. Think about where errors may have crept into your calculation. The modern value for the average radius of the earth is about 6370 km (3950 miles), so there are some sources of error in your calculations. List several errors that might affect your result.

Density

Problem 5

The mass of the earth is 6×10^{27} g. Knowing the radius of the earth, you may compute its volume (which for any sphere is $\frac{4}{3}\pi r^3$) and then its density, in grams per cubic centimeter. To simplify the arithmetic, use exponential notation in your calculations. Show all your calculations.

Problem 6

Most rocks of the earth's crust have densities between 2.5 and 3.0 g/cm³. Given your calculation of the previous problem, can you say that the earth has a uniform density? What conclusions can you draw regarding the interior structure of the earth?

Relief

The relief of the earth's surface seems impressive, particularly if you are struggling up a steep mountain trail. To visualize the surface relief in a more objective manner, compare the relief of the earth with that of an object one can more readily comprehend, such as a pingpong ball.

Problem 7

Mount Everest, the world's highest peak, stands at about 29,000 feet above sea level; the Marianas trench is the deepest part of the ocean floor, about 36,000 feet below sea level. Compute the ratio of maximum surface relief to the radius of the earth.

Problem 8

A typical pingpong ball has a diameter of about 1.5 inches and a surface relief of 0.01 inch. Which is smoother, the earth or the pingpong ball? Show your calculations.

Ocean Basin Topography

The earth is a very smooth surface, but its minor surface irregularities have provided some very important clues to its internal workings. The topography of the continents is too complex to provide much insight, but that of the ocean basins, as revealed by depth soundings, was one of the most important pieces of information in the early development of plate-tectonic theory. Figure 17.2 is a bathymetric contour map of the world's oceans. The contour lines are lines of constant water depth.

Mid-Ocean Ridges. Intuitively one might expect the continental shelf areas at the perimeters of the ocean basins to be rather shallow, but one does not expect to see shallow areas in the middle of the ocean basins. It is perhaps surprising to see not only very large but also linear and continuous shallow areas in all the world's ocean basins—the system of *mid-ocean ridges* and *rises*.

Problem 9

On Figure 17.2 mark with a red pencil line the mid-ocean ridges. Draw a topographic profile across the Mid-Atlantic Ridge from A to A' on a separate sheet of graph paper. Keep this profile for use in later problems in this exercise.

Oceanic Trenches. In addition to the global oceanic ridge system, there is a less regular linear system of especially deep regions of the ocean floor, mostly around the edges of the Pacific basin. These *oceanic trenches* are generally arc-shaped and close to continental masses or chains of islands.

Problem 10

On Figure 17.2 trace with a black line the oceanic trenches. Draw a topographic profile across the Japan Trench from A to A' on a separate sheet of graph paper, using the data given in Figure 18.3. Keep it for later use.

THE EARTH'S HEAT

External and Internal Heat

The energy budget of the earth is of prime importance in understanding the external and internal processes that modify our planet. In human terms the most

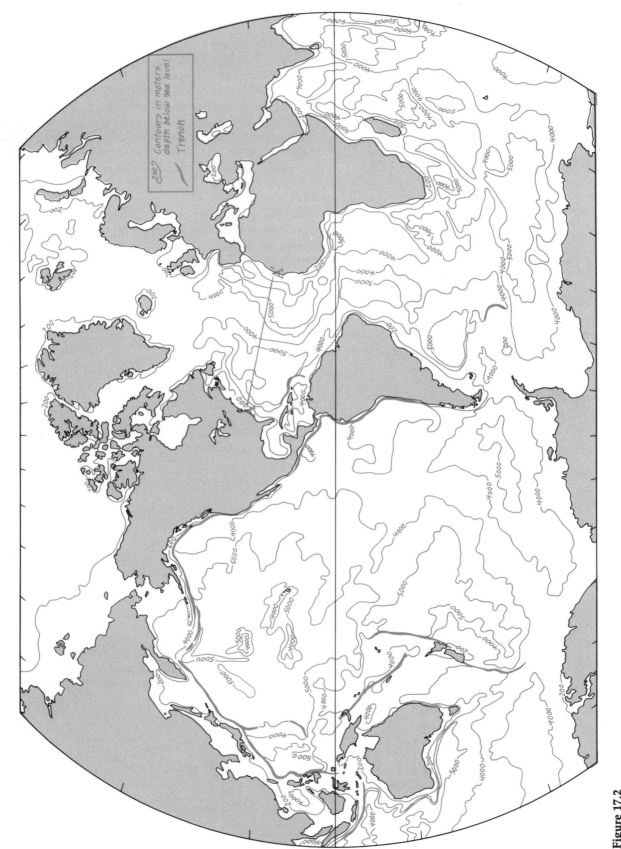

Figure 17.2
Bathymetric contour map of the world's oceans. Water depth is contoured in meters.

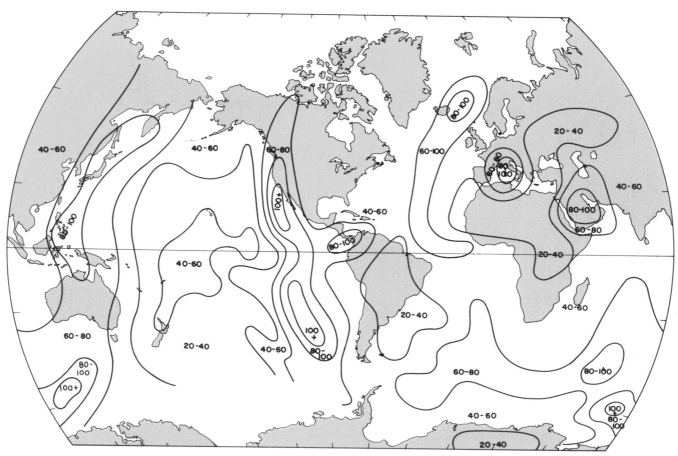

Figure 17.3
Map of global heat flow (milliwatts per square meter).

obvious source of heat is the sun, which supplies about 1.3×10^{23} calories per year to the earth (the calorie of science—the amount of energy needed to heat 1 gram of water by 1°C—not the calorie of diet, which is 1000 times larger). The energy supplied by the sun heats the earth's atmosphere and surface and drives the external or surface processes you studied in Part C. But does the energy supplied by the sun also heat the interior of the earth, thus contributing to mountain-building, earthquakes, and other internal geologic phenomena? The very low conductivity of heat by rock, combined with the large size and high reflectivity of the earth, suggests that it does not.

Problem 11

What evidence from Exercise 4 (Metamorphic Minerals and Rocks) indicates that the interior of the earth is hotter than the crust, thus supporting the idea that the sun is not the source of the earth's internal energy?

Heat Flow

Geologists have measured the flow of heat from the interior of the earth at many places. On the average the heat flowing through an area the size of a football field is roughly equivalent to the heat given off by three 100-watt light bulbs. Figure 17.3 shows a global heat-flow map.

Problem 12

Compare Figure 17.3 with Figure 17.2. Over what features of the ocean basins is the heat flow highest?

The low heat-flow areas on the continents largely correspond to very old and geologically stable *shields,* or *cratons.* The areas of higher heat flow correspond to younger regions of more active crust.

Having measured the heat flow at the surface, and knowing the different thermal properties of oceanic

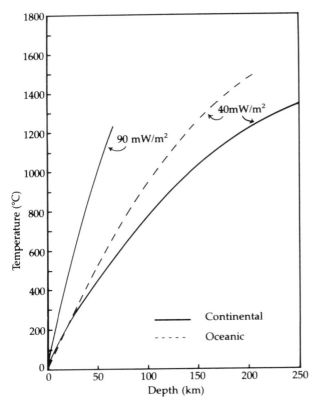

Figure 17.4
Continental and oceanic geotherms.

and continental rocks, geologists can calculate *geo-thermal gradients,* or *geotherms* (variations of temperature with depth), for oceanic and continental regions. Figure 17.4 shows continental geotherms for stable and active regions as solid lines and shows a typical oceanic geotherm as a dashed line.

Problem 13

Data in the table below represent the temperature at which rock in the upper mantle begins to melt. Plot these data on Figure 17.4, and draw a smooth curve between the points. Above this line mantle rocks are partially melted; below it the rocks are solid.

Depth (km)	Melting Point (°C)
0	1100
50	1110
100	1165
150	1240
200	1340
250	1460

Problem 14

What is the most likely physical state, solid or molten, of the lower crust and upper mantle (50–100 km depth) beneath continental areas of young active mountains (high heat flow, >80 milliwatts/square meter)?

Problem 15

What is the most likely physical state of the lower crust and upper mantle beneath the cratons (areas of very low heat flow, <50 mW/m^2)?

Heat-Flow Patterns in the Ocean Basins

It is an important part of geology to compare the different properties of a region whenever possible and to attempt to explain why any observed correlation might exist.

Problem 16

Return to the cross section A–A' across the Mid-Atlantic Ridge that you prepared earlier in this exercise for Problem 9. On this same profile plot values for the heat flow of the ocean floor across the Mid-Atlantic Ridge at A–A' in Figure 17.3. Express the correlation you see in words.

Problem 17

Rocks of the earth's crust (like most other substances) increase in volume when heated. How might this fact account for the correlation you have just found?

Problem 18

Figure 17.5 is a detailed map of the heat flow in the Pacific basin near Japan. The areas of anomalously low heat flow correspond with what topographic feature of the ocean floor?

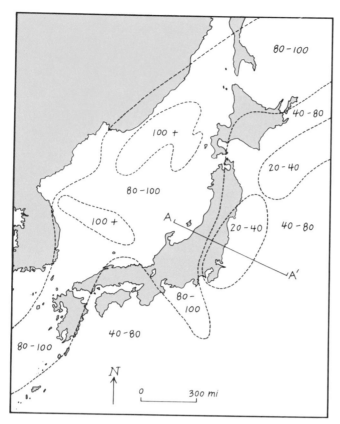

Figure 17.5
Heat-flow map of Japan (milliwatts per square meter).

occurred when the earth's iron core separated from the rest of the earth. Release of gravitational energy as the heavy iron sank toward the center of mass would have heated the whole earth, perhaps to a semimolten state. Gravitational heating, however, was important only very early in the earth's history; the fact that the earth is still hot implies a continuing source of heat energy. This heat source was a mystery until the discovery of radioactivity in 1896. *Radioactivity* is the emission of various particles from atomic nuclei, usually at very high energies. These particles are stopped by collision with surrounding atoms, and their kinetic energy is transformed into heat. Geophysicists estimate that radioactive decay accounts for about half the present heat production of the earth.

Problem 20

The principal mechanism by which heat is transferred from the interior of the earth to its surface is believed to be *convection* in the earth's mantle, a process that involves the flow of hot material from one place to another. Flow usually implies a liquid state. Since we know that the earth's mantle is solid, except for small pockets of liquid just beneath the crust (the asthenosphere), how can convection occur there? (Recall that Exercise 7 concerned the behavior of solids under high pressures and temperatures.)

Problem 19

On your topographic profile across the Japan trench, A–A', plot the heat-flow values shown in Figure 17.5. How good is the correlation between topography and heat flow here?

The Source and Transfer of the Earth's Internal Heat

Heat is associated with the kinetic energy of the atoms and molecules of a substance. That the interior of the earth is hot is well accepted, but what are the sources of this heat? Current views of the formation of the earth, although differing in details, agree that the earth (and other planets) formed by gradual accretion of small bodies under gravitational attraction. As particles began to fall toward the accreting planet, their potential energy was converted to kinetic energy (motion), which was released upon impact to the whole mass of the body as heat. Similar release of heat

---

THE EARTH'S GRAVITY

Like any body with mass, the earth has a gravitational field. Scientists have had a good understanding of the effects of gravity for over 200 years. They can make very precise calculations of satellite orbits, planetary paths, rotations of distant star systems, and many other complex phenomena by using the relatively simple formula for the gravitational attraction between two bodies first stated by Isaac Newton:

$$F = \frac{G\,m_1 m_2}{r^2}$$

where F = the attractive force between the two bodies
G = the gravitational constant, a fundamental constant of the universe
m_1 and m_2 = the masses of the two bodies
r = the distance between centers of mass of the bodies

Curiously, however, there is still no widely accepted theory to explain *why* two masses should attract each other. Gravity is the most familiar of the fundamental forces of nature, yet its exact mechanism remains a mystery.

Gravity Measurements

Geologists measure the gravitational force at various points on the earth's surface in order to detect and analyze deviations from the expected values. These deviations are called *gravity anomalies* and may reflect subsurface conditions that depart from the ordinary. They can be used to study the earth's structure or, for instance, to locate denser than average ore bodies. As an example of how gravity anomalies are detected and measured, consider the following factors.

Latitude Correction. The earth is not precisely spherical in shape but, because of its axial spin, is more nearly ellipsoidal—flattened at the poles and bulging at the equator—so that the equatorial radius r_e is larger than the polar radius r_p.

Problem 21

The entire mass of the earth is assumed to be concentrated at its center, C, as shown in Figure 17.6A. Would the gravitational attraction F be larger at the poles or at the equator of the spheroid? Why? (Recall that $F = G\, m_1 m_2 / r^2$.) See Figure 17.6A.

In order to compare gravity from place to place, we apply a correction to simple gravity measurements to compensate for the change in the earth's radius with latitude. The correction is zero at the poles and positive at the equator. If the earth were a simple spheroid of uniform density, the latitude correction could account for all gravity differences—the anomaly would be zero. The actual situation is not so simple.

Elevation Correction. Another correction to gravity measurements is needed to compensate for the fact that most measurements are not performed on the surface of the spheroid, which is close to sea level, but at some distance above it.

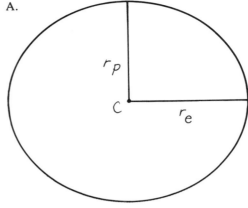

A.

Latitude correction

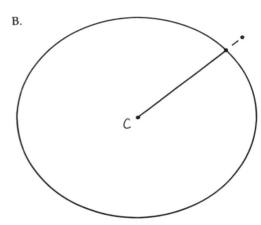

B.

Free-air correction

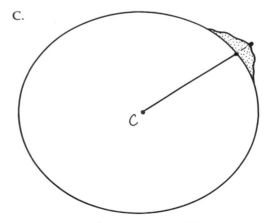

C.

Bouguer correction

Figure 17.6
Gravity measurement and corrections.

Problem 22

Figure 17.6B shows the case of two objects, one at sea level and one at a higher elevation (as if the spheroid had a longer radius). Will the gravitational attraction be greater at high elevations or at low elevations? Why?

The so-called *free-air correction* compensates for the elevation above the spheroid at which gravity is measured. It is a positive number for sites above sea level, because gravity rises as elevation decreases, and negative for those few sites that are below sea level. Any anomaly remaining after the free-air correction is the *free-air anomaly*.

Excess Mass Correction. The free-air correction ignores the mass between the higher elevation and the surface of the spheroid. There is, of course, considerable mass (dotted area in Figure 17.6C) between the elevated site of measurement and the surface of the spheroid. An approximate correction known as the *Bouguer correction* compensates for the extra gravitational attraction of this mass.

Problem 23

Should the Bouguer correction for the extra mass be added to or subtracted from the free-air anomaly?

The *Bouguer anomaly* is given by the measured gravity, minus the theoretical value of gravity at the latitude of measurement, plus the free-air correction, minus the Bouguer correction for estimated excess mass. If the earth were of uniform composition and structure, the Bouguer anomaly would be approximately zero anywhere on the earth. Any deviation from this represents anomalous conditions within the earth.

Isostasy

Surprisingly, the Bouguer anomaly on land is often negative, as if some part of the earth's crust between the elevation of measurement and the surface of the spheroid has no mass! Since that condition obviously cannot exist, the cause must lie in the mass of the rock below the surface of the spheroid (approximately the boundary between the crust and mantle). In some way the excess mass of the crustal rocks above the spheroid must be balanced by rocks of lower than nor-

mal density below the spheroid. The idea that rocks of the crust "float" in the denser substance of the mantle is an example of *isostasy*. Stated differently, the mass of a vertical column of given height and cross section through the mantle, crust, and water of the earth is the same in all places as long as the region is in *isostatic equilibrium*. A familiar analogy to the idea of isostatic equilibrium is the level at which ice cubes of different sizes float in water. The larger cubes, which float higher out of the water, have deeper low-density bases than smaller ice cubes, which float lower out of the water.

Geophysicists conclude from isostasy that high mountain ranges are underlain by deep roots of low-density material. Let us examine observational evidence to support this statement. Because of the physical properties of continental crust, seismic waves travel more slowly in it than they do in the material of the upper mantle. On March 22, 1957, a moderate earthquake occurred near San Francisco. The epicenter and time of origin are known because the earthquake caused surface damage in a small area, and very accurate local instruments timed the onset of the shaking. Thus the epicentral distances to the various seismographic stations shown in Figure 17.7 are also known very accurately. By using travel-time curves for the region, you can predict arrival times for each station very closely. Any substantial disagreement between predicted and observed travel times must be the result of differences in wave velocity due to different media along the wave path.

Problem 24

Using the travel-time curve for *P*-waves given in Figure 17.8, determine the expected travel times to each station listed in the table below, and record these times in column 4. Calculate the differences between predicted and observed travel times, and record these values in column 5.

Problem 25

For those stations showing substantial disagreement (more than 2 seconds), did the *P*-wave travel slower or faster than expected?

Problem 26

Where do these stations lie with respect to the Sierra Nevada and San Francisco? Refer to Figure 17.7.

Problem 27

Is the existence of deep roots below mountains confirmed by your results? Explain.

Another interesting consequence of the concept of isostasy is the *postglacial rebound* of parts of the earth's crust after large thicknesses of glacial ice have melted. The rate of uplift of the crust in Scandinavia since the melting of glacial ice has been as much as 1 cm per year.

Figure 17.7
Seismographic stations in and near northern California.

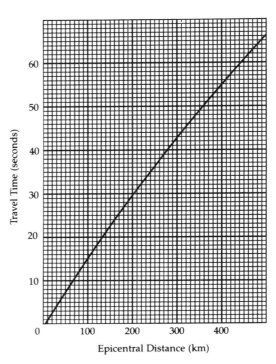

Figure 17.8
P-wave travel-time curve for northern California.

Earthquake Travel Times

Seismographic Station	Epicentral Distance (km)	Observed Travel Time of *P*-Wave (s)	Predicted Travel Time of *P*-Wave	Difference Between Observed and Predicted Times
Lick Observatory	83	14.4	_____	_____
Ukiah, Calif.	175	25.0	_____	_____
Fresno, Calif.	258	37.9	_____	_____
Mineral, Calif.	307	43.3	_____	_____
Reno, Nev.	312	46.2	_____	_____
Yerington, Nev.	325	47.7	_____	_____
Shasta, Calif.	336	45.7	_____	_____
Fallon, Nev.	380	56.2	_____	_____
Arcata, Calif.	382	52.4	_____	_____
Tinemaha, Calif.	383	56.1	_____	_____
China Lake, Calif.	482	67.2	_____	_____

Problem 28

Explain the postglacial rebound (uplift) in terms of isostasy.

Gravity Anomalies

Although a close approach to isostatic equilibrium prevails in many parts of the world, there are regions that clearly depart from isostatic equilibrium. In these areas, no reasonable assumption for the thickness or density of the crust will reduce the Bouguer anomaly to zero. A positive anomaly indicates excess mass; a negative anomaly indicates deficiency of mass. Both types of anomaly may reflect the dynamics beneath the crust that act against isostasy, pulling the crust down or holding it higher than isostatic equilibrium requires. Another cause of gravity anomalies is chemical inhomogeneities in the crust—large bodies of very dense material, for example.

Figure 17.9 shows some selected isostatic gravity anomalies. They are measured in milligals. The gal, named for Galileo, expresses the strength of the gravitational field; 1 gal is an acceleration of 1 cm per second every second.

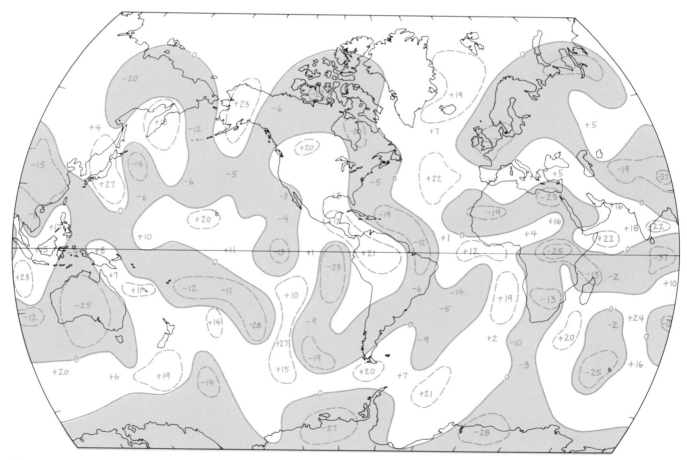

Figure 17.9
Worldwide isostatic gravity anomalies (in milligals).

Problem 29

To what factor would you attribute the strong negative gravity anomalies in the region of Hudson Bay in northern Canada? Recall that negative anomalies indicate deficiency of mass.

Problem 30

To what factor do you attribute the strong positive anomaly over Hawaii, an area of active volcanism? Recall that a positive anomaly indicates excess mass.

Problem 31

As shown in Figure 17.9, all the major oceanic positive anomalies (except Hawaii) are associated with mid-ocean ridges. Why would this be so?

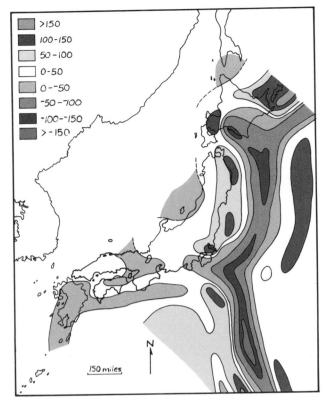

Figure 17.10
Gravity anomalies near Japan.

Problem 32

Figure 17.10 shows the gravity anomalies in the vicinity of Japan in greater detail. To what topographic feature does the large negative anomaly correspond? (Refer to your topographic profile A–A′ across Japan, prepared earlier in this exercise for Problem 10. Why might there be such a mass deficiency?

GEOMAGNETISM

Scientists and navigators have known for four centuries that the earth has a magnetic field. We have studied the properties of this field and used it in navigation for almost as long. However, the source of the geomagnetic field is still not fully understood.

The Nature and Source of the Geomagnetic Field

At first the geomagnetic field was believed to derive from permanently magnetized minerals within the earth. The lines of magnetic force lie roughly in north–south arcs, as if there were a great bar magnet in the earth's core (a dipole field). The traditional iron-filings experiment is worth repeating to refresh the image of the earth's magnetic field.

Problem 33

On a mat or sheet of stiff paper, draw a circle whose diameter is at least twice the length of the bar magnet you will use. Place the paper over the bar magnet so that the magnet lies on the vertical diameter. Sprinkle iron filings onto the paper and tap it gently to allow the filings to orient themselves in the magnetic field of the bar magnet. With a pencil trace on the paper the general orientation of the filings outside the circle you have drawn. Remove the paper from the magnet, and shake the filings back into their container.

If you imagine that your circle is the circumference of the earth and that the top of the bar magnet is the magnetic north pole, the pattern of filings close to the circle illustrates an important property of the earth's magnetic field, its *inclination*, or the dip angle of a compass needle that swings freely in the vertical plane. The inclination varies systematically with latitude.

Problem 34

What is the approximate inclination, in degrees, of a compass needle at the equator?

Problem 35

What is the approximate inclination of a compass needle at the north pole?

Problem 36

What is the approximate inclination of a compass needle at 45° north latitude?

The other important measurement of magnetic orientation is the *declination*, the horizontal angle between magnetic north and geographic north. Geographic and magnetic north are usually within 25° of each other, but they rarely coincide.

Several lines of evidence show that the source of the earth's magnetic field is not a permanent magnet. Permanently magnetized minerals lose their magnetism above a certain temperature, the *Curie point*; most of the planet's interior is at temperatures well above the Curie points of magnetic minerals. Furthermore, permanently magnetized minerals cannot move about rapidly enough to account for the observed changes in the strength and direction of the earth's magnetic field throughout human history. The declination at London, for example, is known to have changed gradually from 11.5° E in 1580 to 24.5° W in 1819.

Evidence from seismic waves shows that part of the earth's core is fluid, and scientists now believe that circulation of an electrical conductor (most likely molten iron and nickel) in the liquid outer core generates the earth's magnetic field. There is little agreement, however, on exactly how this circulation occurs, what energy source drives it, and how this flow gives rise to the magnetic field. It seems clear, however, that the rotation of the earth is essential to the geomagnetic field; planets that do not rotate rapidly do not have magnetic fields associated with them, despite their apparent similarity to earth in other ways.

Magnetic Anomalies

The difference between the observed geomagnetic field and the idealized dipole field is called a *magnetic anomaly*. Linear magnetic anomalies on the seafloor are of great importance to modern geology; they are considered in the exercise on plate tectonics. Other magnetic anomalies seem to fall into two groups of distinctly different sizes.

Small anomalies range up to 300 km in diameter and do not change in position or amplitude with time. Figure 17.11 shows several small magnetic anomalies in northern Michigan that are probably caused by deposits of magnetic iron ores beneath the surface.

Large magnetic anomalies are between 3500 and 9000 km in diameter. These anomalies of continental magnitude change with time in both position and amplitude. Figure 17.12 shows some large magnetic anomalies.

The depth to the source of a magnetic anomaly is estimated by considering the size and the amplitude of the anomaly. An approximate rule of thumb is that the depth to the source is equal to the *half width* of the surface anomaly. If one draws a profile of the anomaly from the magnetic contours (in the same way that topographic profiles are drawn from topographic contours), the half width of the anomaly is the width of the anomaly measured at one-half the amplitude.

Problem 37

Consider the small magnetic anomalies shown in Figure 17.11. What is the approximate depth to the source of these anomalies? Show your work. See Figure 17.11B for an example.

Problem 38

In what part of the earth (core, mantle, or crust) is the source of the small anomalies?

Problem 39

Consider now the large anomalies shown in Figure 17.12. What is the approximate depth to the source of these anomalies? Show your work.

Problem 40

What does this location imply about the origin of the large anomalies?

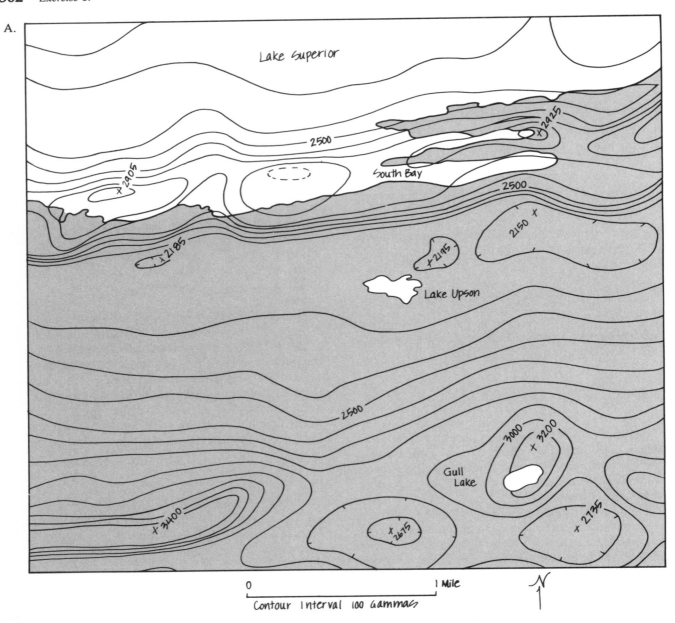

A.

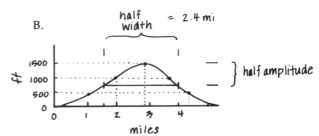

B.

Figure 17.11
A. Magnetic anomalies in northern Michigan. B. An example of finding an anomaly's source depth from its half width.

Paleomagnetism and Polar Wandering

Some rocks acquire at the time of their formation a permanent *(remanent)* magnetization, the direction of which is parallel to the earth's magnetic field at the time and place of formation. Most important is *thermal remanent magnetization,* which is acquired by orientation of magnetic minerals when an igneous rock, commonly basalt, cools through the Curie point of those minerals. Even rocks many millions of years old retain this remanent magnetization, and it is possible to measure the inclination and the declination of the ancient magnetic field that produced the alignment of minerals in the rock.

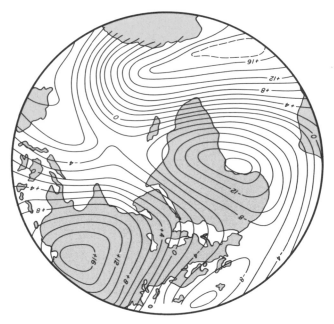

Figure 17.12
Large magnetic anomalies (1000 gammas).

It is an interesting fact that geomagnetic poles determined for Quaternary (very young) rocks cluster closely around the geographic pole, not the present geomagnetic pole, which is 11.5 degrees of latitude from the geographic pole. This evidence strongly suggests that the geomagnetic pole position, averaged over a sufficient length of time, is in fact indistinguishable from the pole of rotation of the earth. This inference in turn supports the idea that earth's rotation is essential to its magnetic field.

Problem 41

Paleomagnetic data from different periods of geologic time show an interesting pattern. Figure 17.13 shows the *apparent polar wandering curve* for North America. The geomagnetic pole position departs significantly from its present position and seems to change systematically with time. There are several simple interpretations of the data shown in Figure 17.13. What are two possible interpretations?

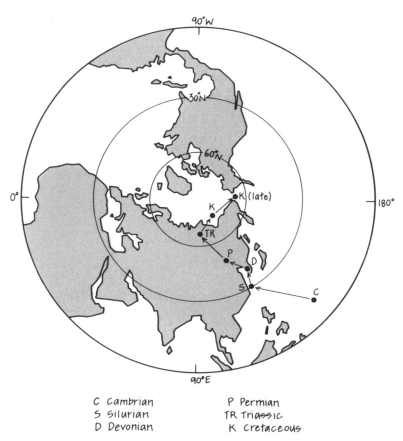

C Cambrian   P Permian
S Silurian   TR Triassic
D Devonian   K Cretaceous

Figure 17.13
Paleomagnetic pole positions for North America.

Plate Tectonics

The eye is blind to what the mind cannot see.

ANON.

I shall consider this paper [first modern paper on plate tectonics] an essay in geopoetry.

HARRY H. HESS

Plate-tectonic theory is the backbone of modern geology. Geologists frame most of their ideas and questions within the theory of plate tectonics, such as where to find oil, how to locate new and vital sources of metal ores, why ice ages come and go, how to account for large-scale extinctions of prehistoric life forms, what geological processes operate on Venus, Mars, and Mercury. Plate-tectonic theory is central to all these questions and many more.

Acceptance of this vital theory by most geologists has come about only within the past 25 years, although the idea of lateral movement of continents has been in the literature for hundreds of years. The theory of continental drift proposed by Alfred Wegener beginning in 1912 came very close to modern theory, but it failed to gain widespread permanent acceptance. Technological developments following World War II provided many observations and measurements previously unobtainable, and geologists of the 1950s and 1960s pieced the data together into plate-tectonic theory.

The first part of this exercise concerns the evidence in support of plate tectonics already at hand by 1960. You will need to return to previous laboratory exercises to extract those data and ideas pertinent to the idea of global plate tectonics.

THE EARLY EVIDENCE FOR PLATE TECTONICS

The great scientific and engineering progress during and after World War II led to the development of instruments used to explore the topography, heat flow,

earthquake activity, and magnetism of the ocean basins, the hitherto unexplored part of our planet. This vast flood of new information from the oceans was crucial to development of the "new geology" by geologists and physicists of the early 1960s.

Mid-Ocean Ridges

Problem 1

Inspect Figure 18.1, the global map of the ocean basins. What seems to be the most prominent topographic feature of the Atlantic basin? Note that this feature extends around Africa and into the Indian and Pacific Oceans.

Problem 2

The top part of Figure 18.2 is a profile view of the Mid-Atlantic Ridge obtained by seismic-reflection (sonar) studies. Ignoring the minor peaks and valleys in the profile, trace a smooth, generalized profile over Figure 18.2. Below that is another profile showing the heat flow from the ocean floor in microcalories per square centimeter per second. Notice the sharp peak in heat flow over the central rift valley. What familiar geologic feature is suggested by both the generalized profile and the peak in heat flow?

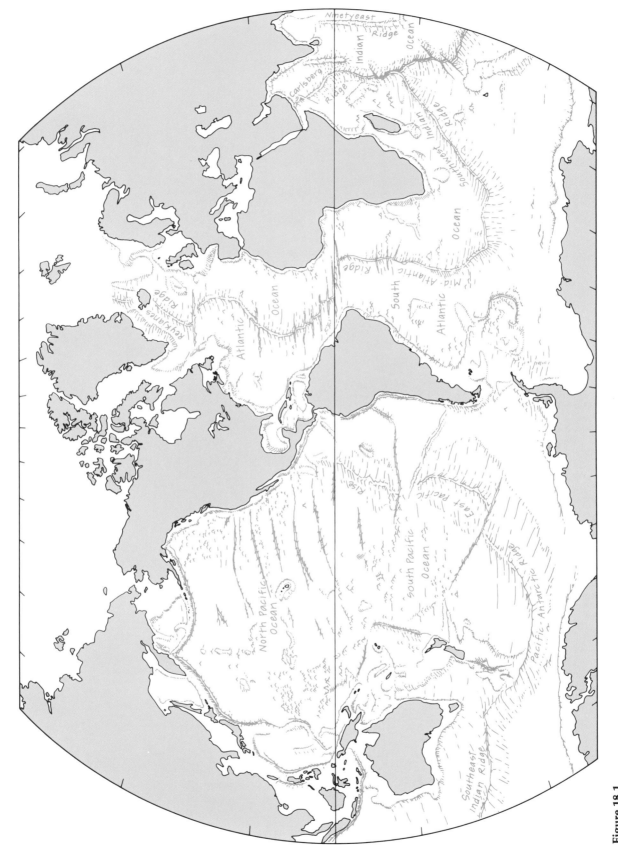

Figure 18.1
Global map of the ocean basins.

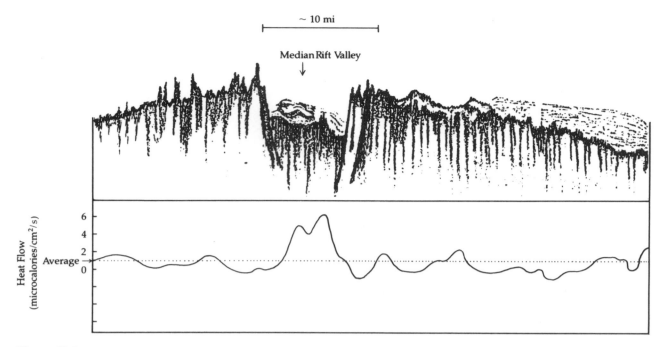

Figure 18.2
Seismic-reflection profile (top) and heat-flow profile (bottom) across the Mid-Atlantic Ridge.

Problem 3

Refer to Figure 2.10 in Exercise 2, Igneous Rocks. Does the global pattern of basaltic volcanism substantiate your answer to the previous question?

Problem 4

Examine the map of global seismicity (Figure 16.12 in Exercise 16, Earthquakes). What focal depth earthquakes are associated with the mid-ocean ridges?

Problem 5

Figure 18.2, the seismic-reflection profile of the Mid-Atlantic Ridge, shows fairly well the dips of the faults that border the median rift valley. Do these faults appear to be normal faults or reverse faults?

Problem 6

Does this type of faulting imply compression or extension of the crust?

Problem 7

From these processes that are acting at the mid-ocean ridges, formulate an overall description of events happening at the mid-ocean ridges.

Oceanic Trenches

Problem 8

Examine again Figure 18.1, the map of the ocean basins. Where do most of the oceanic trenches occur—in which ocean basin and in what relationship to land masses or chains of islands?

Problem 9

Refer to Figure 2.10 again. What type of volcanic activity is associated with the islands and land masses near trenches? Is this the same type of volcanism associated with mid-ocean ridges? Does this suggest that the same or different processes are active near trenches and at mid-ocean ridges?

Problem 10

Refer to Exercise 4 on metamorphism. Notice the correlation between the location of the blueschist belts shown in Figure 4.9 and the oceanic trenches around the Pacific basin. What general conditions of temperature and pressure are necessary to produce blueschist–facies metamorphism?

Problem 11

What do blueschist–facies conditions imply about conditions in the region between the trenches and the volcanic belt? (See Figure 4.8 for a more detailed view of the position of a typical blueschist belt.)

Contour intervals:
Ocean –2000 m
Land 1000 m

N
0   300 mi

Figure 18.3
Topographic map and profile of Japan and nearby Pacific Ocean.

Problem 12

To see more closely what is happening in the oceanic trench regions of the Pacific basin, consider Japan. From Figure 18.3, sketch a topographic profile across the trench and through the arc of volcanoes in Japan from A to A'. Indicate with small arrows above the section the positions of trench, blueschist belt, and andesitic volcanic arc. Use the grid provided below.

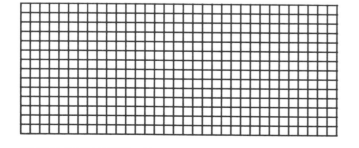

Problem 13

The final piece of evidence is supplied by seismology. Examine the map of worldwide seismicity in Exercise 16, Figure 16.12. Oceanic trenches and adjacent land masses are associated with what focal depth earthquakes?

Problem 14

On your topographic profile across the Japan trench, plot the focal depths of earthquakes to show the trend of the focal depth with respect to trench and volcanic arc. Shallow foci range from the surface to 70 km; intermediate foci are between 70 and 300 km; deep foci are between 300 and 700 km.

Problem 15

First-motion studies indicate that the sense of motion in trench areas is reverse faulting. Place arrows on your diagram to show this motion. Does reverse faulting indicate compression of the crust or extension of the crust?

Problem 16

Recall the discussion in Exercise 17 of heat flow across the Japan trench. On your profile, color areas of higher than average heat flow red, and color those of lower than average heat flow blue.

Problem 17

Summarize the evidence presented with respect to oceanic trenches, and describe what processes are occurring to produce the three parallel structures—trench, blueschist belt, and volcanic arc.

SUPPORTING EVIDENCE FOR PLATE-TECTONIC THEORY

When a theory such as plate tectonics emerges—elegantly simple, yet at the same time a revolution in our understanding—at least a few scientists design research programs to test it. Within the structure of scientific reasoning (inductive reasoning), it is not possible to prove unequivocally that a theory is true; the possibility always remains that contradictory evidence will be found. There are theories, however, such as the atomic theory, relativity, and evolution, that have successfully withstood the onslaught of so much experimentation and observation without the appearance of any strong contradictory evidence that they are no longer questioned by the scientific community. Plate tectonics would appear to be joining the ranks of these successful theories, although many details remain to be resolved. It is instructive, then, to look at some facets of the process by which such a theory becomes widely accepted.

When continental drift (the precursor theory to plate tectonics) was first seriously proposed by the German meteorologist Alfred Wegener, it was a theory whose time had not yet come. Some of the evidence Wegener presented was incomplete and inaccurate, and the physical mechanism he proposed was easily disproved. In any case, the most compelling evidence is found in the Southern Hemisphere, but most geologists lived and worked in the Northern Hemisphere. This is a strong argument for geologists to travel! Detailed exploration of the ocean basins, which did not take place until the 1950s, provided the key to plate-tectonic theory.

The emergence of the present theory of the earth was heralded by the publication of Harry H. Hess's landmark paper "The History of Ocean Basins" in 1962. Hess, an American marine geologist at Princeton University, described an earth in constant evolution. New basaltic oceanic crust, formed by partial melting of upper mantle material, rises to the surface at the linear volcanic system known as the mid-ocean ridges. This new crust is forced aside as still newer material wells up at the ridges in a process named *seafloor spreading*. Because the earth is assumed not to be expanding, crust must be consumed somewhere. Hess said that this process occurred at the system of oceanic deeps or trenches, most of which ring the Pacific basin. It was postulated that cooled oceanic crust dives underneath another slab of crust, which may be either continental (andesitic) or oceanic (basaltic), in the process that today is called *subduction*. Shallow- to deep-focus earthquakes accompany this process, and andesitic volcanism, caused by partial melting of the basaltic oceanic crust, occurs landward of the trench. This theory was simple yet heretical, being at odds with every prevailing theory of a static earth.

Although it was only one of several theories in the geologic literature at the time, enough geoscientists began research to test the idea that strong supporting evidence soon became available. But what constitutes strong supporting evidence for a theory? A scientific observation *already* in the literature, which can be explained satisfactorily only by the theory in question, is perhaps the most valuable kind of supporting evidence. A prediction made on the basis of the theory and subsequently confirmed by observation or experiment is another crucial test of a theory. The new theory soon had the benefit of both kinds of strong supporting evidence.

An Unexplained Observation from the Literature: The Linear Magnetic Anomalies

During the 1950s, with the development of a reliable shipboard device to measure the earth's magnetic field, oceanographic expeditions routinely began to measure the magnetic field over the ocean basins. The results of one such survey conducted by the Scripps Institute of Oceanography were published in 1958 and 1961. The data, surprising and unexplainable in terms of any then-current theory, showed a series of *linear magnetic anomalies* on the seafloor off the coast of Washington state. Dubbed "magnetic stripes," these anomalies were areas where the magnetic field was greater than average (a positive anomaly) or less than average (a negative anomaly). Note that the terms *positive* and *negative* are slightly misleading; the measured magnetic field direction was consistent in all cases with the present geomagnetic field orientation; only the intensity varied, either above or below the average. Figure 18.4 shows the more regular pattern of linear magnetic anomalies measured in 1965 around the Reykjanes Ridge south of Iceland.

The explanation for the magnetic stripes within the context of seafloor spreading was not long in coming. In the early 1960s, many geologists came to accept growing evidence that the direction of the earth's magnetic field spontaneously reverses from time to time; that is, what we call magnetic north abruptly switches to south. In 1963 British geoscientists Fred Vine and Drummond Matthews combined the knowledge of magnetic field reversals, known from measurements made on land, with the idea that new oceanic crust is continually forming at mid-ocean ridges (seafloor spreading) to explain the magnetic stripes. (Canadian geologist Lawrence W. Morley came up with the same idea and submitted a paper in 1963 to two respected journals of scientific research, both of which rejected it as too speculative!) They reasoned that, as new basaltic oceanic crust forms and cools at the volcanically active center of an oceanic ridge, it becomes permanently magnetized in the prevailing direction of the earth's magnetic field. As still newer crust forms at the ridge axis, the cooled material on both sides moves away from the ridge. If the magnetic polarity reverses, the next strip of newly formed crust is magnetized in the opposite direction. As this process continues for millions of years throughout many successive magnetic polarity changes, the entire seafloor comes to consist of bands of basalt magnetized in alternate directions. The linear positive and negative anomalies on either side of a mid-ocean ridge represent strips of oceanic crust with, respectively, greater and less than average magnetization.

Confirmation of the Vine–Matthews–Morley hypothesis came with new measurements that revealed that the magnetic stripes were parallel to the ridge axis and symmetrical on either side of it. Furthermore, the ridge axis was always found to be the site of a large positive anomaly, as one would expect. This interpretation of linear magnetic anomalies in terms of seafloor spreading combined with geomagnetic field reversals proved to be the decisive factor in the acceptance of the new theory by many geoscientists, especially after several years, when the age of the world's ocean floors had been determined and mapped.

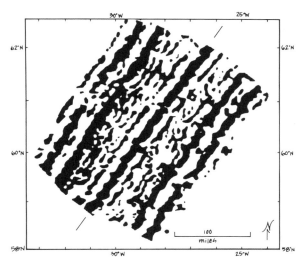

Figure 18.4
Magnetic anomalies over Reykjanes Ridge, North Atlantic. Straight lines indicate axis of ridge and central positive (black) anomaly.

When geologists recover and determine the age of material from the ocean floor over magnetic anomalies, they can calculate rates of seafloor spreading for a particular ridge system. The difficulties inherent in such an undertaking render the resulting time scale only approximate; nonetheless, it may be used to get some idea of the spreading rates. Figure 18.5 shows sample magnetic profiles from the South Atlantic and North Pacific Oceans along with an approximate time scale for the South Atlantic.

Problem 18

If an event that occurred t years ago is now found at a distance x from the ridge axis, the rate of motion is x/t. From the profile in Figure 18.5, calculate the recent (past 40 million years) spreading rate on the Mid-Atlantic Ridge. Show your work. Express your answer in cm per year.

Problem 19

Now correlate the magnetic profile from the North Pacific to that of the South Atlantic by matching patterns, in much the same way that you correlated tree-ring records in Exercise 5. Draw lines between the two profiles that connect the same magnetic event. Using the magnetic time scale, calculate the rate of spreading in the North Pacific.

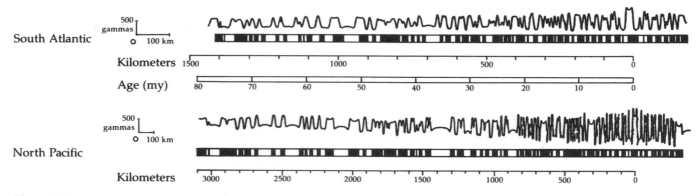

Figure 18.5
Magnetic profiles from several oceanic ridges.

A Prediction from Theory: Transform Faults

In 1965 the Canadian geologist J. Tuzo Wilson spent several months at Cambridge University as a close associate of Harry Hess, Fred Vine, and Drummond Matthews. Shortly afterward he published his classic paper in which he proposed the existence of a new type of fault, based on Hess's theory of seafloor spreading.

It had been well known that mid-ocean ridges such as the Mid-Atlantic Ridge are offset or segmented at many points by huge fracture zones; some offsets are hundreds of kilometers long. These fracture zones, or faults, as well as the mid-ocean ridges they offset, are the sites of numerous shallow-focus earthquakes. Figure 18.6 illustrates ridge segments in the equatorial Atlantic Ocean, the fracture zones connecting them, and epicenters of some recent earthquakes there.

Problem 20

Observe the offset of the ridge along the east–west-striking fault near St. Paul's Rocks. If this were an ordinary strike-slip fault displacing a static feature such as a dike or a fence line, what would be the sense of motion, right- or left-lateral?

Wilson's great insight was to realize that the feature being displaced, the Mid-Atlantic Ridge, is not a static feature according to Hess's theory of seafloor spreading. Rather, the ridge is an active feature where new crust is being created and then carried away to both sides of the ridge. Wilson reasoned that if creation of crust occurs along ridges, the movement of material actually observed along the fault would be opposite to that predicted by conventional geologic reasoning.

Let us see just what Wilson saw. Figure 18.7 shows a ridge crest, r_1, at which oceanic crust is created and moved laterally away from the ridge in the direction of the arrows. Similarly, r_2 is another segment of the same active ridge where new crust also is created and moved away in the direction of the arrows. Fault F offsets the ridge.

Problem 21

Ignoring the obvious displacement of the ridge segments, concentrate instead on the arrows indicating the movement of the newly created oceanic crust. What is the sense of motion of the new oceanic crust between ridge segments r_1 and r_2, right-lateral or left-lateral?

Problem 22

Is this answer consistent with your earlier answer?

Problem 23

Will there be movement along fault F to the right of r_1 or to the left of r_2? Explain your answer.

Wilson saw that the ridge segments themselves were not moving but remained fixed. Only *between* them did relative motion, due to ridge spreading, occur on the fault; beyond the ridges the fault was quiet, both sides moving abreast. The fracture zone would grow ever longer, but only the central portion would have seismic activity.

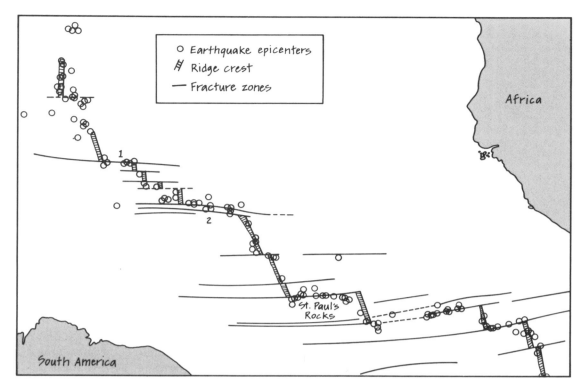

Figure 18.6
Ridge segments and earthquake epicenters along the equatorial part of the Mid-Atlantic Ridge.

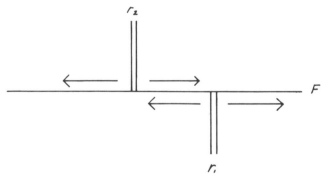

Figure 18.7
Ridge crests and transform faults.

On this kind of fault, the spreading motion of one ridge segment is transformed to lateral (strike-slip) motion on the fault and back to spreading motion on the next ridge. Wilson pointed out that similar transformations of motion would occur with trenches, where shortening instead of spreading takes place. Wilson recognized the existence of this new class of faults, which he called *transform faults*, as a direct corollary of the creation and destruction of crust according to Hess's theory. He wrote, "It is confusing, but true, that the direction of motion along [such a fault] is in the reverse direction to that required to produce the apparent offset." Wilson considered the ridge offsets to be the result of preexisting weaknesses in the rifted continents, in this case Africa and South America. Wilson's prediction, then, was that a new type of fault connects the active ridges and trenches of the world; the motion is confined to the segment of the fault between the active elements and is in the opposite direction to that predicted by conventional geologic reasoning.

Proof soon came from fault-plane or first-motion studies like the one you did in Exercise 16, Earthquakes. By the mid-1960s, accurate and sensitive seismic arrays made possible the detailed study of distant earthquakes such as those near the mid-ocean ridges. Lynn Sykes studied the first motions of these earthquakes; some of his results are shown in Figure 18.8.

Problem 24

Examine again Figure 18.6. Is the motion along the east–west-striking faults indeed confined to the part of the fault between offset ridge crests?

A.

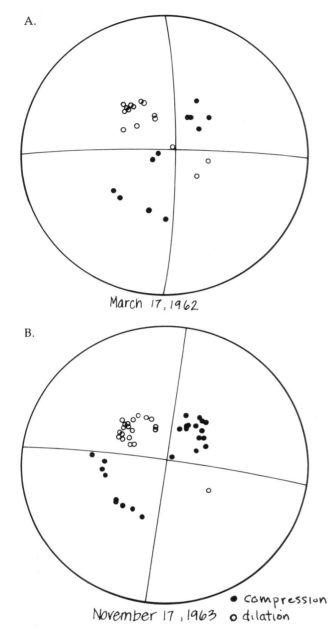

March 17, 1962

B.

November 17, 1963   • compression   ○ dilation

Figure 18.8
First-motion study solutions for the two earthquake events shown in Figure 18.6.

Problem 25

Figure 18.8 shows Sykes's first-motion solutions for events 1 and 2 of Figure 18.6. The two possible nodal planes are drawn as the two lines trending approximately north–south and east–west. Since we know from Figure 18.6 that the faults strike east–west, we can discard the north–south possibilities. Were the earthquakes right-lateral or left-lateral?

Problem 26

Is this motion in the sense predicted by Wilson? Explain your answer.

A PLATE-TECTONIC MODEL

The three attached sheets allow you to construct a simplified working model of the plate motions of the eastern Pacific. The model is crude, first because it neglects certain features of the area (such as the division of the Farallon plate into the Cocos and Nazca plates and the Galapagos and Chile ridges), and second because it is planar, so the motions only approximate those on a sphere. Nevertheless it is realistic in a general way. Here are the steps for constructing it:

1. Take sheet 1 and cut off and discard the stippled part along the line indicated. Cut so the black line remains on the part you keep. Cut out the small elliptical area in the central United States. Reinforce the sheet with tape under the dots labeled COC/AME and PAC/AME.

2. Cut the stippled part from sheet 2 as before, and reinforce with tape below the COC/PAC and PAC/AME dots. Press a thumbtack through the PAC/AME dot from below, and tape over the head of the tack to keep it in place.

3. Fit thumbtacks similarly through the COC/AME and COC/PAC dots on sheet 3, and tape over their heads. Do not cut this sheet.

4. Attach sheet 2 on top of sheet 3 by pressing the COC/PAC dot onto the appropriate tack. Attach sheet 1 by pressing the PAC/AME and COC/AME dots onto their respective thumbtacks, allowing the point of the COC/PAC thumbtack between sheets 2 and 3 to protrude freely through the elliptical hole in sheet 1 .

The model is worked by grasping sheet 2 at X and swinging it gently to the left and right. This movement shows the displacements of the Pacific and American plates relative to a fixed Cocos plate. The allowable motion can be increased by enlarging the elliptical slot in sheet 1 and cutting a narrow slot oriented N45°E through the PAC/AME dot in sheet 1.

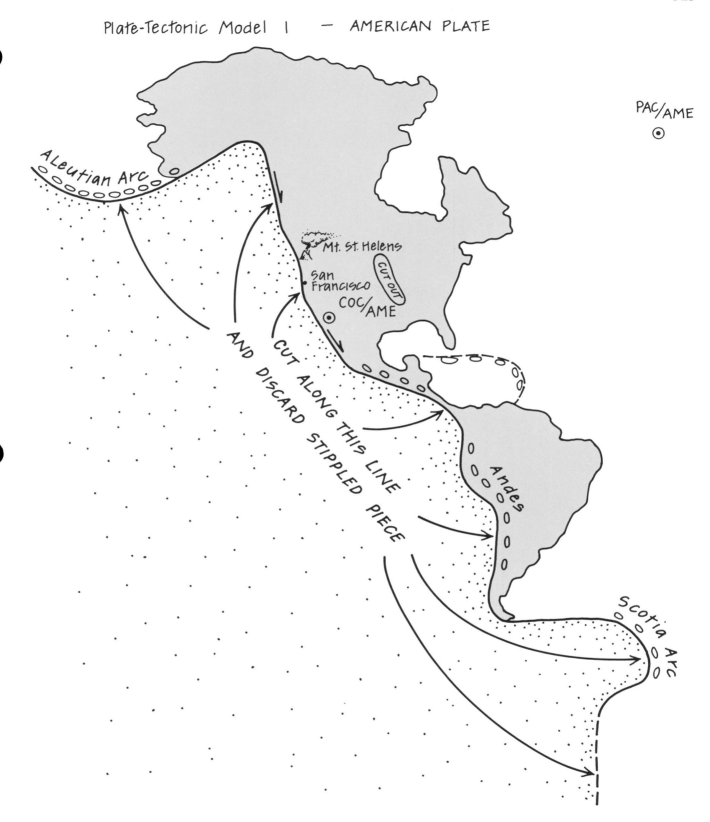

Plate-Tectonic Model 2   —   PACIFIC PLATE

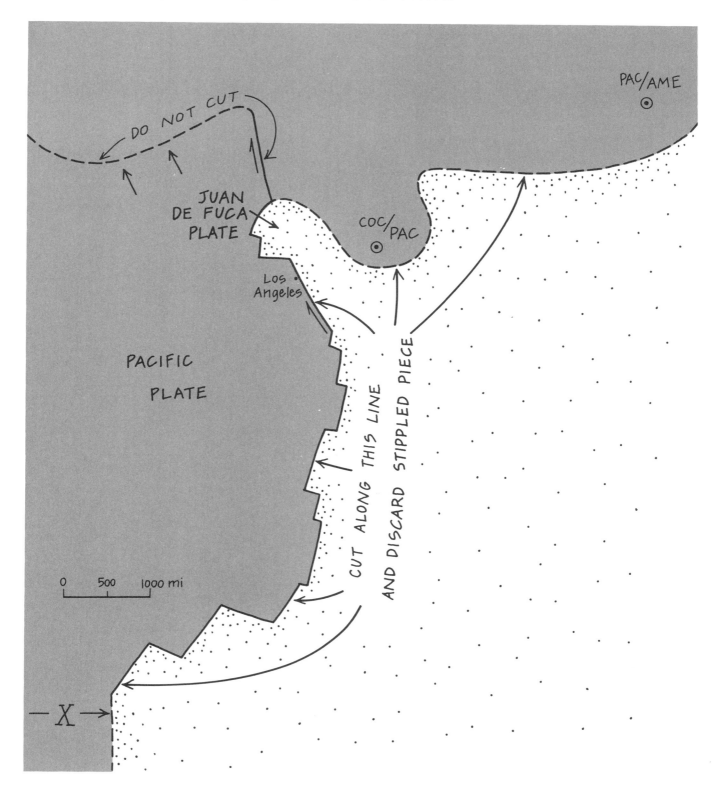

Plate-Tectonic Model 3 — COCOS PLATE
(Do not cut this sheet)

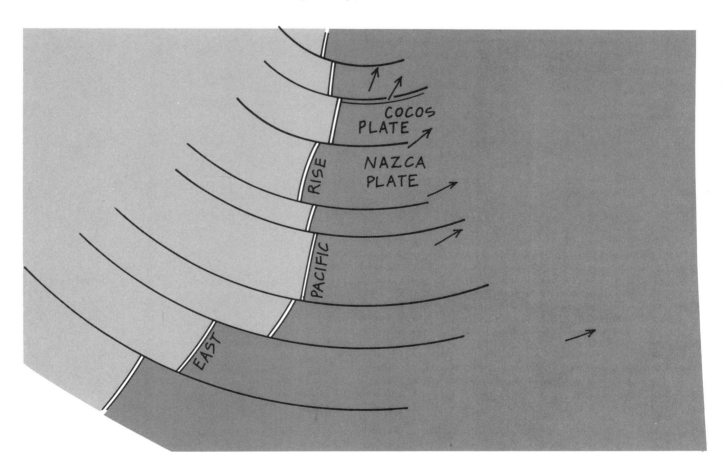

Note the following features as the point X is moved to the left (present-day motion:)

1. The Pacific plate is consumed at the Aleutian trench and the Cocos plate at the Chile-Peru trench.
2. These two plates slide past one another along the San Andreas fault zone and other similar zones farther north.
3. The East Pacific Rise moves beneath North America in the vicinity of the Gulf of California.

Problem 27

List the three types of plate margins shown on this model and some locations representing each type.

The motion of one portion of the earth's surface relative to another portion can be represented by a rotation of that portion about a particular point, a *pole of rotation*. The position of each thumbtack is the position of a pole of rotation of one plate with respect to another. This geometry results in several important features of plate motion.

Problem 28

What is the general shape of the transform faults in the Pacific basin west of South America?

Problem 29

How does the spreading rate vary from north to south along the East Pacific Rise?

The ellipses along the west coast of South America and Mexico and along the Aleutian arc represent andesitic volcanism.

Problem 30

With what type of plate boundary is this activity associated?

Problem 31

What is the reason for the volcanic activity?

Problem 32

Why is there no volcanic activity at present in the region from southern California to the abrupt right-angle bend of the coast in southern Alaska, except for a small region around the Oregon–Washington–southern British Columbia coast?

Problem 33

To what is the activity of Mount St. Helens attributable?

Problem 34

What type of plate margin characterizes the California coast at present?

Los Angeles is west of the main San Andreas fault system. Movement along the fault is right-lateral, as shown by arrows, and is proceeding at an average rate of approximately 3 cm per year. For the following questions make the admittedly uncertain assumption that this motion will continue in the same direction and at the same rate for the next 300 million years or so.

Problem 35

Near what city will Los Angeles be in 27 million years?

Problem 36

What is its ultimate fate, geologically speaking?

The Basin and Range district of Nevada and Utah has extended and widened by 100 percent in the Cenozoic Era by a process of normal faulting. The region is also characterized by an abnormally high heat flow and many areas of hot springs and recent volcanism.

Problem 37

With this evidence in mind and by observing the plate-tectonic model, loosen up your mind and propose a hypothesis to account for the extension of the Basin and Range.

FIELD STUDY: THE EARTH'S CRUST

The Earth's Crust at Present

Examine Figure 18.9, which illustrates the ages of different parts of the oceanic crust.

Problem 38

Mark in red pencil the places where the crust is spreading (divergent plate margins). Show the direction of spreading with arrows.

Problem 39

One may get an approximate idea of the spreading rates of mid-ocean ridges by noting the width of sea-floor that is underlain by rocks of a certain age. Place an F where the rate of spreading is fast and an S where the rate is slow.

Problem 40

What factor might explain the absence of crust that is from 100 to 230 million years old in the East Pacific?

Problem 41

How has the size of the Atlantic and Pacific Oceans changed in a general sense over the past 230 million years?

Problem 42

The east coast of the United States was once adjacent to what area? Mark this area with a green line.

Problem 43

What type of fault probably occurs at

X_1

X_2

X_3

Problem 44

For each location marked by a box on the map, determine the rock most likely to occur in that place.

1.

2.

3.

4.

5.

Earth's Crust During the Past 200 Million Years

Drawings that show the positions of the continental crust at different times since the early Mesozoic Era appear at the bottom of right-hand pages in Exercises 21–22. These drawings represent the frames of a motion picture. If you flip through the pages rapidly from the last page forward, the continents will appear to move as they have moved in the past 200 million years to their present positions, and you will see the breakup of the supercontinent of Pangaea into the continents as we know them today. The numbers on each "frame" of the motion picture are the times (in millions of years before present) at which that continental configuration existed.

Problem 45

At what time did the breakup of Pangaea begin, and where did the rupture first start?

Problem 46

Where was the last point at which Pangaea was joined? At what time was the rupture completed?

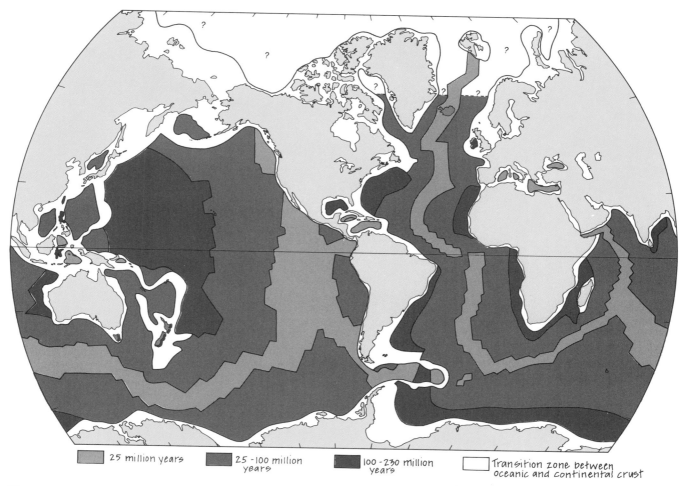

25 million years   25 -100 million years   100 -230 million years   Transition zone between oceanic and continental crust

Figure 18.9
Age of the ocean basins.

Some of Alfred Wegener's strongest evidence for continental drift was taken from the study of fossils. There are correlations of fossil organisms across the present Atlantic Ocean between South America to Africa and between other continental masses that are now separated as well.

Problem 47

Considering the time of the split between Africa and South America and the continuous evolution of organisms throughout geologic time, what would be the age of the youngest fossils that correlate between those two continents?

Other data that supported the idea of continental drift presented by Wegener concerned ancient climates. He cited evidence for glaciation in the Permian Period on such widely separated (at present) continents as southern India, South Africa, South America, and Australia.

Problem 48

From the reconstruction of Pangaea at the beginning of the motion picture, comment on the validity of Wegener's evidence.

A puzzling pattern in the apparent location of the earth's magnetic pole became apparent to geophysicists in the 1950s. It appeared from the magnetic record in iron-rich rocks that the earth's magnetic poles have moved significantly over the course of geologic time, an interpretation that was, needless to say, not well received.

Problem 49

Refer to Exercise 17, Figure 17.13, for the apparent polar wandering curve for North America. How might this puzzle be resolved in terms of modern plate-tectonic theory?

REFERENCES FOR PART D

Books

Bolt, B.A., *Earthquakes—A Primer*. W.H. Freeman, New York, 1978.

Calder, N., *The Restless Earth*. Viking Press, New York, 1972.

Cox, A., ed., *Plate Tectonics and Geomagnetic Reversals*. W.H. Freeman, New York, 1973.

Cox, A., and B. Hart, *Plate Tectonics: How It Works*. Blackwell, London, 1986.

Glen, W., *Continental Drift and Plate Tectonics*. Charles E. Merrill, Columbus, Ohio, 1975.

———, *The Road to Jaramillo*. Stanford University Press, Stanford, Calif., 1982.

Hallam, A., *A Revolution in the Earth Sciences*. Oxford University Press, Oxford, 1973.

———, *Great Geological Controversies*. Oxford University Press, Oxford, 1983.

King, P.B., *The Evolution of North America*. Princeton University Press, Princeton, 1977.

Marvin, U.B., *Continental Drift: The Evolution of a Concept*. Smithsonian Institution Press, Washington, D.C., 1973.

McAlester, A.L., D.L. Eicher, and M.L. Rottman, *The History of the Earth's Crust*. Prentice-Hall, Englewood Cliffs, N.J., 1984.

McPhee, J.S., *Basin and Range; In Suspect Terrain; Rising from the Plains*. Farrar, Straus & Giroux, New York, 1981, 1983, 1986.

Planet Earth Series, *Continents in Collision* and *Earthquakes*. Time-Life Books, Alexandria, Va., 1982.

Press, F., and R. Siever, *Planet Earth—Readings from Scientific American*. W.H. Freeman, New York, 1974.

Richter, C.F., *Elementary Seismology*. W.H. Freeman, New York, 1958.

Sullivan, W., *Continents in Motion*. McGraw-Hill, New York, 1974.

Takeuchi, S., S. Uyeda, and H. Kanamori, *Debate About the Earth*. Freeman Cooper, San Francisco, 1970.

Uyeda, S., *The New View of the Earth*. W.H. Freeman, New York, 1978.

Wilson, J.T., ed., *Continents Adrift and Continents Aground*. W.H. Freeman, New York, 1976.

Wyllie, P.J., *The Way the Earth Works*. John Wiley & Sons, New York, 1976.

Yanev, P., *Peace of Mind in Earthquake Country*. Chronicle Books, San Francisco, 1974.

U.S. Government Publications

U.S. Geological Survey Information Pamphlets are available free from

Distribution Branch
U.S. Geological Survey
Box 25286, Federal Center
Denver, CO 80225

Active Faults of California
Earthquakes, 1985.
Plate Tectonics and Man, 1980.
The San Andreas Fault, 1985.

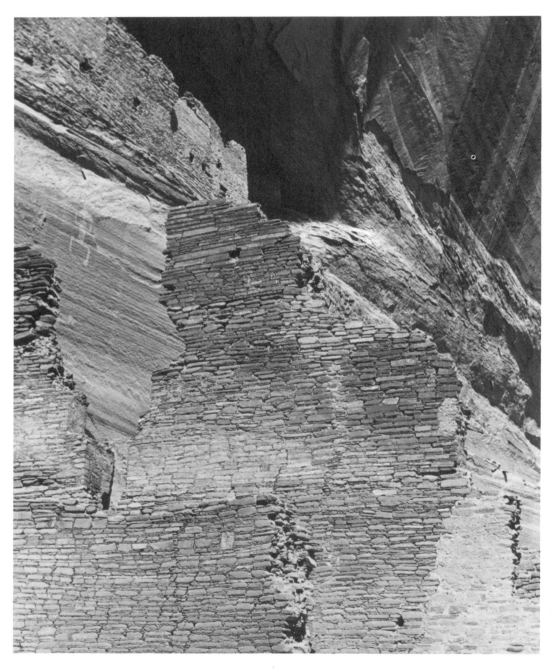

Petroglyph at White House Ruin, Canyon de Chelly National Monument, Arizona. It is believed that the Anasazi, the ancient ones, abruptly deserted advanced settlements such as this one around A.D. 1200 due to changing environmental and geological conditions.

Geology, Space, and Civilization

You have explored many features of the earth in Parts A through D of this book. No study of our earth would be complete, however, without a view of the earth in space or a look at how civilization influences and is influenced by geologic processes and events. Exercise 19 explores some of the planetary bodies of our solar system. This exercise is an excellent review of the earlier exercises; the planets and their moons display an amazingly diverse array of materials, surface processes, and interior conditions.

Exercises 20, 21, and 22 examine the interactions of geology and civilization. In many ways our history is one of ever-expanding use of the earth's materials, from stone to iron to energy resources. This progres-

sion, and the effects it has had on civilization, is the subject of Exercise 20. Exercise 21 deals with the interaction of humans and processes occurring on the earth's surface—how surface processes affect people and, conversely, how human activities affect surface processes. Perhaps the most dramatic effects of earth processes on civilization occur as a result of the earth's internal energy. Earthquakes and volcanic eruptions have destroyed entire cities and, in rare cases, whole cultures. Exercise 22 explores the effects of earthquakes and volcanoes on civilization and the progress that has been made in understanding the dangers and predicting when these events may occur.

Planetary Geology

We are all transitory inhabitants of the third planet revolving around an ordinary, somewhat smaller-than-average star in an average galaxy which contains over two billion similar stars and undoubtedly many billions of other planets.

BRUCE MURRAY

How little we knew, until recently, of even the tiny portion of the galaxy we call the solar system! Starting in the 1960s, we began to get our first close-up glimpses of other worlds. In the early stages of space exploration, knowledge of the geologic processes occurring on earth helped scientists to interpret what they saw and photographed on other bodies of the solar system. Now, however, what we are learning about other planets and moons is helping us to better understand geologic processes that occur on earth. The gravity, tectonic activity, or atmosphere of the earth, for example, cannot be manipulated by experiment, but the landforms of the moon or Mars show how surface processes operate in the absence of an atmosphere or under a smaller gravitational field. Other planets and moons also furnish valuable clues for the interpretation of earth's early history, which has been almost totally obscured by the constant destruction and renewal of almost every part of the earth's crust.

The presence of a magnetic field around Mercury may help elucidate the poorly understood terrestrial field. The very old plagioclase-rich rocks of the moon may shed light on the early crustal evolution of the earth. The ancient cratered surfaces of the rocky inner planets testify that the earth itself must have undergone a similar period, soon after its accretion, of intense bombardment by extraterrestrial material. The dense, corrosive, and hot atmosphere of Venus may even warn of impending dangers to our own planet from the greenhouse effect. Far from being an esoteric impractical field of study, then, planetary geology can contribute much to our understanding of the physical world in which we live.

MATERIALS THAT MAKE UP THE PLANETS

Igneous Rocks

The only planetary body other than the earth from which we have samples is the moon. The samples returned by the U.S. manned lunar missions and the Soviet unmanned missions have helped us to understand the internal evolution of the moon and, by inference, possibly the early stages of terrestrial evolution. There are no rocks on the moon that correspond to the andesites and granites of the earth's continental masses. Most of the major rock types found on the moon are mafic. Apparently the relatively small size of the moon resulted in rather rapid cooling; there was insufficient energy left to drive the plate tectonic processes that chemically "distill" the more felsic rocks from their mafic precursors on the earth.

Perhaps the most significant lunar rocks in terms of inferences about the early crust of the earth are rocks consisting almost entirely of the calcium-rich plagioclase anorthite ($CaAl_2Si_2O_8$). Such rocks are called *anorthosites.* Anorthosites are rather rare on earth, but they are overwhelmingly the most abundant lunar rock. Several lines of evidence suggest, in fact, that the entire lunar crust, 50 to 100 km thick, is anorthositic. These are the oldest lunar rocks as well (4 billion years or older), and they predate any rocks found thus far on earth.

To understand the implications of the lunar anorthosites for crustal evolution and differentiation of the earth, recall from Exercise 2 the minerals that crystallize earliest in the normal sequence of igneous minerals.

Problem 1

Considering both the discontinuous and the continuous sides of the Bowen reaction series (Exercise 2, Figure 2.2), which two or three minerals would crystallize first (at the highest temperatures) from a mafic silicate melt as it cools?

Problem 2

Of these minerals, which would be richest in iron and magnesium and therefore densest? Which of these minerals would be the least dense?

Some scientists have suggested that the moon accreted from small particles in orbit around the earth. These particles would have arrived with considerable speed (kinetic energy), which would have been converted to heat upon impact. If lunar accretion proceeded more rapidly than this heat could be dissipated into space, it is likely that at least the outermost several hundred kilometers of the moon melted. The outer surface of the moon would have been covered with an ocean of white-hot molten rock.

Problem 3

From this model of early lunar history and your answers to the previous questions, show how an anorthositic crust might have formed.

Problem 4

Cite a terrestrial example of magmatic differentiation (see Exercise 2) and relate it to the model you have just proposed.

Figure 19.1
Rocks from Utopia Planitia, Mars. (Viking Lander 2 image.)

Problem 5

Assuming that the early history of the earth also involved large-scale heating and melting, why do we see no large areas of terrestrial anorthositic crust like that of the moon?

Although no samples of Martian rocks are available, the Viking landers from 1976 to 1984 transmitted excellent images of the surface of Mars and performed many chemical tests and physical observations. Figure 19.1 shows a typical rock from Utopia Planitia (Viking Lander 2).

Problem 6

What is its texture, and how did this rock form?

Problem 7

The Martian atmospheric pressure is only 0.006 bar; that of the earth is 1 bar. Why, then, might this type of rock texture be more abundant on Mars than on earth?

Surely one of the most curious objects in our solar system is Io, one of the four major moons of Jupiter. Io is about the same size and density (3.35 g/cm³) as our moon and, like the moon, is devoid of water, but there the similarity ends. Because of its brilliant orange-red color and because it has a tenuous atmosphere of SO_2, Io was thought to have a surface rich in sulfur. Unusual as this idea seems, the images transmitted from Voyager 1 in 1979 revealed an even more dramatic surprise. The surface of Io lacks the impact craters associated with every other moon and planet, suggesting that some process currently reworks the surface. Figure 19.2 shows this process in action. The bluish-white plume above the surface of Io is a volcanic eruption of a type probably unique in our solar system. The material ejected is not a silicate but appears to be molten sulfur. The major source of internal heat within Io is not the decay of radioactive substances, as it is on earth, but powerful tidal forces generated within Io by its strong gravitational interaction with nearby Jupiter and the adjacent moon Europa. Sulfur melted by these forces moves upward toward the surface because it is hotter and therefore less dense than its surroundings. It is thought that large reservoirs of relatively cool liquid SO_2 exist near the surface of Io. When ascending molten sulfur encounters the cooler liquid SO_2, violent eruptions result.

Problem 8

What liquid on the earth plays a role analogous to that of SO_2 on Io in causing violent volcanic explosions? What is the texture of rocks produced in such explosive eruptions?

Problem 9

Use your imagination (also recall the way igneous rocks are named) and suggest a rock name for these volcanic rocks of Io.

Figure 19.2
Volcanic plumes on Io as seen from Voyager 1.

Sedimentary Rocks

Clastic sedimentary rocks are formed when other rocks are weathered and the resulting loose sediments are transported, deposited, and cemented together (this was the subject of Exercise 3). Weathering and transport by wind, water, and ice imply the existence of an atmosphere and hydrosphere.

Problem 10

Would you expect sedimentary rocks to be common on other planets? Explain your reasoning.

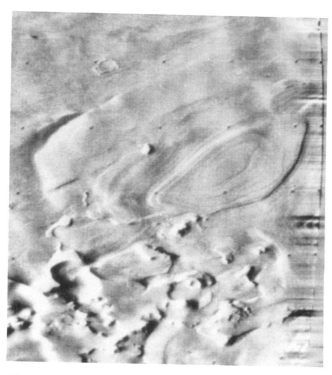

Figure 19.3
Layered terrain in the south polar region of Mars. (Mariner 9 image.)

Figure 19.3 shows a Mariner 9 image of layered terrain in the south polar region of Mars. Apparently sediments were deposited, cemented, and then subsequently eroded to expose a benchlike topography. This area seems to be one of the rare known occurrences of sedimentary rocks on other planets.

Problem 11

Agents that commonly cement terrestrial sedimentary rocks are calcite, silica, iron oxides, and clay minerals. Given the low Martian temperatures (currently below $-80°C$) and the presence of water ice in the polar caps of Mars, can you suggest what agent might cement the layered deposits?

Metamorphic Rocks

The majority of terrestrial metamorphic rocks are formed in regional metamorphism under conditions of deep burial and accompanying high temperature and high pressure and are exposed at the earth's surface only after long periods of erosion have removed great thicknesses of overlying rocks. (This subject was covered in Exercise 4.)

Problem 12

Would you expect to see rocks formed by regional metamorphism exposed on the surfaces of other planets and moons? Explain your answer.

TIME, CORRELATION, AND MAPS

Two processes operating at various times over the past 4.6 billion years have been largely or completely responsible for shaping the present surfaces of the earthlike planets and moons (with the notable exception of earth itself). One of these processes is volcanism; the second and in some cases more important process is meteorite impact. Although the earth bears a few dozen isolated scars of meteorite impact, most of these features have long been obliterated by weathering, erosion, and crustal deformation. The impact of a meteorite excavates a depression in the surface, raises a rim of debris, and scatters other debris (*ejecta*) in a more or less radial pattern around the impact crater, as shown in Figure 19.4. Meteorite impact is a rock-forming process. The high pressures (and somewhat elevated temperatures) generated by shock waves spreading out from an impact cements fragments together and resets the radiometric "clocks" in the excavated material. The crater floor, the rim, the ejecta blanket, and the rays are all new rock material, usually a breccia. The relative ages of rocks produced during different impacts can be determined using the principle of superposition (covered in Exercise 5) along with other criteria.

Problem 13

Compare the shape of the top crater in Figure 19.4 with the feature that formed in the raindrop impact experiment of Exercise 10.

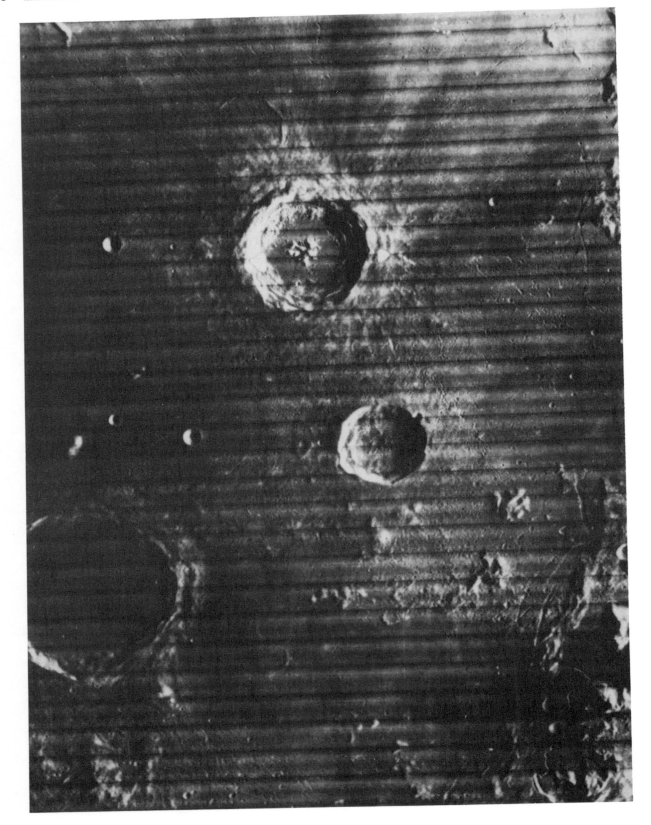

Figure 19.4
Three lunar craters.

Problem 14

Discuss several criteria for determining relative ages of impact crater deposits. Consider:

• overlapping deposits (superposition)

• freshness or sharpness of the rim

• small-crater density over the main feature

• definability of the ejecta blanket and ray material

Problem 15

Using your criteria for relative age, number the three large craters in Figure 19.4 according to age, 1 for oldest, 3 for youngest.

Figure 19.5 shows a photograph of the Imbrium Basin, a large multiringed impact basin excavated by a very large meteorite impact and later filled with flood basalts (dark area at upper right).

Problem 16

Locate the craters A, B, C, D, E, and F on the photograph. Which of these craters has well-defined ray systems?

Problem 17

Which craters have well-defined ejecta blankets?

Problem 18

Which craters have sharp rims but subdued ejecta blankets?

Problem 19

Classify the craters into three groups, 1 for oldest, 3 for youngest.

Problem 20

What can you infer about the relative ages of the Imbrium Basin, the flood basalts that fill low areas, and the craters D and A?

SURFACE PROCESSES

Modifications of the surface of the earth by chemical and physical weathering, mass movement, moving water, wind, and ice are among the most extensively studied geological phenomena. This section considers the roles of these surface processes on other planets and moons. In general, there has been only very minor breakdown of surface materials by chemical weathering (see Exercise 9) on the other planets and moons. Without an atmosphere there is no transport by water, ice, or wind. Nevertheless, some surface processes are greatly enhanced on planets with tenuous or no atmospheres. Impact cratering is much more important on airless bodies because even tiny particles can penetrate to the surface instead of burning up in the atmosphere. Impacts can physically weather (and transport) surface material. You considered the effect of impact cratering in the previous section of this exercise.

Mass Movement

Gravity-induced movement of loose surface material (see Exercise 9) is common on the earthlike planets, and these mass movements are generally quite large as compared with those on the earth.

Problem 21

Study the three large craters in Figure 19.4. Where has mass movement occurred? What kind of movement (slump, slide, flow, fall) seems to have occurred?

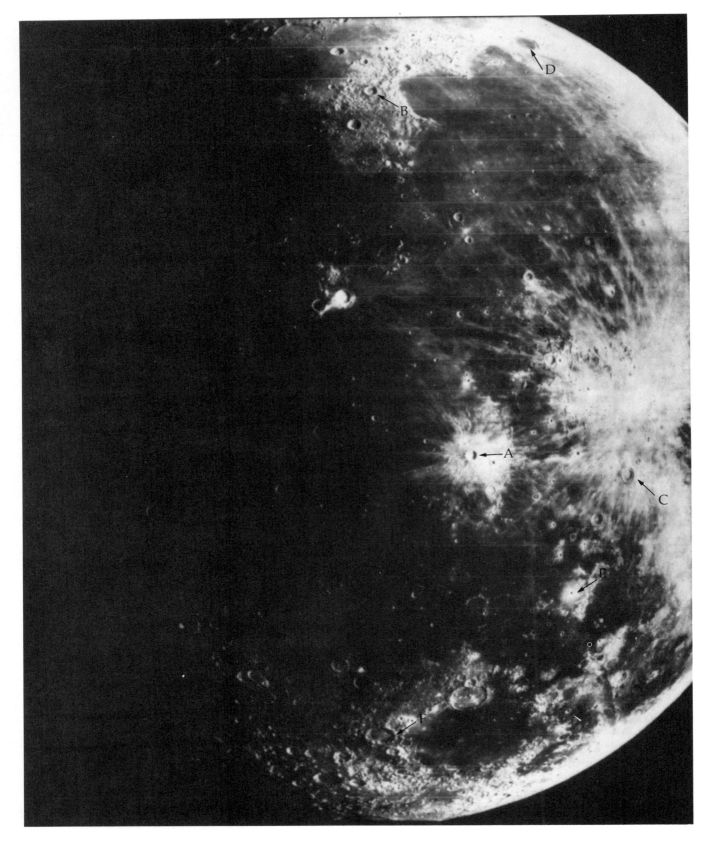

Figure 19.5
Photograph of Imbrium Basin.

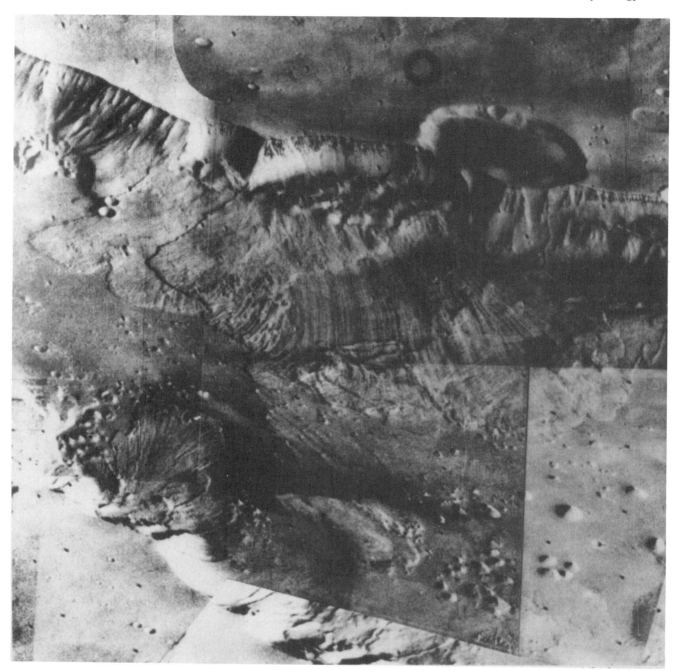

Figure 19.6
Valles Marineris, Mars. Area shown is about 70 km
by 150 km. (Mosaic of Viking Orbiter 1 images.)

Problem 22

Figure 19.6 shows Valles Marineris on Mars. The area
shown is 70 km (43 miles) wide. What appears to be
the main mechanism by which the canyon enlarges
itself?

Problem 23

Trace on Figure 19.6 the outlines of the three most
recent movements you see. How would you characterize the top part of the large movement of the opposite wall of the canyon? The bottom part?

Problem 24

Consider the size of the area in Figure 19.6 and compare the Martian slides with the rather large slides in the Rio Grande Gorge south of Taos, New Mexico (Figure 10.6).

Fluvial Processes

Of all the earthlike planets and moons, Mars is the only one other than earth to show signs of fluvial activity (see Exercise 12). One of the most surprising results of the Mariner and Viking missions was the discovery of what appear to be channels, valleys, and other features of fluvial origin on Mars. These features differ markedly in some respects from terrestrial analogs, a disparity that is not surprising, given the tenuous nature of the present Martian atmosphere and the difference in the mass of the two planets.

Problem 25

The Martian gravity is roughly one-third that of earth, so a given particle of sediment weighs less on Mars than on earth. What difference would that make in the settling time of a particle traveling as suspended load in a Martian stream?

Problem 26

On earth, silt and clay are commonly carried as suspended load in moving water, and sand and pebbles are carried as bed load. What size of particles might have been carried as suspended load in Martian channels? As bed load?

Figure 19.7 shows Parana Vallis, a comparatively well-developed valley network on Mars, and Nirgal Vallis, a somewhat less-developed valley. The areas shown are roughly 250 km long.

Problem 27

Select three or four prominent trunk valleys, and outline the drainage basins of each (see Exercise 12). Compare the resulting map with the terrestrial examples. Do adjacent drainage basins on Mars border continuously along their perimeters (that is, do they share common ridge lines) as they do in the well-developed terrestrial valley systems?

Problem 28

What does this observation suggest about the length of time that liquid water persisted on Mars?

Problem 29

Are most of the craters younger or older than the valley network?

Problem 30

Since the period of heavy meteorite bombardment on Mars ended about 4 billion years ago, would you say that the erosional event (and hence the period of denser atmosphere) represented by the Martian valleys was early or late in the planet's history?

Ice and Groundwater

It seems likely that there is water ice in the polar caps of Mars and between detrital grains in the wind-blown layered deposits near the polar regions. Another line of evidence suggests that ice may be even more widespread on Mars. Figure 19.8 shows the head of an outflow channel. These features are large channels originating in jumbled chaotic terrain. They lack the tributaries that characterize the valley networks of Mars. The very localized source of such large quantities of water poses some interesting problems.

Observe the chaotic terrain immediately adjacent to the canyon (right side). The large blocks of material detached from the headwall appear to have been emplaced by slumps. On earth, large slump features such as these have usually been lubricated by substantial quantities of water. There is no obvious source of surface runoff of water in this area of Mars, but substantial quantities of water may exist as groundwater.

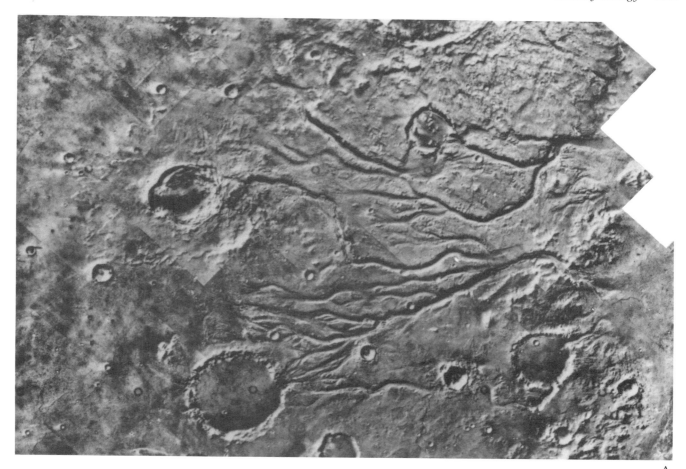

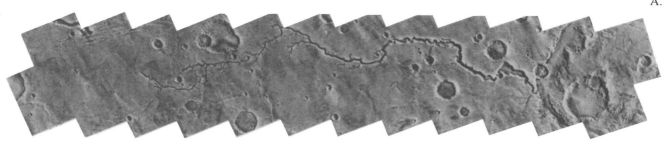

Figure 19.7
A. Parana Vallis. B. Nirgal Vallis. (Viking Orbiter 1
images.)

A.

B.

Problem 31

Recalling the cold (rarely warmer than $-80°C$) surface
temperatures of Mars, speculate on what common solid
impermeable substance might serve to confine the
groundwater and protect it from evaporation.

Problem 32

What events on Mars might generate enough heat to
trigger the release of liquid groundwater from a confined aquifer to form the outflow channels?

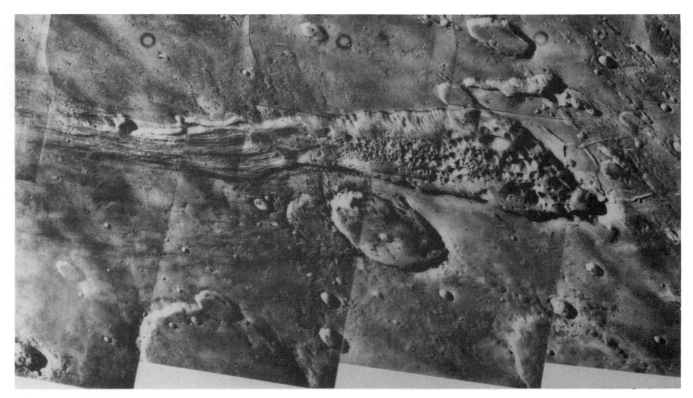

Figure 19.8
Capri Chasma, an outflow channel on Mars. Chaotic terrain at the right. Area
shown is about 300 km by 300 km. (Mosaic of Viking Orbiter images.)

Wind

To compare the relative effectiveness of wind and water
as agents of erosion, one must know the momentum
of the transported particles that do the actual eroding.
The momentum of a given size of particle depends on
its mass and its velocity. Wind speeds typically reach
higher values than water speeds, but because of the
lower density and viscosity of air, greater wind speeds
are required to set particles of a given size into motion.
(See Exercise 15.) Mars experiences very high wind
velocities (in excess of 100 mph); wind velocities on
earth are normally much lower.

Problem 33

Figure 19.9 shows several dune fields on Mars. Examine the largest field. At the top and bottom are some
barchan dunes. What wind direction (assuming north
is up) is indicated by the alignment of the barchans?

Problem 34

What type of dune predominates in the center of the
large dune field?

Problem 35

What do these two types of dunes suggest about the
sand supply in this area?

INTERNAL PROCESSES

Moonquakes

Quakes—that is, sudden ground movements—require
a supply of internal energy. On earth the source of
this energy is the hot molten core, which transfers
heat energy to the mantle, which then transfers heat

Figure 19.9.
Large dune field on Mars imaged by Viking orbiter.

energy to the surface by convection and causes actual movement of material in the upper mantle. It is probably the movement of mantle material that causes crustal plates to move. That movement, in turn, causes most earthquakes.

Seismographs have been placed on the moon and on Mars. The resulting data show that both bodies are seismically quiet internally.

Problem 36

Seismographic data on the moon reveal it to be solid throughout. From a consideration of the nature of *P*- and *S*-waves and the media through which each may travel, what seismic evidence might suggest that the moon contains no molten material similar to the molten outer core of the earth? (See Exercise 16.)

Problem 37

If the moon is solid, what is the most likely source for the small, infrequent moonquakes that have been recorded?

Geophysics: Gravity, Isostasy, Heat, Magnetism

The bodies of the solar system have provided planetary scientists with contrasts and surprises. The surprising variety of surface features that have been observed on the planets and moons of the solar system are the result of the variety of different internal compositions and conditions. The sulfur volcanoes of Io are caused by its unique chemical composition. The moon's cold, solid bulk is due to its small size and largely silicate composition.

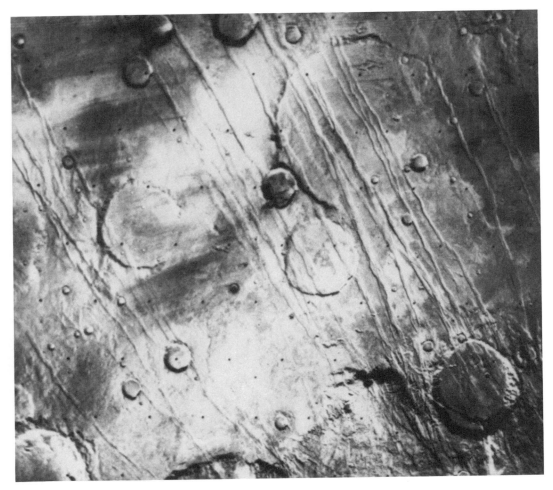

Figure 19.10
Normal faulting, Syria Rise, Mars.

Gravity and Isostasy. Examination of the shape of Mars reveals a pronounced bulge, the Syria Rise, which stands about 7 km (over 20,000 feet) higher than the average elevation. Gravity studies of the bulge reveal no underlying "root" of low-density rock similar to that of the mountain ranges of earth. The center of the bulge is criss-crossed with normal faults and down-dropped blocks. (See Figure 19.10.) Moreover, this is the area of the Tharsis shields—volcanoes of probable basaltic composition.

Problem 38

Suggest an explanation (or two) to account for these features of the Syria Rise.

Heat and Isostasy. Some of the larger planets and moons have hot interiors, whereas most of the small ones are cold throughout. The planet Mercury and Tethys, a moon of Saturn, are both thought to be cold, solid bodies.

Problem 39

Discovery Scarp on Mercury, a feature 500 km long, seems to be a thrust fault (indicating crustal compression). There are no extensional faults on Mercury; the planet is now a cold, solid body with a large iron core covered with a thin crust of silicate material. Given that Mercury was once molten, how do you account for the compressional features of Mercury's crust?

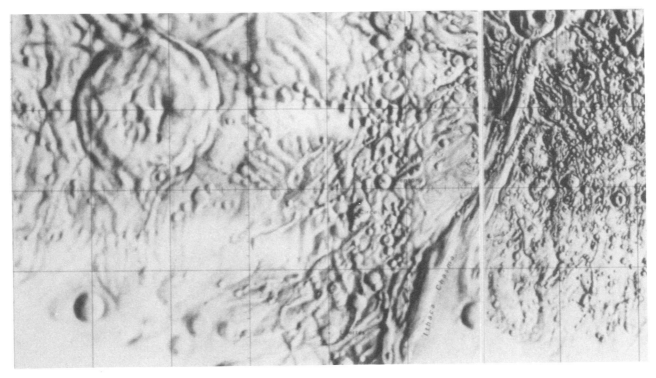

Figure 19.11
Surface of Tethys. (Map prepared from Voyager 1 and 2 images.)

Problem 40

The surface of Tethys, shown in Figure 19.11, displays a great canyon system (shown here running NNE–SSW) bounded by normal faults that extend halfway around the moon. It forms 5 to 10 percent of the surface area of Tethys. Is this feature compressional or is it extensional?

Problem 41

The density of Tethys is 1.2 g/cm^3, which suggests that much of the moon consists of water ice. In the absence of an exterior heat source, Tethys would be expected to lose heat from its surface and freeze from the outside inward. Assuming this model is correct, explain the expansion features found on the surface of Tethys.

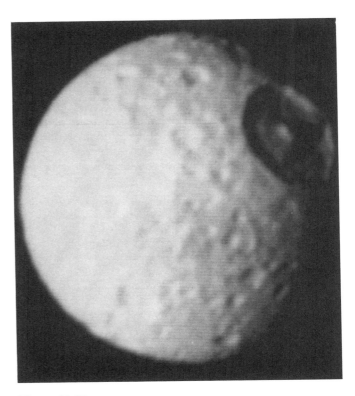

Figure 19.12
Mimas. (Voyager image.)

Figure 19.12 shows Mimas, another moon of Saturn. Contrast the very large but relatively flat-looking impact crater on Tethys (Figure 19.11) to the large crater with obvious high relief on Mimas.

Problem 42

If the interior of such bodies behaves in a plastic manner—a situation more typical of warm solids than of cold ones—isostatic rebound would tend to smooth out features of great relief. What were the relative temperatures of Tethys and Mimas at the time of the impact cratering shown in Figures 19.11 and 19.12?

Magnetism. Of all the earthlike planets—Mercury, Venus, earth, Mars, and moon—earth displays by far the most substantial magnetic field. The causes of the geomagnetic field are not well understood. The consensus of geophysicists is that circulating currents in the earth's iron core act as a dynamo and give rise to the magnetic field. Some of the physical parameters of these bodies are shown in the table below.

Problem 43

Compare Venus and earth. Their diameters and densities are almost identical. What parameter seems to be essential to the presence of a magnetic field?

Problem 44

Compare Mars and earth, whose densities differ markedly. What difference in bulk chemical composition of the two planets may this difference reflect, and why would this factor affect the magnetic field?

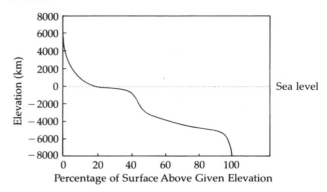

A. Earth

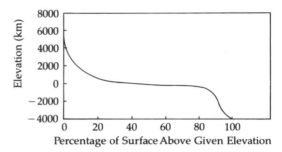

B. Venus

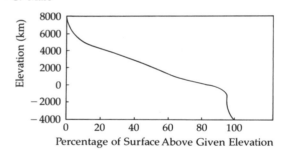

C. Mars

Figure 19.13.
Hypsographic curves of the surfaces of earth, Venus, and Mars.

Physical Properties of Some Solar System Bodies

Property	Mercury	Venus	Earth	Mars	Moon
Diameter (km)	4880	12,100	12,756	6794	3476
Density (g/cm³)	5.4	5.3	5.5	3.9	3.3
Axial rotation (days, hours, minutes)	59 d	243 d	23 h, 56 m	24 h, 37 m	27 d, 7 h, 41 m
Magnetic field (as a fraction of earth's)	0.01	0.00005	1	0.004	0.0001

Problem 45

How would the diameter of the body, which governs the rate at which it cools and solidifies, affect development of a magnetic field?

Plate Tectonics

Earth, the planet dominated by plate tectonics, is characterized by constant production and destruction of a relatively thin basaltic oceanic crust, while the continental crust, too thick and buoyant to be destroyed, rides higher on the underlying mantle. One rather simple signature of this process is the distribution of the earth's average crustal elevation, which reflects the existence of the two distinct kinds of crust. (See Figure 13.2.) The continents stand roughly 4.5 km higher than the ocean basins, and there are only minor deviations from these two dominant elevations. Recently, radar images that show the crustal elevations of the surfaces of Mars and Venus have become available for comparison with earth. Hypsographic curves for the three planets are shown in Figure 19.13. Notice how clearly the two types of terrestrial crust are revealed.

Problem 46

Compare the three curves. Do the surfaces of Venus and Mars resemble that of earth in any of the features that are suggestive of plate tectonics? What are the similarities or differences?

Geology and Civilization: Minerals and Energy

The geologist can't escape dedication to history.

RENÉ DUBOS

One important factor that distinguishes humans from any other inhabitant of the earth is their knowledge and use of the earth's mineral and energy resources, which has gradually become an absolute reliance on those resources. Current archeological evidence indicates that humans used stone tools over 2 million years ago. Since that time humans have continued to utilize the earth's resources for both beneficial and destructive purposes. In many ways, human history from the Stone Age to the Space Age is one of increasing knowledge and use of rocks and minerals. Concrete, steel, glass, metals, plastics, gasoline, and silicon chips are just some of the important products we get from the earth's mineral and energy resources.

MINERAL RESOURCES

One of the most important archeological sites for early humans is Olduvai Gorge in Tanzania, discovered in the early 1960s by Mary Leakey. This site reveals the oldest association of stone tools and hominid (human-like) bones. Stone tools discovered there are of distinctive rock types and in some cases were transported considerable distances from their sources. Obviously these early humans recognized that certain rocks were more suited to their purpose than others. The excavation of the 1.8-million-year-old "Bed I" at Olduvai yielded 2275 stone artifacts. Most of the stone fragments were flakes, produced in sharpening other stones for use as choppers and knives. The early culture at Olduvai used both basalt and quartzite for tools.

Early Hominids

Problem 1

Examine the geologic map of Olduvai Gorge in Figure 20.1. Note the location of the primary excavation of Bed I, where basalt and quartzite tools were found. What is the nearest source of basalt and quartzite?

Problem 2

How far were these particular rocks transported?

Problem 3

In what way might the journey to a distant outcrop shared by other groups of hominids have affected civilization?

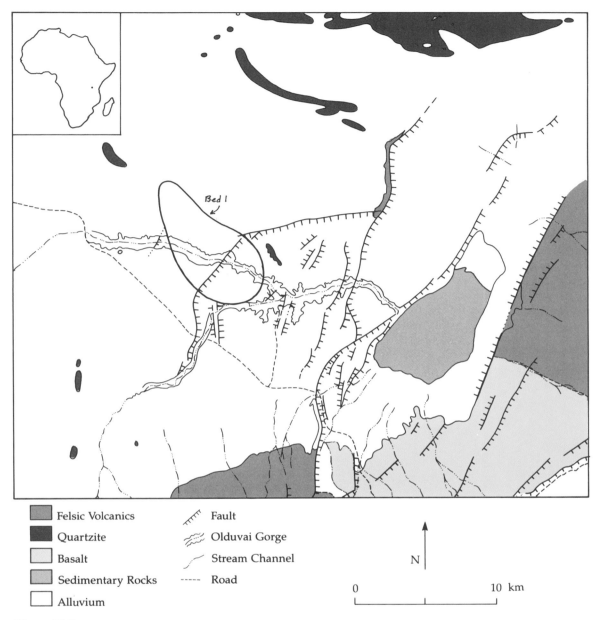

Figure 20.1
Geologic map of Olduvai Gorge, Tanzania.

Problem 4

At Olduvai Gorge and all other early hominid sites, quartzite and other quartz-rich rocks such as chert, flint, and jasper were used to produce blade-shaped tools. What simple physical properties of quartz would have made it useful to early man?

Ancient Civilizations

A major turning point in human history was the discovery, production, and use of metals. Stone Age people recognized the distinctive features of metals and selectively gathered them, but they used metals only as ornaments and decoration. The first widespread use of metals was by the Sumerians in the Near East, who lived along the Tigris and Euphrates rivers 7000 years ago. In addition to the first extensive use of metals, the Sumerians are credited with many other major advances in human history, including the first written language, early work in mathematics and science, and the development of the city–state society. Sumerian culture centered around large cities in Mesopotamia between the Tigris and Euphrates Rivers. The major cities of Ur, Uruk, and Kish had populations near 50,000. Irrigated fields outside the cities provided food for the inhabitants. Ships and caravans transported metals and precious gems to the cities, where craftsmen worked them into tools and decorations. As early as 4000 B.C., the Sumerians were keeping records by using pictographs on clay tablets. By 3000 B.C., they had a full syllabic alphabet in the form of a wedge-shaped script (cuneiform).

Although Neolithic man had used metals, it was not until the Sumerian culture that fire was used to concentrate metals from rock. Copper was the first and most important metal smelted by the Sumerians. By 1000 B.C., the Near East civilizations had produced all the major metals (copper, bronze, iron, gold, and silver) that man would use until the twentieth century.

Problem 5

Although not a metal, clay has played a very important role in the development of civilization. What physical characteristics of clay made it useful for recording the first written language?

Problem 6

In what other ways is clay useful to humans?

Problem 7

Examine the samples of copper, bronze, and iron provided by your instructor. What physical characteristics of these metals make them useful?

Problem 8

Your sample of copper ore is similar to ores found near Mesopotamia (azurite, malachite, or similar minerals). What simple physical properties make this copper ore easy to recognize?

Problem 9

Examine the location of the metal ore deposits in the Middle East and Europe shown in Figure 20.2. Where did the Sumerians probably find the ore for the metals they used?

Problem 10

Many of the blue and green ores of copper are carbonates and phosphates that form by weathering in arid climates. Suggest some reasons for the fact that the Sumerians smelted and utilized metals about a thousand years earlier than European cultures.

One of the best-known early cultures was the Egyptian civilization, which began as small separate agricultural communities along the Nile River. These communities were first united under the Pharaoh Menes around 3100 B.C. During the First Dynasty, great achievements in architecture, painting, and sculpture set a standard for all future dynasties. The Egyptians had a written language that had developed into a sophisticated literature by 2300 B.C.

The study and utilization of metals by the Egyptians apparently began with the migration of a group of Sumerians into northern Egypt around 3500 B.C. As in Sumeria, copper was the first metal smelted by the Egyptians. Although copper was extremely useful, gold was the foundation for the prosperity of the Egyptian culture. As an example, letters in the form of clay tablets from the ruins of Akhenaten at Tell-el-Armana mention the trade of Egyptian gold for slaves from Assyria and Babylonia. The exhaustion of Egyptian gold resources contributed to the decline of Egyptian culture, and Egypt was finally conquered by the Persians in 525 B.C.

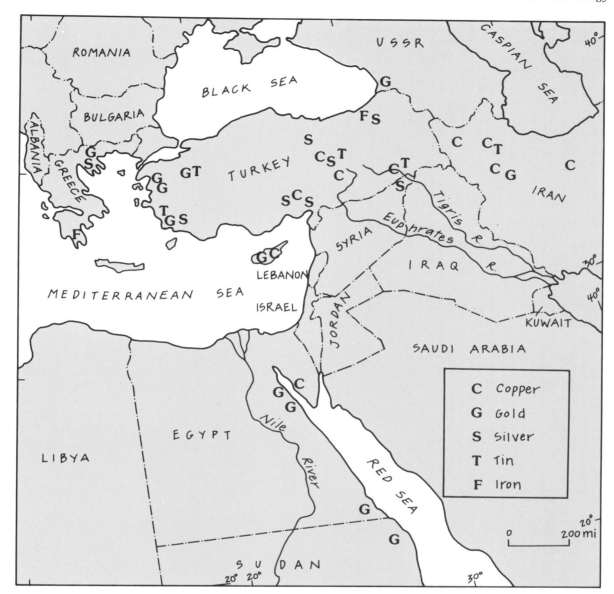

Figure 20.2
Ore deposits of the ancient Middle East.

Problem 11

Examine the location of the metal ore deposits in the Middle East, as shown in Figure 20.2. Where did the Egyptians find their copper? Their gold?

Problem 12

Artifacts of silver and tin have been found in Egyptian ruins. Where might these metals have come from?

The huge pyramids and monuments of ancient Egypt were built by various Pharaohs in the belief that everlasting life could be gained from preservation of material objects. The pyramids, still standing today after almost 5000 years, are the world's first massive stone monuments. One of the grandest of them, the Great Pyramid, is over 470 feet high and contains over 2 million blocks of limestone and granite. Herodotus, a Greek scholar, stated that over 100,000 men labored 20 years to build the Great Pyramid.

Problem 13

Which type of rock would be easier to carve with copper tools, granite or limestone? Why?

Problem 14

Copper saws embedded with the mineral corundum have been found at ancient quarry sites in Egypt. Recall the Mohs scale of hardness (page 9). Why was corundum used by the Egyptians?

Problem 15

Alabaster gypsum was a stone commonly used by the Egyptians for carved vases, jars, shrines, and statues. Examine the sample of alabaster provided by your instructor, and describe its hardness, luster, and color. What qualities make alabaster gypsum such an excellent material to carve?

One of the greatest ancient treasures ever discovered was the tomb of King Tutankhamen. The tomb of the eighteen-year-old Pharaoh was filled with over 5000 pieces of gold jewelry. The coffin was solid gold and weighed 2500 pounds. Although Tut was not an important Pharaoh, King Tut's tomb is an outstanding example of the wealth of the Egyptian civilization.

Problem 16

If gold sells for $500 per ounce, what is the value of the gold in Tutankhamen's coffin? (For simplicity, we assume the ounce of everyday life, 16 to the pound, or 28.35 grams. In fact, gold and other precious metals are sold by the troy ounce, an archaic unit equal to 31.1 grams.)

Modern Civilization

Perhaps the most significant step in modern civilization was the use of rock materials for energy. The British in the late eighteenth century initiated the extensive use of coal and petroleum, which has changed human civilization forever. Where once a thousand men were needed for a task, it now takes only a few workers and a machine. It was in the tin mines at Cornwall, England, once worked by Roman slaves, that this era began. Here the steam engine, developed by Thomas Newcomen in 1705 and powered by wood, was used to pump water out of the lower levels of the mines. Later improved by James Watt, the steam engine, with coal as its source of energy, powered the Industrial Revolution. Today petroleum is our primary energy resource, and it is the basic resource for the production of fertilizers, plastics, and other synthetics.

Problem 17

Examine Figure 20.3, which shows the power output of basic machines. About how many kilowatts of power can a single human produce? A horse?

Problem 18

To illustrate the dependence of the American standard of living on a cheap energy source, make the following calculation. The average person in America today uses about 10 kilowatt hours of energy per day. Men and women on farms typically worked 12 hours a day before the widespread use of power tools and electricity. How many of these 12-hour "man-days" does 10 kilowatt hours correspond to?

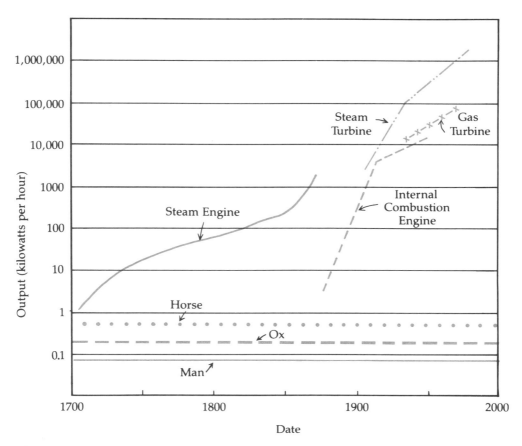

Figure 20.3
Power output of basic machines over recent centuries.

Problem 19

Iron and steel, by virtue of their greater strength and hardness, are superior to copper for use in machinery and large works of construction. Iron, however, requires more energy to produce from its ores than copper. This is one reason why iron was not used extensively by ancient civilizations. Examine the geologic map of England, Figure 20.4. What geologic features of England influenced the development of modern industrial society?

Problem 20

Examine Figure 20.5, which shows the energy produced from various sources in the United States. What was the primary source of energy in 1850? In 1900? In 1950?

Resources and Wars

World War II was one of the most devastating events in human history. In total over 45 million people died in a struggle that was truly worldwide in scope. Millions of tons of metals were used to manufacture the thousands of tanks, ships, and planes used in battle. Powered by oil, mobile armies moved across Europe, Africa, and Asia. Never before had the earth's resources been utilized in such enormous quantities in such a short period of time. The outcome of World War II depended heavily on use of metal and energy resources.

Problem 21

Examine Figure 20.6, which shows the major ore deposits and oilfields of Europe and western Asia that were known at the time of World War II. Did Germany have large resources of metals or oil?

Figure 20.4
Ore map of Great Britain.

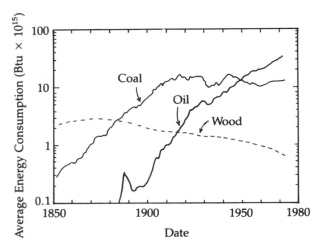

Figure 20.5
Primary energy sources of the United States, 1850–1975.

Problem 22

The most important metals of modern warfare are iron (for steel), copper, and aluminum. Gold is also essential as a currency to pay for supplies and soldiers. Figure 20.6 shows the expansion of Germany between 1935 and 1945. Why were Poland and Czechoslovakia crucial to the German war plans?

Problem 23

The partition of the region of Alsace–Lorraine between Germany and France at the conclusion of World War I was a source of aggravation to Germany. What economic factor made this region so attractive to Germany?

Problem 24

Adolf Hitler, Germany's leader, had negotiated a treaty in 1939 with Joseph Stalin of the Soviet Union in which he promised not to invade Russia. In defiance of this treaty, Hitler invaded Russia in 1941. Part of the German army drove north and east toward Moscow, while most of the attack was directed southeast through Stalingrad. What vital commodity induced Hitler to undertake this ultimately disastrous invasion? (Refer to Figure 20.6.) Where was this commodity found?

Problem 25

Germany's ally, Italy, invaded North Africa. While one objective was surely the Suez Canal, Hitler was also looking farther east. When the Italian offensive stalled, Hitler sent troops to Africa. What commodity was so essential to the Axis war effort that Hitler was willing to expend German muscle to get it? Where was he heading when he invaded northeast Africa?

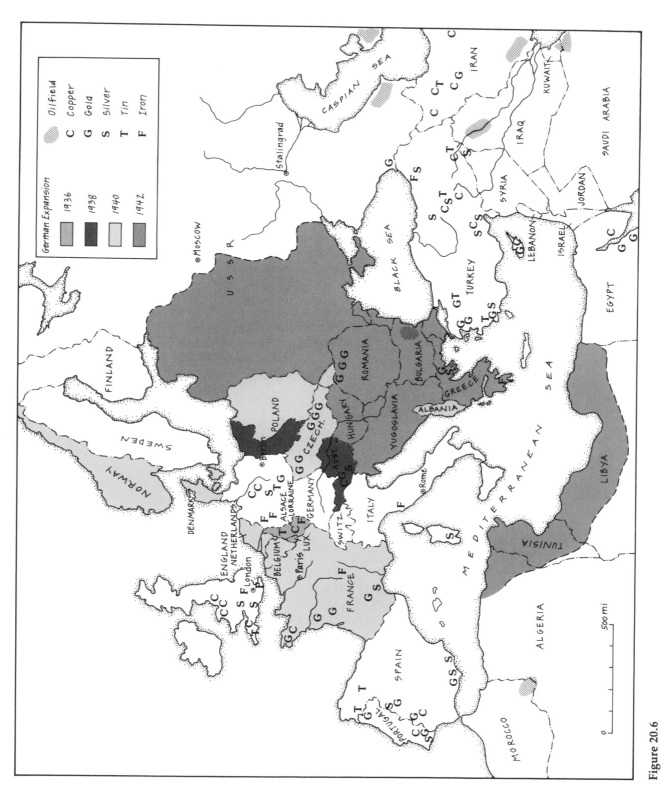

Figure 20.6
Ore deposits and oilfields of Europe and the Near East known at the time of World War II. Germany's territorial expansion between 1935 and 1945 is indicated by the shaded area.

Problem 26

Both of these efforts failed, with the result that during the Battle of the Bulge German tanks halted on the battlefield due to lack of fuel. Ironically, since World War II Britain and Norway have discovered large oil reserves in the North Sea. If Hitler had been able to exploit these fields, how might the course of the war have been different?

Problem 27

Until the middle of 1941 the United States was supplying Japan with 80 percent of its petroleum and the scrap metal used for steel production. The United States ceased the export of petroleum and scrap metal at that time due to hostile acts by the Japanese. Figure 20.7 shows Japanese expansion during World War II. Relate the pattern of Japanese expansion to metal and petroleum resources in the Pacific. Why was Indonesia so important to Japan?

ENERGY RESOURCES AND STRUCTURAL GEOLOGY

We live in a unique time. During the past 50 years modern civilization has produced and used more energy than in all previous human history. Despite the dwindling of basic energy resources, however, the world's appetite for energy is still accelerating. In 50 more years, our present energy sources will be scarce, and the ways we use energy will have to be very different.

The *fossil fuels* are the most widely used energy resource. These are carbon compounds from which energy is derived by burning, that is, the rapid oxidation of reduced carbon to CO_2. Reduced carbon may occur in the free state as elemental carbon, the major constituent in coal, or in various combinations with hydrogen (hydrocarbons), as in petroleum. Energy is released when reduced carbon is oxidized by burning or by decay (the same chemical process). Some agent must have supplied energy, therefore, in order to achieve the reduced state of carbon. The sunlight of past geologic ages is the source of this energy; it pro-

duced the plant and animal growth necessary for the formation of coal and petroleum. Thus coal and petroleum are, in effect, fossilized solar energy stored by previously living organisms. Clearly this is a nonrenewable energy resource, unless we are able to wait millions of years for its replenishment!

Coal

Coal forms when vast numbers of land plants are buried before oxidation (decay) can take place. This condition prevails in swamps, where the abundant vegetation dies and accumulates in shallow, stagnant water. When the free oxygen in the water is entirely consumed in decay, preservation of carbon begins. As the preserved plant material gets buried by additional plant material or by clastic sediments, it undergoes compression and heating at depth, which drive off gases and water. The accumulation of 1 foot of plant material may take 100 years or more. One foot of moderate-grade coal requires deposition of from 25 to 50 feet of plant material. Evidently swamp and marsh conditions were extensive during certain periods in the past as coal beds are typically tens of feet thick and hundreds of miles wide.

Problem 28

Assume that deposition of plant material in a swamp environment occurs at the rate of 1 foot each 100 years and that compaction reduces the original thickness of material to 1/35th of its original thickness. How much time is required for deposition of enough plant material for a coal seam 10 feet thick? Show your calculations.

Petroleum

Petroleum forms by accumulation of marine organisms in seawater that lacks free oxygen. Typical settings include coastal basins, where a large number of organisms live and where the supply of oxygen in the seafloor sediments is limited. After death, organisms settle through the water and accumulate with silt and clay on the quiet seafloor. The process of decay is arrested by lack of oxygen, and the resulting rock is a carbon-rich black mudstone. The deposit is subsequently buried, compacted, and heated under pressure at depth. Although the exact process by which the organic material is converted to petroleum is poorly understood, it can generally be said that chemical reactions slowly transform the organic material into liquid and gaseous hydrocarbons. The increasing

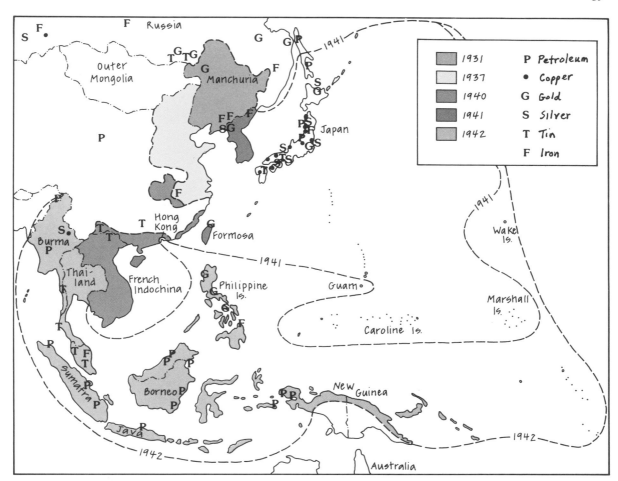

Figure 20.7·
Japanese expansion during World War II.

pressure of burial forces the mobile hydrocarbons out of the mudstone. Being of low density, they migrate upward along cracks and through interconnecting void spaces in the rock. Sandstones and fractured limestones are common conduits of hydrocarbons. Petroleum accumulates when it reaches an impermeable barrier, often shale or mudstone; these end points of migration (*traps*) are where geologists look for petroleum.

The path of oil migration is controlled to a great extent by the orientation of the rock layers through which it migrates. In tilted beds, hydrocarbons flow upward parallel to the dip. Petroleum in folded strata accumulates near the crests of anticlines if an impermeable layer lies above permeable sandstone or limestone. Faults may also influence the migration of petroleum and cause it to be trapped against impermeable zones of rock.

Natural gas, petroleum, and water separate by density at a trap. As illustrated in Figure 20.8, natural gas lies above the petroleum, which lies above the water.

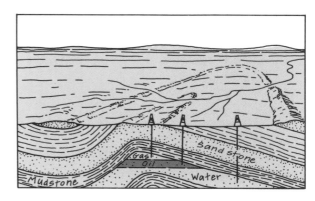

Figure 20.8
Water, petroleum, and gas in a stratigraphic trap.

The fluids saturate the rock, filling small void spaces that are rarely over 1 cm^3 in volume. When a drill punctures the trap, the gas, petroleum, and water may flow toward the zone of low pressure created by the conduit to the surface. More commonly, however, petroleum must be pumped to the surface.

Tools for Finding Petroleum. Just as topographic contour lines help describe the shape of the earth's surface, lines of equal elevation—structure contours—can be drawn for surfaces underground (the top surface of a rock unit, for example) to describe their shape. (See Exercise 7.) Structure contours are estimated from surface outcrops of beds or from cores of rock brought up from drill holes.

Problem 29

The drill-hole data shown in Figure 20.9 are from the Bruner oilfield near Tulsa, Oklahoma. Draw 10-foot structure contours from the drill-hole data. (Elevations are below sea level.) What is the geologic structure (syncline, unconformity, etc.) in this oilfield?

Problem 30

Figure 20.10 shows a structure contour map of the Glenn Pool in Oklahoma. The contoured surface is the upper contact of the Wilcox Sandstone, a petroleum-producing unit. Color in red pencil the areas where you would drill for oil or gas.

As you might expect, structure contour maps are powerful tools for petroleum development. In areas that lack drill holes, subsurface projections must be made from surface map data. The Hewitt field in southern Oklahoma, a major oil producer, was found in 1916 by William J. Mallard from surface mapping of the area.

Problem 31

Imagine that you are a struggling wildcat driller and have drummed up enough money for one drill hole. From Mallard's original map (Figure 20.11) and your knowledge of structural geology, can you find the Hewitt field? Locate the best site for your exploration hole.

Problem 32

The petroleum at the Hewitt field is found in Permian sandstones at a depth of approximately 800 feet. Subsequent drilling established that an angular unconformity separates the oil-producing Permian sandstones from the overlying Cretaceous sedimentary rocks. Was Mallard truly skillful or just lucky? Why?

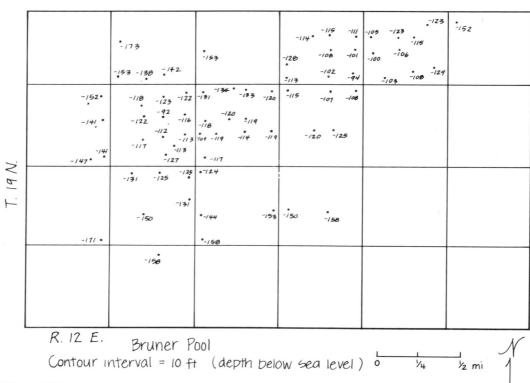

Figure 20.9
Drill-hole data from the Bruner oilfield near Tulsa, Oklahoma, showing depth to the top contact of the Oswego Formation.

Problem 33

Examine the depths to an impermeable mudstone, established by drill holes, as shown in Figure 20.12. If you thought there might be petroleum trapped beneath the mudstone, where would you drill for it? Show two or three proposed drill sites, and explain the reasons for your selections.

World Use of Petroleum. In 1956, geologist M. King Hubbert correctly predicted that production of oil in the United States would peak around 1972 and decline thereafter, creating an oil shortage. Hubbert estimated the total world reserves of oil at 1952 billion barrels. In the years since 1956, about 550 billion barrels of oil have been produced. Show all your calculations for the following questions.

Problem 34

What has been the average annual rate of oil consumption in the world since 1956?

Problem 35

A barrel of oil is 42 gallons. What is the average annual rate of consumption in gallons?

Problem 36

About how much oil remains to be produced, according to Hubbert's estimate?

Problem 37

Assuming that the average annual rate of oil consumption remains constant, how long will the oil last?

Problem 38

Do you think that constant rate of consumption is a valid assumption?

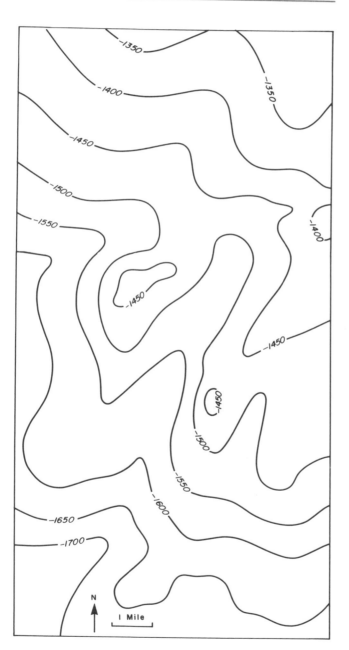

Figure 20.10
Structure contour map of the Glenn Pool, Oklahoma.

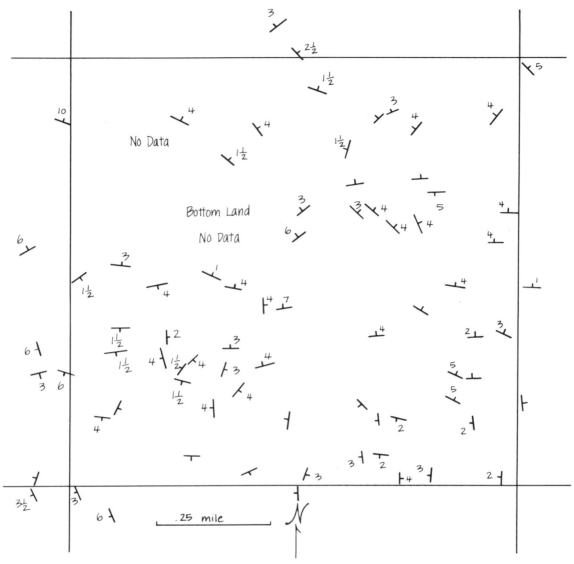

Figure 20.11
W. J. Mallard's map of the Hewitt Field area, Oklahoma.

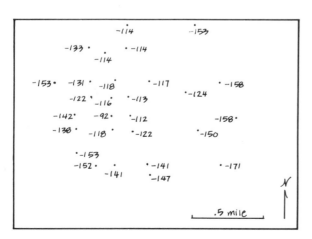

Figure 20.12
Drill-hole depths to an impermeable mudstone layer.

Problem 39

In the United States, daily oil consumption was 15.1 million barrels in 1983. The population of the United States in 1983 was about 230 million. What was the average daily rate of oil consumption per person in the United States?

Problem 40

What was the average annual rate of oil consumption per person in the United States?

Problem 41

If the population of the world (5 billion people) were to consume oil at the rate of the average American, how long would the remaining oil last?

Problem 42

Write a short paragraph commenting on the implications of your calculations in the previous question.

Geology and Civilization: The Surface Environment

"That's the sort of thing that draws people into geology," Park said. "Geologists go into the field because of love of the earth and of the out-of-doors."
"The irony is that they go into the wilderness and change it," Brower said.

JOHN MCPHEE, *Encounters with the Archdruid*

One way organisms survive is to exploit a vacant ecological niche; through variation and natural selection, some characteristics of the species change over time in order to enhance its survivability in that niche. We humans, with our great mental capacity, take a more aggressive approach to our surroundings. Rather than passively adapting to new conditions, we often actively modify the environment so that it is more suitable for us. Some people would argue that many modifications of the physical environment are dangerous. For example, houses are built today on sandy beaches exposed to the open ocean, across active fault zones, or on steep hillsides that are prone to sliding. This exercise addresses several case histories of futile attempts to thwart natural processes for human purposes.

Geology has at times influenced the course of prehistory and history. This exercise will also explore some interesting examples of the interaction of human history and geologic circumstances and phenomena.

WEATHERING

The structures humans build undergo chemical and physical weathering, just as rocks and minerals do. Weathering decreases the quality, usefulness, and safety of homes, commercial buildings, and roads. Annual losses due to weathering cost billions of dollars. One may easily observe the effects of weathering in monuments, statues, and the remains of historic buildings.

Problem 1

Although cemeteries are not places where geology comes immediately to mind, they provide natural laboratories for calculating how fast different materials weather under different climatic conditions. Examine Figure 21.1, which illustrates headstones of different compositions. Which material weathers fastest, and which weathers slowest? Considering the nature of the materials and the time each has been exposed to the elements, formulate an explanation for the different rates of weathering observed.

Problem 2

Assuming that you wish to be remembered by posterity, what material would you want your headstone to be—limestone, granite, or a silica-cemented quartz sandstone? Why?

Limestone

Figure 21.1
Headstones of different rock types.

Quartizite

Problem 3

Mount Rushmore in South Dakota and Stone Mountain in Georgia are large monuments carved in the local granitic bedrock. Consult the maps of global temperature and rainfall in Exercise 9. How would the difference in climates of South Dakota and Georgia affect physical and chemical weathering? Which monument will probably survive longer?

SLOPE PROCESSES

Landslides are of vital concern to humans because of their widespread distribution and destructive potential. It is estimated that, in the United States alone, landslides cause a billion dollars in damage and an average of 25 deaths per year. The monetary costs are due directly to actual physical damage to buildings and roadways and indirectly to lower productivity in agriculture and forestry. Most of the costs and deaths occur in urban areas, where the population density is high and space is limited. A city can reduce the poten-

tial losses by landslide only through proper geologic evaluation and zoning ordinances.

In early January, 1982, a major storm system dropped record amounts of rain over the San Francisco Bay Area. Rainfall varied locally but was as high as 24 inches during a two-day interval. (The average rainfall each year in this area is about 20 inches.) The storm triggered numerous landslides, causing an estimated $300 million in damage and 24 deaths. This study examines the Glenwood subdivision in San Rafael, California, where several debris avalanches occurred on the morning of January 4. The locations of the initial failures of slopes are shown on the topographic map in Figure 21.2, and many are visible in the aerial photographs taken on January 6 (Figure 21.4).

Problem 4

Using a yellow pencil, mark the gullies in the hills around the Glenwood subdivision on the topographic map. A debris avalanche, like a stream, takes the shortest path down a hillslope. This direction is, of course, the steepest gradient, which is the path perpendicular to the contour lines. For each point of initial failure, draw in red pencil the path the flow would take down the hillslope. Confirm these paths by locating the avalanches on the aerial photographs where possible.

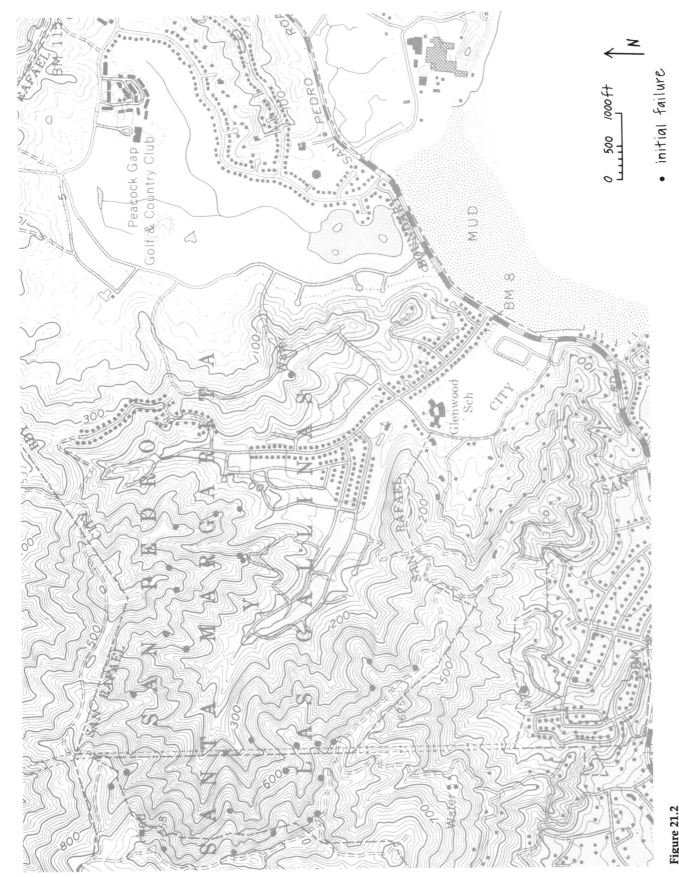

Figure 21.2
Slope failures in the Glenwood subdivision, San Rafael, California.

Problem 5

For several of the debris avalanches, determine the slope of the hillside at the point of failure. This measurement is made by determining the elevation change within a 100-foot horizontal distance measured along a line perpendicular to the hillslope. This procedure is illustrated in Figure 21.3. Use the chart in Figure 21.3 to convert percent slope into degrees, and list your results below.

Elevation Change	Slope (percent)	Slope (degrees)

At what slope (degrees) do most slope failures occur?

Problem 6

Discuss some possible reasons that failures do not occur on gentler or steeper slopes.

Problem 7

Suppose you are a planning commissioner studying the future development of the Glenwood subdivision. What sites should be avoided in planning buildings? On the topographic map shade the hazardous areas orange and the safe areas green.

GROUNDWATER

The groundwater aquifers that are used for home and industrial purposes are generally continuous under a large surface area. As a result, many landowners can potentially influence a given groundwater reservoir. The natural permeability of the aquifer causes problems because any one landowner can easily influence a neighbor's groundwater use. The two major problems that can result are selective depletion of the aquifer and contamination of the aquifer.

The following case study was done at Hershey, Pennsylvania, during the late 1940s. Hershey is in a small valley underlaid by limestone in the Appalachian Mountains. The valley has numerous springs and shallow wells that provide water for the local farms and the (at that time) world's largest chocolate factory (see Figure 21.6). One reason for the location of the Hershey chocolate factory was its proximity to an excellent water source, Derry Spring.

A mile and a half northeast of the chocolate factory, the limestone bedrock was mined from a quarry operated by the Annville Stone Company. The limestone in this area is bedded and fractured. Many of the natural springs are located at the intersections between

Conversion of Slope Percent to Degrees

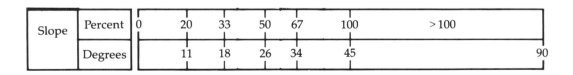

Slope	Percent	0	20	33	50	67	100	>100	
	Degrees		11	18	26	34	45		90

Vertical distance = 90 ft

Horizontal distance = 100 ft

Slope = 90 percent

Figure 21.3
Measurement of slope from map.

Figure 21.4
Aerial photographs of the Glenwood subdivision.

$\nwarrow N$ A.

crosscutting joints. Because of the high groundwater table, the mining operation required the pumping of 3500 gallons of water per minute to keep the quarry site dry. For several years the mining operation, the chocolate factory, and local residents existed in harmony.

In May, 1949, Annville Stone Company began a deeper and larger mining program that increased the pumping of the quarry to 6500 gallons of water per minute. During the summer of 1949, many springs and wells dried up. Throughout the valley, water levels in farm wells dropped, and many residents had to haul water from other sources. During July, 1949, sinkholes began to develop on the surface, and in the following months over 100 sinkholes formed, ranging in depth from 2 to 10 feet. Although 1949 was not an abnormally dry year, the groundwater table dropped

dramatically. The change in the water table elevation between October, 1948, and November, 1949, is contoured in Figure 21.5.

Problem 8

Based on the data, describe what happened to the groundwater table in the Hershey Valley during the summer of 1949.

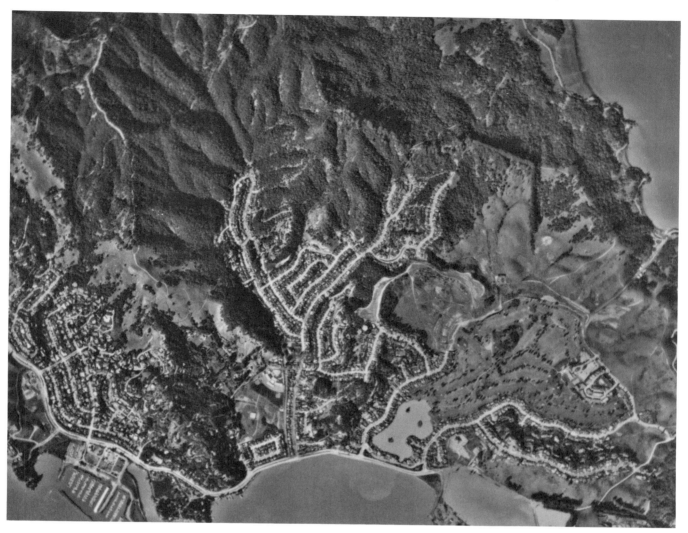

Figure 21.4
Aerial photographs of the Glenwood subdivision.

B.

Problem 9

To what would you attribute the changes in the groundwater system?

Beginning in December, 1949, Hershey Chocolate Corporation began to pump water into the limestone aquifer in order to recharge Derry Spring. Soon afterward, a rise in the water table was observed in several wells. This rise required an increase in the pumping at the Annville quarry from 6500 to 8000 gallons per minute. The extra pumping increased the cost of the new operation, and in early 1950 Annville Stone Company asked for a court injunction to stop the artificial recharge of the aquifer by Hershey Chocolate Corporation.

Problem 10

You are the presiding judge. How would you decide the case? Defend your decision.

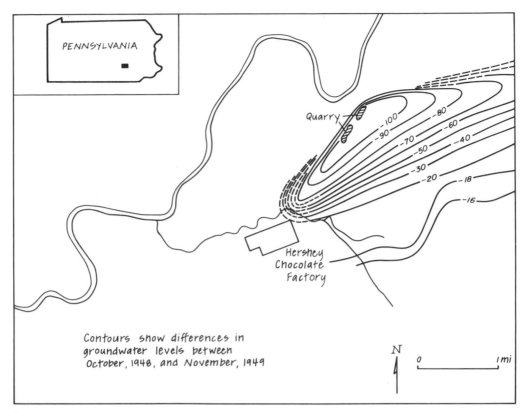

Figure 21.5
Change in water table elevation near Hershey, 1949.

RIVERS AND STREAMS

Rivers are literally the staff of life for many cultures; they supply water, a means of transportation, and the fertile soils that develop over river deposits in floodplains. Because humans settle where there is abundant water and rich farmland, many large population centers are located near major rivers, making river flooding one of the most dangerous surface hazards. In the United States alone, flood damages each year cost taxpayers over $3 billion. Probably the worst flood disaster in history occurred in China in 1931 when over 3,000,000 lives were lost in the floods of the Huang Ho (Yellow) River.

Damage caused by rivers is due partially to the unpredictable nature of flooding. At best, our knowledge is limited to statistical averages based on less than two centuries of historical records. For most rivers, minor floods occur once every two years. The average time between floods of a certain size is the *recurrence interval*. A large flood, for example, may have a recurrence interval of perhaps 20 or 50 years, depending on the river and its floodplain. The larger the flood, the longer the recurrence interval.

Examine Figure 21.7, the topographic map of the

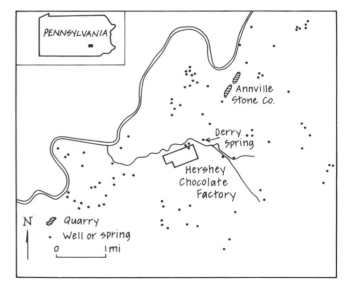

Figure 21.6
Area surrounding Hershey, Pennsylvania.

Kansas River at Topeka, Kansas. The graph in Figure 21.8 shows the flood levels between 1903 and 1956 measured at the gaging station on the Topeka Avenue Bridge.

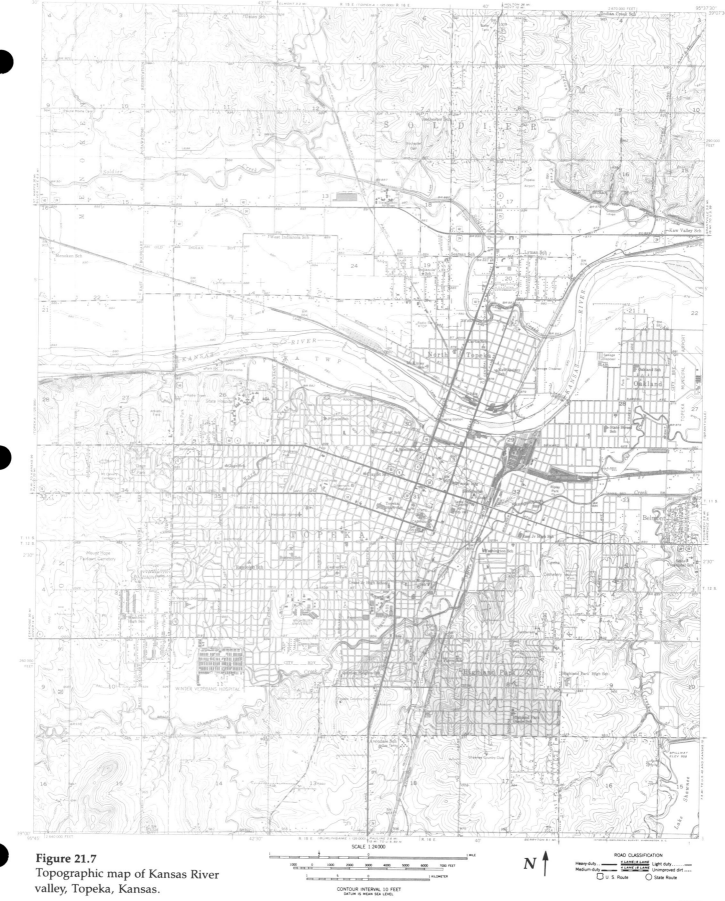

Figure 21.7
Topographic map of Kansas River
valley, Topeka, Kansas.

SCALE 1:24 000

CONTOUR INTERVAL 10 FEET
DATUM IS MEAN SEA LEVEL

N

ROAD CLASSIFICATION

Heavy-duty
Medium-duty

Light duty
Unimproved dirt

U. S. Route
State Route

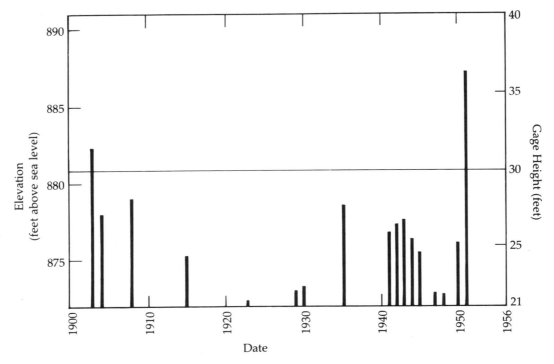

Figure 21.8
Flood levels at Topeka, 1903–1956.

Problem 11

How many floods occurred during this 54-year period?

Problem 12

What is the recurrence interval for major floods? For smaller floods?

Problem 13

What is the highest flood during this 54-year period?

Problem 14

At this water level, determine how far the flood waters extended across the floodplain. Draw in red pencil on the map the extent of the flooded area.

Although rivers themselves have natural levees along their banks in many areas, humans have raised these levees and built dams to further protect the floodplain from inundation. Artificial levees confine the stream to its channel during high water levels, and dams across tributaries help to control excessive discharges. Artificial flood control is not new. The Egyptians living along the Nile 5000 years ago built dams to control flooding. Today most of the world's major rivers have been tamed to a degree by flood-control practices. The benefits of levees and dams in flood control are obvious; not so obvious are the effects of a dam on the equilibrium of a river system. Consider the longitudinal profile of a stream at grade (that is, the discharge and gradient of the stream are just sufficient to carry the sediment load from the drainage basin). Figure 21.9 shows a typical longitudinal profile. The short vertical line represents a dam built across the stream.

Problem 15

As the stream enters the reservoir impounded behind the dam, its velocity decreases. What will happen to the sediments carried by the stream?

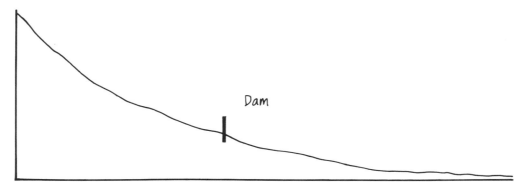

Figure 21.9
Longitudinal profile of a stream.

Problem 16

Water discharged from the lake behind the dam carries no sediment load, yet it is moving fast enough to transport solid material. As a result the stream erodes material until it returns to equilibrium. Where, specifically, does this erosion occur? Draw on Figure 21.9 the new profile of the river after deposition and erosion in response to the dam.

Problem 17

What are some implications of this erosion for towns on the river bank downstream from the dam?

Some flood control measures include straightening and paving sections of river bed through populated areas in order to speed the discharge and dissipate potential floodwaters more rapidly.

Problem 18

If the channel is straightened, what is the effect on the stream gradient? (Gradient was discussed in Exercise 12.) What is the effect on the flow velocity?

Problem 19

How would channel straightening affect the amount of sediment that can be carried by that stretch of river (its capacity) and therefore its erosive power in the vicinity of the cutoff?

Problem 20

In view of your answers to the previous questions, what are some potential problems in this approach to flood control?

It has been suggested that levees be constructed along both banks of the Mississippi River from Illinois to the Gulf of Mexico to control flooding and that dams be constructed on tributaries for the same purpose. These measures would surely help alleviate flooding, but not all the effects of such a plan are beneficial. The delta of the Mississippi River system has already been affected by the levees and dams built along the river.

Problem 21

Examine the maps of the Mississippi delta in 1956 and 1978 in Figure 21.10. How has the size of the delta changed in this 22-year interval?

A.

B.

Figure 21.10
Mississippi delta, 1956 (A) and 1978 (B).

Problem 22

Consider the effects of dams and levees on the sediment load carried by the river, and formulate an explanation for the change in the delta.

Problem 23

Missouri lies on the west bank of the Mississippi River, and across the river on the east bank is Illinois. Before the court ruling that fixed state boundaries, a meander cutoff in the appropriate place could have transferred property in Missouri to Illinois, or vice versa. What interesting political situation might such a cutoff have caused before the Civil War ended slavery?

Another facet of river dynamics is the political effect when a river changes its course. Mark Twain in *Life on the Mississippi* had this to say: "A cutoff plays havoc with boundary lines and jurisdictions: for instance a man is living in the State of Mississippi today, a cutoff occurs tonight, and tomorrow the man finds himself and his land over on the other side of the river, within the boundaries and subject to the laws of the State of Louisiana!"

This situation has been changed by court rulings that permanently fixed the state boundaries in the middle of the prevailing channel at that time, regardless of any subsequent changes in the course of the river. Thus portions of the state of Mississippi now lie on both east and west banks of the river, due to recent changes in the river channel.

Rivers change their courses on a much larger scale as well as by local meander cutoffs. The lower Mississippi River is a good example. It threatens to divert its entire flow into the Atchafalaya River in times of flood. Such a course change would wreak havoc on Gulf Coast shipping since it would leave the major ports of New Orleans and Baton Rouge high and dry. On the other hand, this threatened course change has been suggested as a way to control flooding in the lower Mississippi. With the advent of levees along the upper river, floodwaters reach the lower river unabated and with even higher flood crests than before the levees were built. Flooding on the lower river has become increasingly difficult to control.

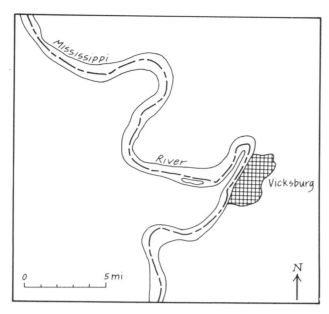

Figure 21.11
Mississippi River near Vicksburg, Mississippi, in 1865.

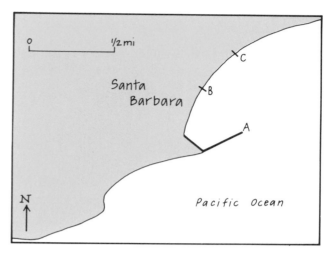

Figure 21.12
Coastline near Santa Barbara, California.

Problem 24

How might the Atchafalaya River be used to control flooding in Baton Rouge or New Orleans?

During the Civil War the Confederate guns at Vicksburg, Mississippi, were a major threat to Union shipping on the river. General Ulysses S. Grant proposed a rather interesting solution to the problem.

Problem 25

Look at the map of Vicksburg as it was in 1865 (Figure 21.11). Are you as imaginative as Grant? Knowing what you do about river dynamics, what might you attempt to do in order to allow ships to bypass the murderous guns at Vicksburg?

SHORELINE PROCESSES

Waves striking a shoreline obliquely create a current moving parallel to the shore, a longshore current. These currents can move enormous amounts of sediment down the shore (longshore drift). When a structure interferes with this transport process, it disrupts the natural equilibrium of erosion and deposition. As an example, consider the breakwater built near Santa Barbara, California, to shelter pleasure boats (Figure 21.12). The direction of longshore drift is east.

Problem 26

Indicate on Figure 21.12 the direction of longshore drift and the orientation of wave fronts that would cause a longshore current in that direction.

Problem 27

Show the refraction of wave fronts around the breakwater by sketching the resulting wave fronts in the bay.

As the waves round the end of the breakwater, some of their energy is dissipated over a long stretch of shoreline, A–B, while the rest of the energy is focused at point C.

Problem 28

What will happen to the sediment carried by the long-shore current as the wave energy drops? Where will the material be deposited?

Problem 29

What happens at point C, where the wave energy is focused?

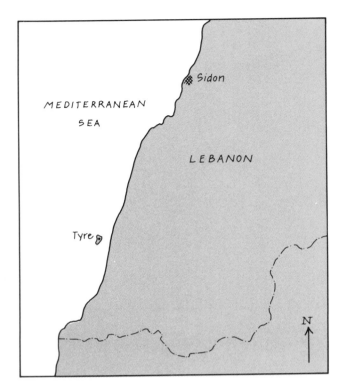

Figure 21.13
Tyre, east side of Mediterranean Sea.

Problem 30

Clearly the problem is too much sand in one place and removal of existing sand in another place. Since the entire shoreline is lined with expensive homes, what might be done to restore the shoreline to its former equilibrium state? Removing the breakwater, while sensible, is not one of your options.

An interesting example of sediment deposition along a shoreline and how it affected history is the ancient city of Tyre, along the eastern shore of the Mediterranean Sea in present-day Lebanon. Before the time of Alexander the Great, Tyre was a fortified island and was virtually invulnerable to invasion from the land. Through the years, wave energy directed toward the shore carried sand around Tyre to the quiet water behind the island, where the wave energy dropped, and sand was deposited.

Problem 31

Indicate by red dots on Figure 21.13 where sand would be deposited. What would happen to the island?

By the time Alexander the Great arrived on the scene in 332 B.C., the island of Tyre was almost connected to the mainland by a long sand spit (a *tombolo*). The opportunistic Alexander merely finished the process by building a causeway, and his army captured Tyre after a seven-month siege.

Hurricanes

Extremely dangerous conditions result when abnormally severe storms impinge on coastal areas. Since 1900, hurricanes in the Atlantic Ocean, Caribbean Sea, and Gulf of Mexico have killed an estimated 50,000 people and have caused over $12 billion in damages in the United States. Most of the damage and deaths arise from walls of water that are swept ashore by the high winds and "pulled" ashore by low atmospheric

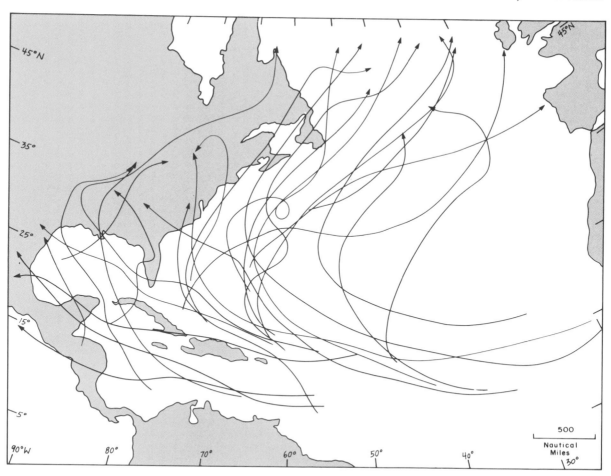

Figure 21.14
Paths of some early September hurricanes in the North Atlantic, 1901–1963.

pressures that cause the surface of the sea to bulge upward near the center of the storm. Such a *storm surge* may raise sea level by as much as 25 feet. In addition, waves tens of feet high pound the shoreline with devastating force, often changing the shoreline permanently.

A hurricane is a vast whirling storm whose winds revolve around a central *eye*. The eye is a clear calm area of extremely low barometric pressure. The difference in pressure between the eye and the outer regions of the storm is great, and violent winds swirl around the center at speeds up to 200 miles an hour. The rotation direction is counterclockwise in the Northern Hemisphere and clockwise in the Southern Hemisphere. The combination of high winds, great waves, and heavy rains in a hurricane causes devastation along shorelines and inland areas.

Examine Figure 21.14, which shows the paths and frequency of hurricanes in the North Atlantic during early September, the height of hurricane season.

Problem 32

Near what latitude do most hurricanes originate, and in what directions do they travel?

A vivid account of a hurricane that struck the Florida Keys on September 2, 1935, was recorded by J. C. Duane, an observer for the National Weather Bureau, who was living on Long Key at the time. Figure 21.15 shows an aerial photograph of Long Key. It is estimated that wind velocities in this storm were from 150 to 200 miles per hour. Destruction was practically complete over a path 30 miles wide. The storm center traveled from southeast to northwest across Florida.

Figure 21.15
Aerial photograph, Long Key, Florida.

Sept. 2, 1935

2:00 P.M.: Atmospheric pressure falling; heavy sea swell and high tide; heavy rain squalls continue; wind from the N or NNE.

3:00 P.M.: Ocean swells changed; large waves now rolling from SE, somewhat against winds, which are still N or NNE.

4:00 P.M.: Wind still N, pressure falling faster; rain continues.

5:00 P.M.: Wind N, hurricane force; swells from SE.

6:00 P.M.: Barometric pressure still falling; heavy rains; wind still N, hurricane force and increasing; water rising on north side of island.

6:45 P.M.: Pressure still falling; wind backing to NW, increasing, heavy timbers flying, 6″ × 8″ beam blown through observer's house.

7:00 P.M.: Now in main building, which is shaking with every blast and being wrecked by flying timbers; water piling up on north side.

9:00 P.M.: Pressure falling very fast.

9:20 P.M.: Pressure very low and still falling, winds stop—only a slight breeze now—sky is clear, stars visible. About the middle of the lull, which lasted 55 minutes, the sea began to rise very fast from the ocean side (SE) of camp. Water lifted the cottage from its foundation and it floated.

10:02 P.M.: Pressure still falling; wind beginning to blow from SSW.

10:15 P.M.: First major wind blast from SSW, full force. House is now breaking up; winds seem stronger than at any time during the storm. Pressure at lowest point. I was blown outside and caught onto some coconut branches, hung on for dear life; struck by some object and knocked unconscious.

Sept. 3, 1935

2:25 A.M.: Became conscious in tree and found I was lodged about 20 feet above the ground. The cottage had been blown back on the island, from whence the sea had receded and left it with all people safe.

Problem 33

Why did the water accumulate on the north shore of the island in the beginning of the storm?

Problem 34

What happened around 9:30 P.M. on September 2?

Problem 35

When did the storm surge hit the shoreline?

Problem 36

Why did the wind direction change during the storm?

GLACIERS

Although glaciers are not normally regarded as a hazard to humans, they might represent the ultimate disaster to us all. A sudden melting of glacial ice would have dramatic effects on sea level. It has been estimated that an average surface temperature change of only 2.5°C (4.5°F) would be sufficient to start melting the polar ice caps.

One way that man could significantly alter the earth's surface temperature is by the continued excessive burning of fossil fuels, which could increase the carbon dioxide content of the atmosphere enough to trigger a runaway *greenhouse effect*. Carbon dioxide in the earth's atmosphere is analogous to the glass in a greenhouse. Energy from the sun passing through the atmosphere strikes the earth and is reemitted in the infrared part of the spectrum (heat). Infrared rays are absorbed by carbon dioxide (as well as by the greenhouse window) and, instead of being radiated away to outer space, are retained in the earth's lower atmosphere. As the proportion of carbon dioxide in the atmosphere increases, more heat is held at the surface of the earth, which might become hot enough to melt significant quantities of ice.

Let us examine some possible effects of this scenario.

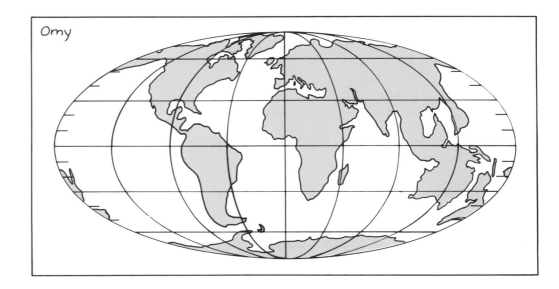
Omy

Problem 37

The surface area of the oceans is about 140,000,000 square miles. What volume of water would change sea level by 1 foot? (1 mile = 5280 feet)

Problem 38

The surface area of Antarctica, which contains about 90 percent of the world's glacial ice, is about 5,000,000 square miles. The average thickness of the continental ice sheet is estimated to be about 7000 feet. If this ice were to melt, what would be the rise in sea level as the meltwater entered the oceans? (Assume that the densities of ice and water are equal, an approximation that will not greatly affect your general conclusion.) Show your calculations.

Problem 39

The following table gives the average elevation (in feet) of some of the world's major cities. How would melting of the Antarctic ice cap affect these cities?

City	Elevation
London	245
Paris	300
Berlin	110
Rome	95
Moscow	394
Beijing	600
Tokyo	30
Sydney	25
Washington, D.C.	25
New York	55
Chicago	595
Los Angeles	340

Problem 40

Return to the topographic map of the United States, Color Map 5. Draw the new coastline formed by melting of the Antarctic ice on Figure 9.2.

In considering the effects of a glacial advance, it is instructive and interesting to use an example from prehistory. An increase in ice volume means a decrease in the volume of ocean water, hence a lower sea level. Figure 21.16 shows the distribution of early archeological sites across Asia and North America. The estimated ages of the sites are given by the map legend.

Problem 41

What is the range in the age of the sites?

Problem 42

What trends do you notice in the ages of the sites occupied by early humans across Asia?

Problem 43

Examine the extent of glaciation and the estimated shorelines of Asia and North America shown on the map. What implication do these ancient shorelines and the geographic progression of hominid occupation in Asia have for the history of man in North America?

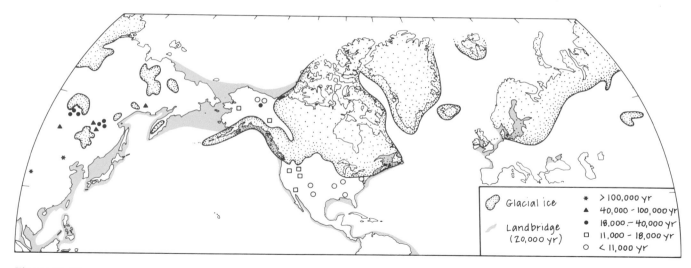

Figure 21.16
Distribution of early hominid sites, Asia and North America.

WIND AND ARID REGIONS

In contrast to water, wind is less commonly a hazard to civilization. Although high winds associated with hurricanes and tornadoes can be very destructive, they do not usually affect a large area. Winds, however, can greatly influence human civilization in their role of carrying precipitation over the land. Over half of the world's population relies on seasonal monsoon winds to bring rain to parts of Africa and Asia that would otherwise be arid. Fluctuations in atmospheric circulation that alter the paths of the monsoon winds can have devastating effects, bringing drought, then famine and death. The tragedy in Ethiopia is the most recent example of famine caused by absence of monsoon rains.

Problem 44

Compare the global population distribution (Figure 21.17) with the pattern of precipitation (Figure 9.4). Is there a correlation between precipitation and population density? Explain.

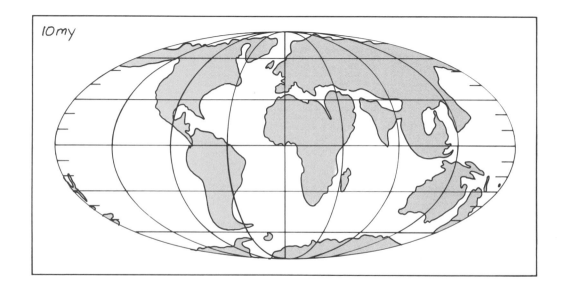

10my

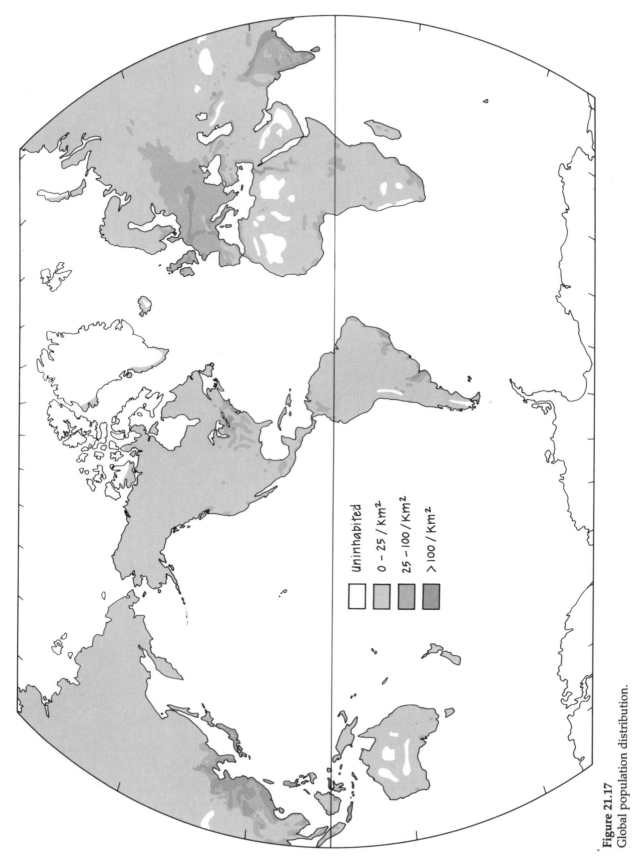

Figure 21.17
Global population distribution.

Legend:
- Uninhabited
- 0 – 25 / Km²
- 25 – 100 / Km²
- >100 / Km²

Problem 45

Examine India in both of the figures. How does population density change with rainfall?

There are many examples in human history of cultures attempting to cope with diminishing water supplies. For example, in Pre-Columbian America, the Anasazi and the Hohokam groups flourished in parts of New Mexico, Arizona, southern Colorado, and Utah. They made pottery and baskets, constructed elaborate irrigation systems, and built complex multistory dwellings. Around A.D. 1200 these people abruptly deserted their settlements.

Problem 46

Return to Exercise 5, Figure 5.1. From the tree-ring data, formulate one possible reason for the mysterious disappearance of these people.

Problem 47

The 1930s were the "Dust Bowl" days of the American Midwest and Southwest. Millions of tons of valuable topsoil were literally blown away by the wind. What do the tree-ring data suggest about the climate during the 1930s in this area?

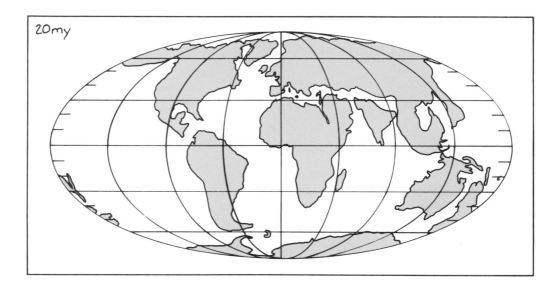

20my

Geology and Civilization: Internal Processes

Civilization exists by geologic consent, subject to change without notice.

WILL DURANT

None of us is immune to geologic hazards. Entire civilizations have been destroyed by geologic catastrophes. Yet, perhaps because of our skills in isolating ourselves from other forces of nature, most people remain ignorant of the ways that geologic processes interact with man, for better or for worse. National governments daily decide issues of great geologic impact—construction of dams, building of extensive networks of highways, utilization of natural resources, appropriation of funds for rebuilding after a disaster—with very little appreciation of the geologic factors involved. On a smaller scale, communities often fail to plan intelligently for new building and development by taking into account the geology of their surroundings. Some locations are simply unsuitable for anything but open space and parks. Few citizens are aware of the potential for geologic disaster when buying property or when voting on public issues that concern land use and development. This exercise is intended to develop that awareness by presenting a series of geologic situations involving volcanic activity and earthquakes for you to evaluate.

VOLCANOES

Hazards to human population from volcanic activity may take several forms. The most obvious danger is from flows of lava, although the 1980 eruption of Mount St. Helens demonstrated that volcanic ash falls, mud flows, and explosive blasts of superheated air, steam, and ash may do much more damage.

Problem 1

From your knowledge of the characteristics of mafic and felsic magmas, which lava type is more likely to flow quietly rather than explode?

Problem 2

Look back at Exercise 2, Figure 2.10. In what general areas of the world would you find volcanoes of this type?

Hawaii

Figure 22.1 shows the island of Hawaii, a land mass formed by five large volcanoes, the largest of which is Mauna Loa. Historic lava flows are shown on the map.

Problem 3

Notice that most of the flows issuing from the north flank of Mauna Loa converge on the island's largest city, Hilo. What is the reason for this flow pattern?

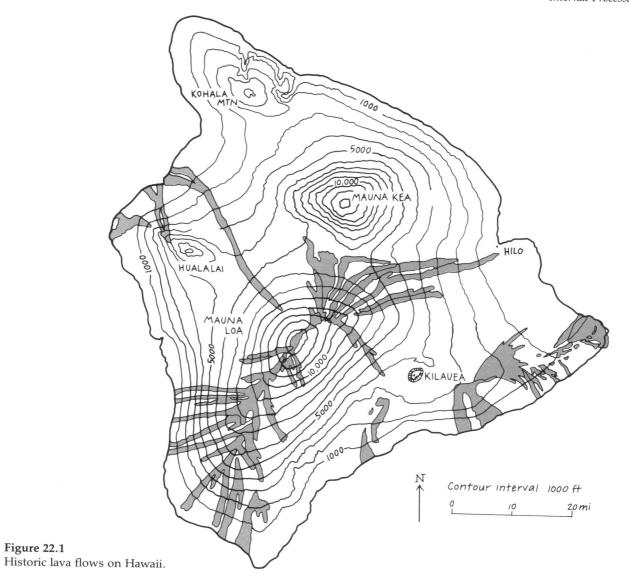

Figure 22.1
Historic lava flows on Hawaii.

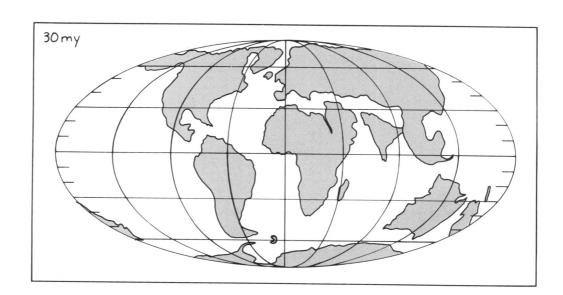

Problem 4

Considering the rather fluid nature of the Hawaiian lavas, suggest a few methods of minimizing the risk to Hilo from lava flows. Be imaginative! Evaluate the potential for success of your suggestions.

Kilauea is probably the most intensively studied volcano in the world, and its eruptions can be predicted quite accurately. *Tiltmeters* placed on the flanks of the volcano can detect changes in slope smaller than 1 part per million (equivalent to lifting one end of a kilometer-long board by 1 millimeter). As magma slowly rises through the earth's crust and fills shallow magma chambers under the vent of the volcano, the broad slopes of the volcano gradually tilt outward, away from the vent. When an eruption drains magma from the chamber, the tilt of the volcano's flanks abruptly returns to near its initial value. This sequence is illustrated in Figure 22.2A.

Figure 22.2B shows a 20-year tiltmeter record of Kilauea. It is clear from the very variable average baseline that there is no critical angle of tilt—no exact level of inflation—that seems to precede an eruption. The abrupt deflations of the volcano, however, have always followed eruptions.

Problem 5

From the periods of gradual tilting (inflation) followed by sudden deflation, state when six major eruptions have occurred in the time interval shown.

Volcanic activity is usually preceded by seismic activity. There are probably several reasons for these "volcanic earthquakes," most of which are rather small (magnitude 3 or less).

Problem 6

Consider what happens to the volcanic structure as the temperature, volume, and pressure of the rocks in the volcanic edifice change prior to an eruption, and give an explanation for the small earthquakes that usually precede volcanic eruptions.

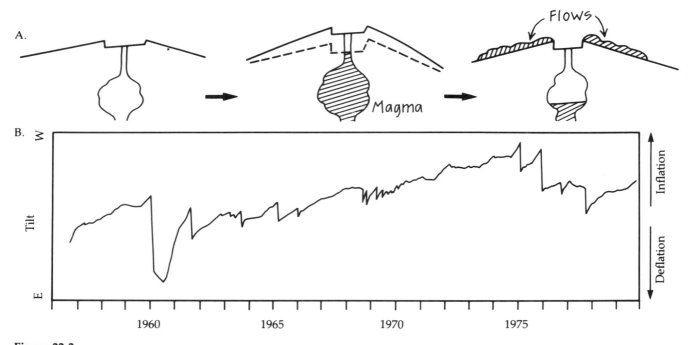

Figure 22.2
A. Tilting of shield volcano with influx of magma. B. Tiltmeter record of Kilauea.

Probably the most consistent precursor signal of volcanic eruptions at Kilauea is an unusual kind of seismic activity called *harmonic tremor* or *spasmodic tremor.* It is a more or less continuous vibration of the ground for many hours or even days at a time. Harmonic tremor may be caused by the escape of volatile material in the low pressure near-surface environment or possibly from the turbulent underground flow of magma. It is usually associated with eruption of fresh magma from a volcanic vent. Figure 22.3A shows a normal seismogram recorded at the Hawaiian Volcano Observatory at Kilauea; Figure 22.3B shows harmonic tremor recorded during an eruption of Kilauea in 1977.

Problem 7

In what ways does harmonic tremor seem to differ from normal seismic activity?

Mount St. Helens

Not every volcano is as benign as the volcanoes of Hawaii. In fact, most volcanic activity on the continents is of a rather different type.

Problem 8

Refer to Figure 2.10 again. What is the composition of the volcanic rocks most often found on the continents?

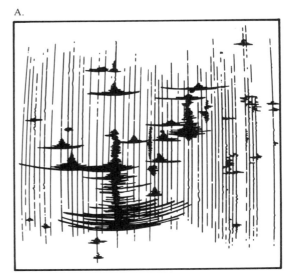

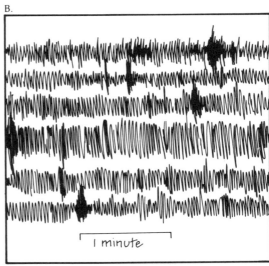

Figure 22.3
A. Normal earthquake activity at Kilauea.
B. Harmonic tremor at Kilauea.

40my

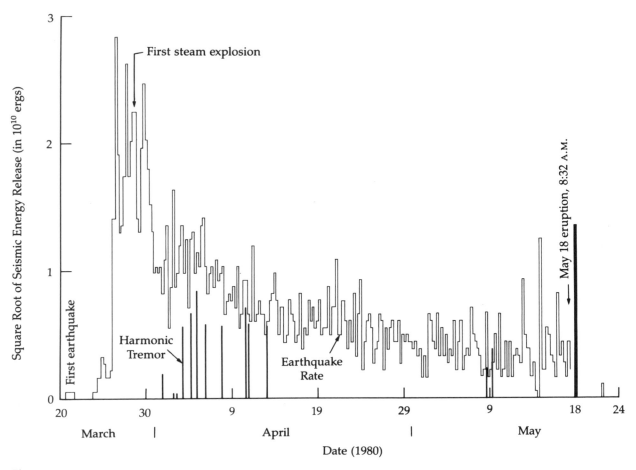

Figure 22.4
Earthquake rate and harmonic tremor at Mount St. Helens, 1980.

Problem 9
Is this type of volcanic activity likely to be a quiescent flow or explosive?

Problem 10
Among the many seismic events recorded near Mount St. Helens were the three shown in Figure 22.5. Which of the three events appears to be of the greatest interest in an area of volcanic hazard? Why?

As an example of continental volcanic eruptions, consider the eruption of Mount St. Helens, Washington, in 1980. This eruption was certainly one of the most thoroughly investigated and photographed volcanic events of all time, though by no means the most destructive. The first clue that Mount St. Helens was stirring to life after a dormancy of over 100 years was furnished by the seismic network of the U.S. Geological Survey, which reported abnormally frequent earthquakes there beginning on March 20, 1980 (see Figure 22.4).

The onset of the major eruption of Mount St. Helens, at 8:32 A.M. on May 18, 1980, has been described by several eyewitnesses, two of whom, geologists Keith and Dorothy Stoffel, were flying in a small aircraft at low altitude directly over the summit crater when the eruption began. They escaped only by diving at full throttle and turning south, away from the main thrust of the eruption. Accounts of the sequence of events in the first few moments of the eruption agree remark-

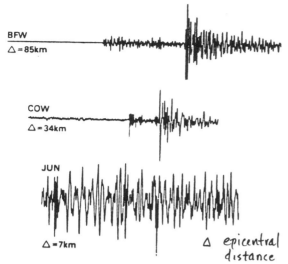

Figure 22.5
Three sample seismograms, Mount St. Helens.

ably well. At 8:32 an earthquake of magnitude 5+ shook the mountain and apparently triggered an instability of the north slope of the volcano. (Geologists had previously determined by laser ranging that the north side had been rapidly and continuously bulging outward over a period of several weeks.) The whole north side of the summit crater began to move downhill as a huge slide. Seconds later there was a gigantic explosion with a strong horizontal blast, due to the sudden release of confining pressure on the hot water-saturated rock within the volcano. As newly exposed rock was blown apart by the hot water and gases contained in it, more rock was exposed and was also blown out; a series of these explosions worked its way back into the core of the mountain during the first 10 minutes after the initial earthquake-triggered slide. It is believed that most of the destruction of the mountain occurred during this period.

Problem 11

The eyewitness accounts suggest that the outburst of May 18 was triggered by earthquake and subsequent slope failure of the north flank, not by magma forcing its way to the surface. What evidence in the seismic record shown in Figure 22.4 supports this explanation of events?

Problem 12

The strong harmonic tremor observed on May 18 started at 11:40 A.M., about 3 hours after the initial explosion. At 12:17 P.M. the color of the eruption column changed from dark gray to light gray, and shortly afterward the first of many pyroclastic flows came down the north flank. What change in the nature of the volcanic event around noon on May 18 is suggested by these three events?

In all, the eruption of Mount St. Helens claimed 65 lives. Early evacuation of a major summer resort just beginning its busy season undoubtedly averted a major catastrophe. Experts estimate that 30,000 lives could have been lost without advance warning of the eruption by the U.S. Geological Survey. Early warning, for example, enabled the Pacific Power and Light Company to lower the level of Swift Reservoir south of the volcano in anticipation of mudflows, which might have pushed huge quantities of water over the dam and led

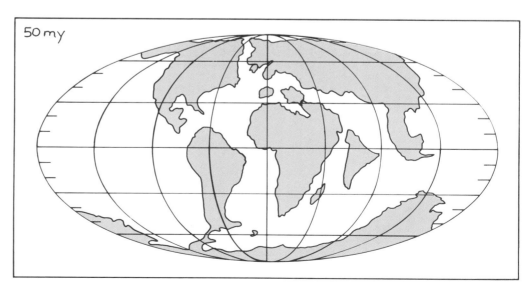

to heavy flooding downstream. Although flows on the south side of Mount St. Helens never materialized, the situation would have been much different had the reservoirs been on the north side of the peak.

Mono Lake, California

The Mono Lake–Long Valley area of eastern California has been the site of highly silicic (felsic) volcanic eruptions for the past several million years. It was therefore of great interest to geologists when seismic activity in the region began to increase in mid-1978, culminating in an unprecedented (in historic times) series of earthquakes, including four of Richter magnitude 6 within a 48-hour period in May, 1980. Since that time there have been other earthquake swarms accompanied by harmonic tremor in a small region near the resort community of Mammoth Lakes. Figure 22.6 shows the general area, with the site of earthquake epicenters indicated by stripes.

Problem 13

Consider the information you might wish to have if you were a geologist assigned to investigate the possibility of a volcanic eruption in this area:

1. Recall the physical changes observed at Kilauea and Mount St. Helens that seemed to herald eruptive phases of those volcanoes.

2. Notice the locality on the map called Casa Diablo Hot Springs, and consider the implications for hot springs of a subterranean influx of fresh magma.

3. Keep in mind not just the epicenters of the earthquakes that normally precede volcanic eruptions but also the depths of the hypocenters and the progression of these focal depths as magma works its way upward.

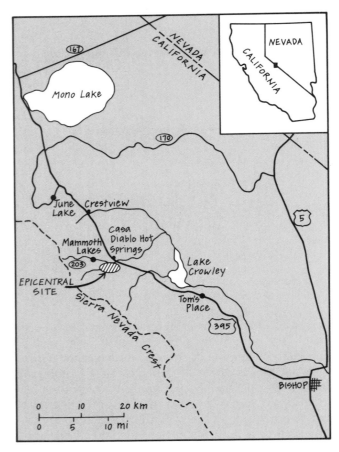

Figure 22.6
Mono Lake–Long Valley area, California.

Bearing in mind all these ideas, design a field research program that you, as chief geologist, would implement in order to determine whether or not a serious volcanic hazard exists in the Mono Lake–Long Valley area. Outline at least four major types of observations that are imperative for your program, and describe the implications of the data you would get.

The epicentral site shown in Figure 22.6 is inferred to be the most likely vent area for possible eruptions. Many previous events in this region were explosive silicic eruptions accompanied by pyroclastic flows and clouds of hot ash, lava flows, and, at some locations, mudflows and floods. Investigations of the geologic record suggest that these hazards exist within a 20-km zone around the possible vent zone. Additionally, areas within 35 km may be subject to ash accumulations of 20 cm (8 in.) or more.

Problem 14

As part of your hazards assessment program, map both of these zones on Figure 22.6 and label them. Since Los Angeles derives much of its water supply from Mono Lake and Lake Crowley, what additional hazards would you note?

Problem 15

Early in 1982, the U.S. Geological Survey did, in fact, issue a volcanic hazard warning for this area of California. This act was accompanied by vigorous protests from local residents, property owners, and real estate brokers, because of the resulting drop in tourist revenues and property values. What are your thoughts on the social and economic implications of such a geologic hazard warning? Should such warnings be issued? Who should issue them?

EARTHQUAKES

Earthquakes pose probably the greatest single threat to life and property of all the natural disasters. When a great earthquake (magnitude 8 or above) occurs in the densely populated regions of California, for example, as many as 100,000 lives may be lost, and property damage will be tens of billions of dollars. An earthquake in Tangshan, China, in 1976 killed some 700,000 people in just a few moments.

Many of the hazards due to earthquakes are quite obvious; others are more subtle. The most obvious hazard during an earthquake is the collapse of structures due directly to the shaking of the ground. Loss of property and life may also result from landslides or from flooding over dams or seismic sea waves induced by an earthquake. In some areas these indirect effects may be more devastating than the direct effects of surface faulting or ground shaking.

Surface Faulting

One misconception entertained by many people is that the ground may split asunder during an earthquake and swallow up people or structures. Of all the hazards that do exist, this one is confined almost entirely to imagination. There is only one documented case of a person being swallowed up by the earth. Although the wrenching or buckling of the ground in surface faulting is a clear hazard to structures directly on the fault line, its effects are not nearly as widespread as other earthquake hazards.

Ground Shaking

The size of an earthquake is most often described in terms of its magnitude on the Richter scale, which is related to the energy released by the earthquake. This

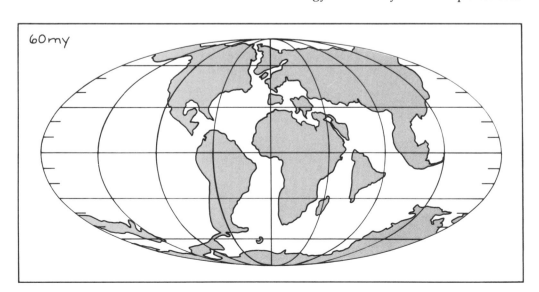

60 my

scale is purely objective; the Richter magnitude of an earthquake is a quantity that depends solely on the reaction of a standard seismograph on bedrock and is independent of distance from the epicenter or any subjective factors. A given earthquake has only one magnitude, regardless of where it was measured.

Quite a different way of describing an earthquake is to measure its *intensity*. Intensity scales measure the effects of an earthquake rather than the energy released. They have been in use for many years since intensity was the only way to evaluate the relative size of an earthquake before accurate recording seismographs were developed. The *Modified Mercalli Scale* is the intensity scale most commonly used in the United States. It is based on reports of ground and building damage and on interviews with people in different parts of the affected area. Intensities are given as Roman numerals from I to XII. Some typical descriptions of damage used in the Modified Mercalli Scale and the approximate magnitudes that produce these effects near the epicenter are shown in the table below.

Because so many factors contribute to the effects of earthquakes, intensities vary considerably throughout the affected area. A given earthquake cannot be assigned a single intensity number; in fact, large differences in intensity commonly occur only hundreds of feet apart. The various reported intensities are plotted on maps, and *isoseismal lines* are drawn to connect points of equal intensity, in the same way that contour lines connect points of equal elevation on topographic maps. Figure 22.7 shows an intensity map of San Francisco for the 1906 earthquake. Notice the wide range of intensities within very short distances in several areas of the city—along 16th Street, for example,

and in the northeast corner of the city. Since the areas of violent and weak intensities are the same distances from the fault (the northwest-trending zone of intensity XI in the southwest corner of the map), some factor other than distance to the fault must be responsible for the wide discrepancy in intensities.

Problem 16

Figure 22.8 is a geologic map showing the same area as the intensity map of Figure 22.7. What correlation, if any, do you see between the two maps?

Problem 17

Figure 22.9 shows three seismograms (at the same magnification) of the same event, an explosion in Nevada, recorded at different sites in San Francisco. What is the major difference in the three records?

Selected Descriptions from the Modified Mercalli Intensity Scale

Modified Mercalli Number	Approximate Richter Magnitude	Description of Effects
III	3	Felt by some people who are indoors, but it may not be recognized as an earthquake. Hanging objects swing.
VI	5	Felt by everyone, and many people are frightened and run outdoors. Walking is difficult. Windows, dishes, and glassware are broken; liquids spill; pictures are knocked from walls. Poorly built buildings damaged; plaster cracks.
VIII	6–7	General fright. Steering of cars difficult. Damage slight in well-built buildings, but poorly built or designed buildings experience partial collapse. Numerous chimneys fall. Cracks appear in wet ground or slopes.
XI	8	Few unreinforced masonry buildings remain standing. Other structures heavily damaged. Broad fissures, slumps, and slides in soft or wet soils. Underground pipelines out of service; rails severely bent.

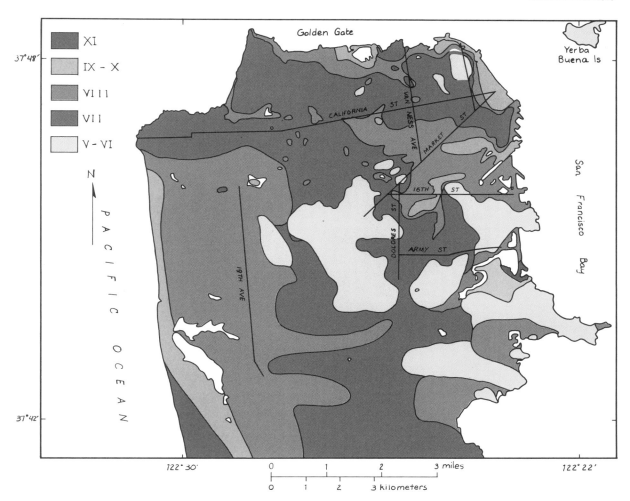

Figure 22.7
Intensity distribution of the April 18, 1906, earthquake in San Francisco.

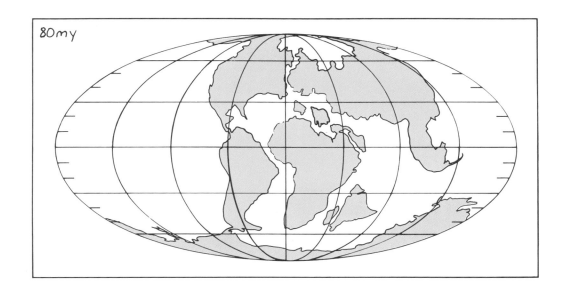

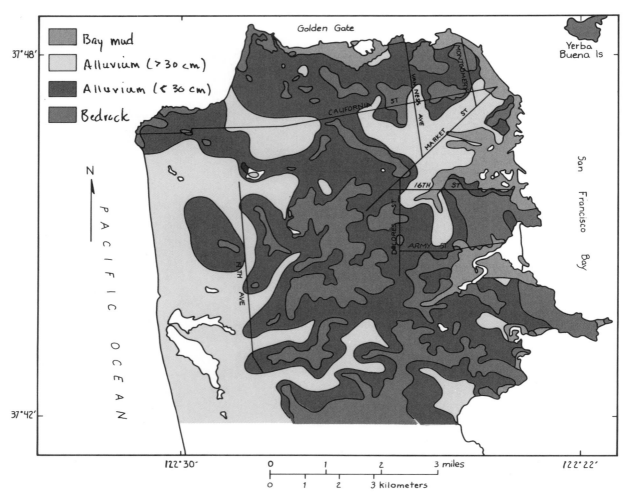

Figure 22.8
Generalized geologic map of San Francisco.

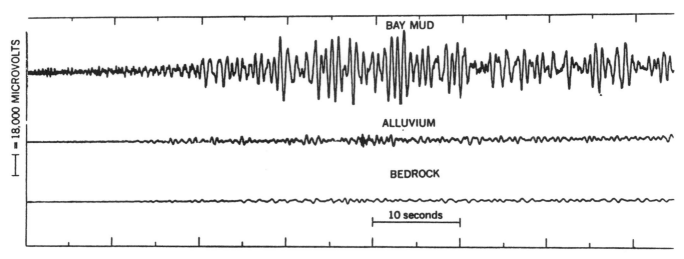

Figure 22.9
Seismograms recorded at three different locations in San Francisco from the same event.

Problem 18

How is the difference in the three seismograms related to the different intensities mapped in Figure 22.7? Is your answer to Problem 16 confirmed by these data?

It is important to realize the vital role played by building design and construction in earthquake country. A most striking example of this role is the comparison of damage caused by the San Fernando (California) earthquake of 1971 (magnitude 6.5) with that caused by the 1972 earthquake in Managua, Nicaragua (magnitude 6.2). Although the energy released in the two shocks was nearly the same, and both cities are built partially on recent alluvium, most of Managua was leveled and 4000 people died, whereas in San Fernando a few buildings collapsed and 64 people died. This difference is due primarily to the very different construction practices in the two areas. Most buildings in the San Fernando area are either wood-frame construction or steel-reinforced concrete, both of which have the strength and flexibility to withstand the large movements inflicted by horizontal ground motion. In Managua, like many areas of the world where lumber is scarce or advanced engineering practices are rare, the most common building materials are adobe or

unreinforced masonry. The widespread collapse of buildings and large death tolls in moderate earthquakes directly reflect these unsound building and design practices.

Problem 19

What other earthquake-prone areas of the world can you think of that employ adobe or masonry construction?

Landslides

Landslides triggered by earthquakes are sometimes as devastating as the ground shaking. The Peruvian earthquake of 1970 triggered a massive ice and rock avalanche from the peaks of the Andes that swept 2 miles down a valley at speeds in excess of 300 km per hour and killed 18,000 people—as many as were killed by all the other effects of the earthquake combined. The massive 1964 earthquake in Anchorage, Alaska (magnitude 8.4), triggered a destructive landslide at Turnagain Heights that destroyed 75 expensive homes. The area of the slide had previously been identified as unstable in a geologic report, but residential development proceeded regardless, in an area that should have been zoned for open space or parkland.

Clearly landslides present a very real potential hazard, but because many complex and related factors contribute to landslide generation, the prediction of slope failure at a specific site is difficult.

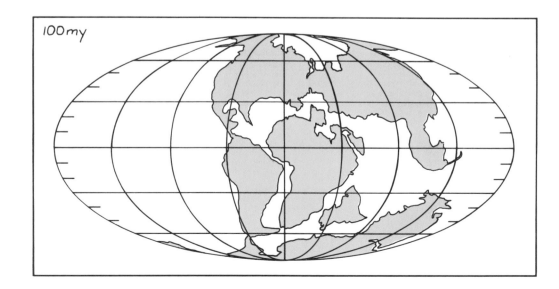

100 my

Problem 20

For each factor in the list below, briefly discuss the influence it might have on landslide potential:

1. Steepness of slope

2. Geologic conditions (bedrock versus unconsolidated sediments)

3. Structural properties of the rock (dip of bedding planes, degree of fracturing, etc.)

4. Vegetation

5. Water content of slope material

6. Proximity to fault

7. Human activities (residential development, regrading of slopes, paving of large areas, etc.)

Figure 22.10 shows aerial photographs of part of the San Francisco peninsula where the San Andreas fault enters the Pacific Ocean. San Francisco is to the left.

Problem 21

How might the fault have influenced the large landslide in the ocean cliffs?

A.

B.

Figure 22.10
Trace of San Andreas fault in Daly City, south of San Francisco, in 1982 (A) and 1956 (B).

Problem 22

Had you been in charge of land use planning for Daly City, the community shown in the photographs, what would you have done differently?

Problem 24

These lakes were enlarged some time ago by old earth-fill dams that actually straddle the active fault trace. What steps would you undertake to lessen the hazard from these dams?

Flooding

Earthquake-induced flooding may be due to several factors; we shall examine only flooding caused by dam failure and seismic sea waves.

Since many dams are located upstream from populated areas, they constitute a major potential for earthquake disaster. The failure of a dam during an earthquake, or, more commonly, the massive displacement of water over the dam, could destroy more buildings and claim more lives than the faulting and ground shaking.

Figure 22.11 shows part of the San Francisco Bay area (San Mateo County). The segment of the San Andreas fault shown in Figure 22.10 is at the northwest corner of the map.

Problem 23

Notice the long narrow lakes, San Andreas Lake and Crystal Springs Reservoir, and the long narrow valley that extends between them. Map the trend of these features, using a red line on Figure 22.11. This linear feature is the active trace of the San Andreas fault, which may be followed in just such a manner for hundreds of miles through California.

Tsunamis or *seismic sea waves* are traveling ocean waves of extremely long wavelength and high speed. Their height in the open ocean is rather small, a few feet at most, which renders them imperceptible to ships at sea. When the waves enter shallow water, their velocity decreases, their height increases to as much as 30 meters (100 feet), and they become a menace to low-lying areas near the coast.

Seismic sea waves are usually the product of an undersea earthquake or landslide, but they may also be started by volcanic eruptions. The most infamous seismic sea wave of historic times was caused by the explosion of the Indonesian volcano Krakatoa in 1883. It devastated the coasts of nearby Java and Sumatra with waves 100 feet high and killed 35,000 people. A major earthquake anywhere in the Pacific basin poses a tsunami threat to the Hawaiian Islands. For this reason, seismic sea wave detectors have been installed along coastlines to detect the waves and give at least several hours of warning so that coastal settlements may be evacuated. The roughly circular lines contoured around Hawaii in Figure 22.12 are times (in hours) for the seismic sea waves generated at any given point to reach Hawaii.

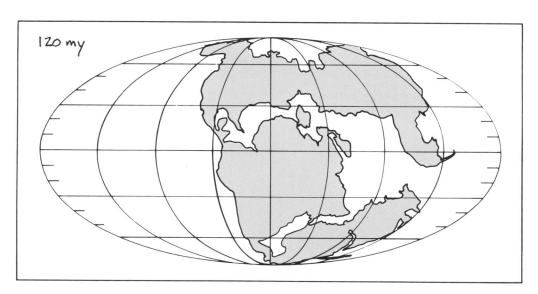

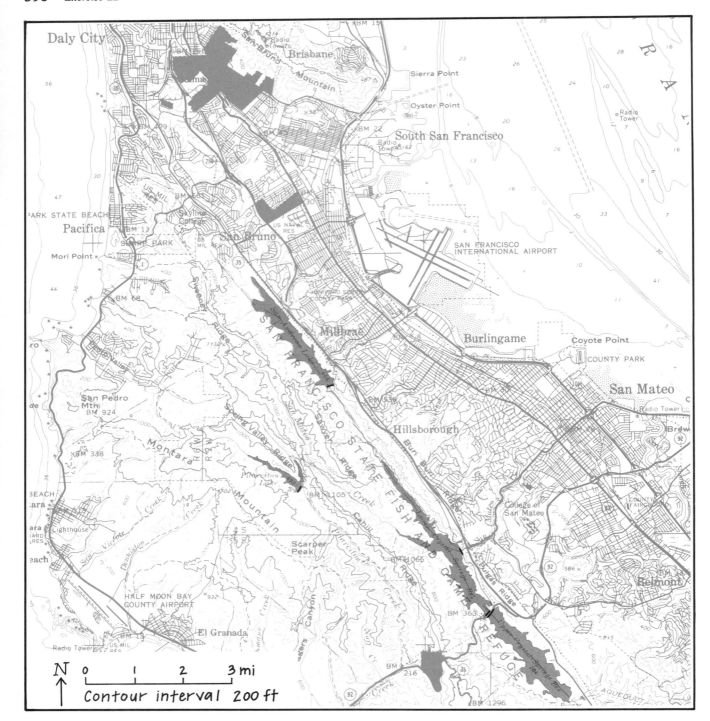

Figure 22.11
Part of San Mateo County, California, showing the trace of the San Andreas fault.

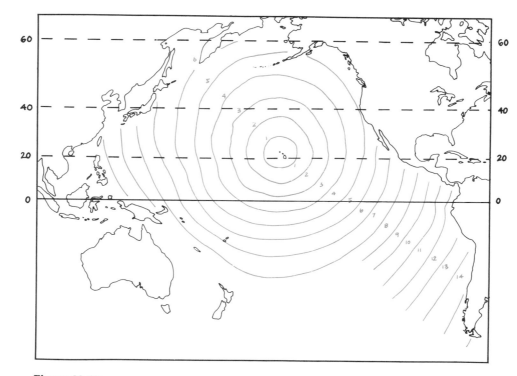

Figure 22.12
Seismic sea wave travel times to Hawaii (in hours) from points in the Pacific basin.

Problem 25

What was the approximate average speed in miles per hour of the tsunami generated by the great Alaskan earthquake of 1964? (Assume an epicenter due north of Hilo on the Alaskan coast.) Latitudes are given at the edges of the map. Ten degrees of latitude equals about 700 miles in the area shown. Show your calculations.

Problem 26

If the interval between wave crests at Hilo was 30 minutes (a typical value), what was the distance from crest to crest of the individual seismic sea waves in the wave train?

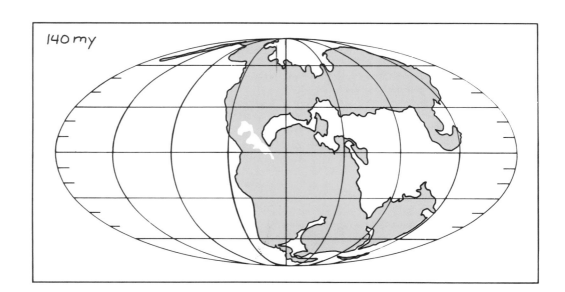

Prediction of Earthquakes

Many precursor signals have been studied for use in earthquake prediction, but the relationship between physical changes in the crustal rocks and the triggering of earthquakes is very complex. Despite a few successful predictions (most notably, the Chinese prediction in February, 1975, of the large Haicheng earthquake 5 hours before it occurred), the failures outweigh the successes.

Problem 27

Consider the social and economic effects of successful (*and* unsuccessful) short-term predictions in a society such as ours. In your opinion, should such predictions be made public? By whom should they be made? Should financial compensation be made to businesses or property owners for the adverse economic effects of such predictions?

For the long strike-slip San Andreas fault system, as well as other fault systems around the Pacific Basin, the only prediction method at present is the *seismic gap* method. This method uses the historical record of major earthquakes in certain segments of the fault system as well as evidence for active fault creep in other parts of the system. Segments of the fault that have undergone neither abrupt earthquake displacement nor quiet creep in historic times are probably due for a major earthquake. Clearly this method will never be the ideal forecasting device (which would predict both the magnitude and the time of the event, within a few days), but it will enable communities and individuals to make long-term preparations for a sizable earthquake.

Figure Figure 22.13 is a fault map of California on which are indicated the extent of historic fault breaks and their dates and the location of fault creep slippage along major faults in the San Andreas system.

Problem 28

Where would you expect the next major earthquake in the San Andreas system? Indicate the location on the map with the number 1.

Problem 29

List two other areas, 2 and 3, that may be due for a major earthquake, in the order you would expect them to occur.

Problem 30

What is the approximate relation between the earthquake magnitude and the length of the rupture along the fault?

HISTORICAL DEVASTATION BY VOLCANOES AND EARTHQUAKES

"Civilization exists by consent of geology, subject to change without notice." The quote at the beginning of this exercise is nowhere more appropriate than it is here. There is strong geologic and archeological evidence to suggest that, on rare occasions, whole cultures have been destroyed by catastrophic volcanic eruptions accompanied by earthquakes or seismic sea waves. Several of these events predate recorded history but are preserved in folk legends. Others, such as the eruption of Vesuvius in A.D. 79 that destroyed Pompeii, were described in vivid detail by witnesses. Intermediate between these two extremes, both in terms of the partial preservation of historic accounts and our knowledge of the time period in which it occurred, is an example from the Mediterranean region.

The legend of Atlantis, the continent which disappeared beneath the sea, along with its flourishing civilization, is a persistent myth among Western civilizations. Despite its name, the mythical continent is thought not to have been in the mid-Atlantic. The archeological and geological facts are these: During the late Bronze Age a highly prosperous and artistically advanced civilization, called by archeologists the Late Minoan, inhabited Crete and other nearby islands in the Aegean Sea. Around 1400 B.C., this flourishing civilization, which dominated the Mediterranean, underwent a sudden and catastrophic decline from which it never recovered. The traditional view has been that the Minoans were conquered and destroyed by invaders. If so, these invaders somehow defeated the most advanced civilization of its time at a single blow and then vanished without a trace and without ever occupying the conquered sites. Recently another theory has been advanced that ascribes the collapse of the Minoan culture to natural, although extraordinary, causes.

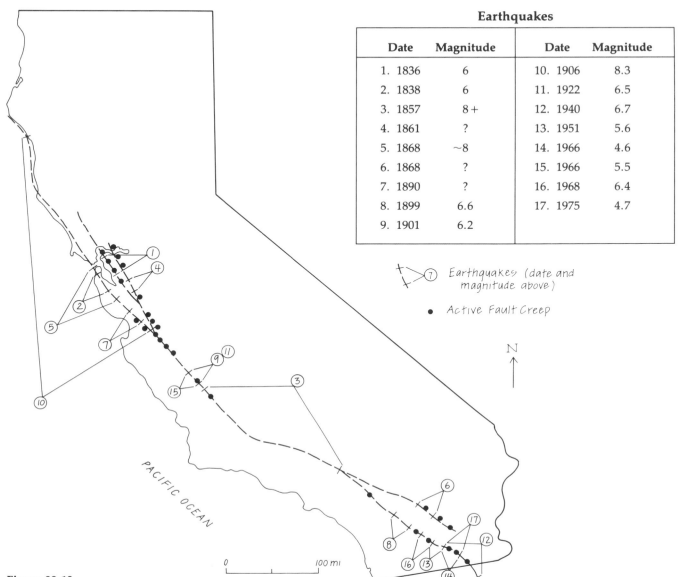

Earthquakes

	Date	Magnitude		Date	Magnitude
1.	1836	6	10.	1906	8.3
2.	1838	6	11.	1922	6.5
3.	1857	8+	12.	1940	6.7
4.	1861	?	13.	1951	5.6
5.	1868	~8	14.	1966	4.6
6.	1868	?	15.	1966	5.5
7.	1890	?	16.	1968	6.4
8.	1899	6.6	17.	1975	4.7
9.	1901	6.2			

Earthquakes (date and magnitude above)

• Active Fault Creep

N

PACIFIC OCEAN

0   100 mi

Figure 22.13
Simplified fault map of California showing historic fault breaks and fault creep areas.

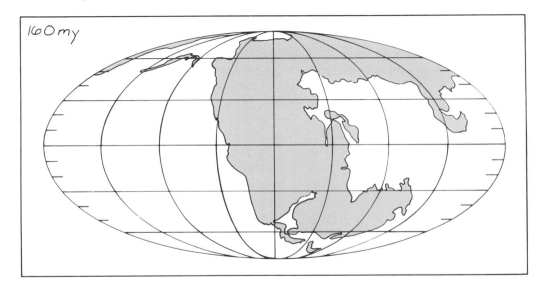

160 my

Figure 22.15 shows the eastern Mediterranean, including Crete and nearby islands. One of these island groups, the Santorin Islands, is shown in more detail in Figure 22.14.

Problem 31

Draw a cross-profile along line A–A' in Figure 22.14, showing the submarine as well as subaerial structure. Use the coordinates provided. Consider both the profile and the map views. What geologic feature does the structure of the Santorin Islands suggest?

Problem 32

These islands seem to be all that remains of a very large volcano. It was probably four times the size of Krakatoa, which produced the devastating tsunami that killed 35,000 people in 1883. What is the origin of Nea Kammeni?

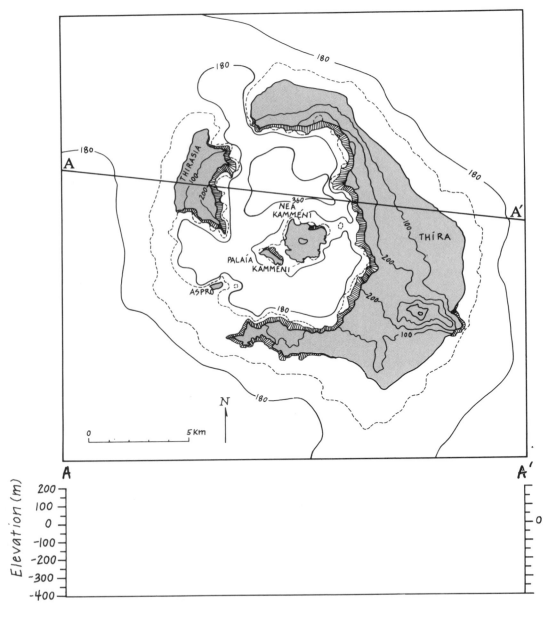

Figure 22.14
Topography and bathymetry of the Santorin Islands.

Figure 22.15
Crete, Santorin Islands, and the eastern Mediterranean.

Intensive archeological work reveals a more detailed chronology for the destruction of the Minoan culture. Around 1450 to 1400 B.C., most of the Minoan sites in Crete were destroyed. The principal agent of destruction was burning, with some structural collapse. At the same time all the Minoan sites in the Santorin Islands were totally destroyed and buried under 5 to 7 meters of volcanic ash. These sites were not reoccupied. The Palace at Knossos on Crete, less severely damaged, was restored and reoccupied for a brief time, estimated at between 20 and 70 years, as were a few of the less important sites. In about 1380 B.C., all the Minoan settlements on Crete were destroyed and covered by a layer of volcanic ash that was about 20 cm deep in the south of Crete and appreciably deeper in the north. The Santorin Islands were buried under a layer of ash from 18 to 24 meters thick. The Minoan fleet, anchored in harbors around Crete and other eastern Mediterranean harbors, was totally destroyed. This cataclysm marked the end of Minoan culture. Minoan sites were not reoccupied for perhaps a century, and then only on a very small scale, with no sign of the former dominance of the area.

Problem 33

In what three major ways might volcanic action in the Santorin Islands and the geologic phenomena commonly associated with volcanic activity affect Crete?

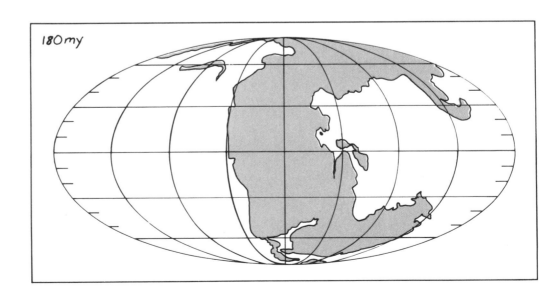

Problem 34

Write a short scenario for the destruction of Minoan civilization (Atlantis?) that accounts for the detailed chronology given above.

More speculative than the theory ascribing the decline of Minoan culture to cataclysmic volcanic events is the idea advanced by some scholars that the Exodus of the Jews from Egypt took place at the same time as the destruction of Minoan Crete. Certainly the time scale does not preclude this possibility, as the date generally given for the Exodus is around 1400 B.C. Consider passages from the Book of Exodus, Chapters 13 and 14:

"And the Lord went before them by day in a pillar of cloud to lead them; and by night in a pillar of fire, to give them light …"

"… and the pillar of cloud brought on darkness and early nightfall …"

"And the children of Israel went into the midst of the sea upon the dry ground: and the waters were a wall unto them on their right hand and on their left."

"And the waters returned and covered the chariots, and the horsemen, and all the host of the Pharaoh that came into the sea after them …"

Problem 35

Does your scenario for the destruction of Minoan Crete contain any elements that might account for these descriptions from Exodus? Explain.

REFERENCES FOR PART E

Books

American Geological Institute, *Environmental Geology, AGI Short Course Lecture Notes,* American Geological Institute, Falls Church, Va., 1970.

Brobst, Donald A., and Walden P. Pratt, eds, *United States Mineral Resources,* Geological Survey Professional Paper 820, U.S. Government Printing Office, Washington, D.C., 1973.

Cargo, David N., and Bob F. Mallory, *Man and His Geologic Environment,* 2nd ed., Addison-Wesley, Reading, Mass., 1977.

Gallant, Roy A., *Our Universe,* The National Geographic Society, Washington, D.C., 1980.

Hays, W. W., ed., "Facing Geologic and Hydrologic Hazards, Earth-Science Considerations," U.S. Geological Survey Professional Paper 1240-B, U.S. Government Printing Office, Washington, D.C., 1981.

Jensen, M. L., and A. M. Bateman, *Economic Mineral Deposits,* 3rd ed., John Wiley & Sons, New York.

Menard, H.W., *Geology, Resources, and Society.* W. H. Freeman, New York, 1974.

Murray, B., *The Earthlike Planets,* W. H. Freeman & Co., New York, 1981.

———, *The Planets: Readings from Scientific American,* W. H. Freeman, New York, 1983.

Mutch, T.A., *Geology of the Moon: A Stratigraphic View,* 2nd ed., Princeton University Press, Princeton, N.J., 1976.

Mutch, T.A., et al., *Geology of Mars,* Princeton University Press, Princeton, N.J., 1976.

Park, C. F., *Affluence in Jeopardy: Minerals and the Political Economy,* Freeman Cooper, San Francisco, 1968.

———, *Earthbound: Minerals, Energy, and Man's Future,* Freeman Cooper, San Francisco, 1975.

———, and R. A. MacDiarmid, *Ore Deposits,* 3rd ed., W. H. Freeman, New York, 1975.

Planet Earth Series: *Solar System, Gemstones, Noble Metals,* Time-Life Books, Alexandria, Va., 1982.

Press, F., and R. Siever, eds., *Planet Earth: Readings from Scientific American,* W. H. Freeman, New York, 1975.

Skinner, B. J., *Earth Resources,* 2nd ed., Prentice-Hall, Englewood Cliffs, N.J., 1976.

Smoluchowski, R., *The Solar System: The Sun, Planets, and Life,* Scientific American Books, New York, 1983.

Watkins, Joel S., Michael L. Bottino, and Marie Morisawa, *Our Geological Environment,* W. B. Saunders, Philadelphia, 1975.

Young, Keith, *The Paradox of Earth and Man,* Houghton Mifflin, Boston, 1975.

U.S. Government Publications

U.S. Geological Survey Information Pamphlets are available free from

Distribution Branch
U.S. Geological Survey
Box 25286, Federal Center
Denver, CO 80225

Earth Sciences and the Urban Environment, 1980.
Engineering Geology, 1981.
Exclusive Economic Zone: An Exciting New Frontier, nd.
Gold, 1981.
Monitoring Active Volcanoes, 1984.
NASQAN: Measuring the Quality of America's Streams, 1980.
NAWDEX: A Key to Finding Water Data, 1980.
Natural Steam for Power, 1983.
Planetary Exploration and Understanding the Earth, 1980.
Prospecting for Gold in the United States, 1986.
Rain: A Water Resource, 1980.
Replenishing Non-Renewable Mineral Resources, 1981.
Role of Earth Sciences in the Disposal of Radioactive Waste, 1980.
Safety and Survival in an Earthquake, 1986.
Severity of an Earthquake, 1985.
Use and Conservation of Minerals, 1981.
Water Dowsing, 1980.
Water in the Urban Environment: Erosion and Sediment, 1976.
Water in the Urban Environment: Real-Estate Lakes, 1976.
Water Use in the United States, 1976.

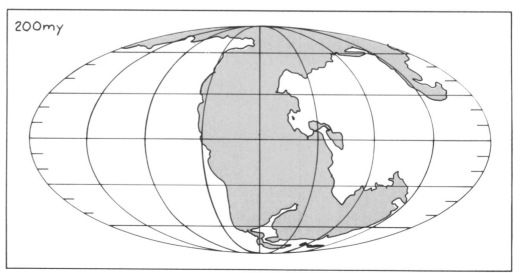

Table of Mineral Indentification

This table is divided into two main groups, minerals with metallic luster and minerals with nonmetallic luster. The minerals with nonmetallic luster are further divided into light- and dark-colored minerals. Each of these subgroups is divided into two sections, those minerals harder than a knife blade (5½) and those softer. Within each group, minerals are listed in order of decreasing hardness. The most diagnostic properties for each mineral are indicated by bold face type.

Metallic Luster

Name and Formula	Crystal Group and Habit	Cleavage/Fracture	Hardness	Color	Streak	Other
Pyrite FeS$_2$	Isometric **Cubic crystals**, Granular masses	none	6–6½	**brass yellow**	greenish-black	
Magnetite Fe$_3$O$_4$	Isometric equant (octahedral crystals)	none	6	dark gray to black	black	**magnetic**
Hematite Fe$_2$O$_3$	Trigonal thick plates, granular masses	none	5–6½	steel gray, black, brown	**reddish-brown**	
Chalcopyrite CuFeS$_2$	Tetragonal		3½–4	**golden yellow,** may tarnish		
Sphalerite ZnS	Isometric equant	**4 perfect**	3½–4	brown, black, yellow	brown to light yellow	luster may be **resinous**
Galena PbS	Isometric **cubic crystals**	**perfect cubic**	2½	**silver**	gray	
Graphite	Hexagonal	1 perfect	**1**	steel gray to black	black	greasy feel, marks paper easily

(Continued)

Nonmetallic Luster, Light Color, Harder Than a Knife Blade

Name and Formula	Crystal Group and Habit	Cleavage/ Fracture	Hardness	Color	Other
Corundum Al_2O_3	Hexagonal **commonly rough barrel shapes,** granular	none (basal parting)	9	gray, variable	red var. is ruby; blue is sapphire
Topaz $Al_2[SiO_4](OH,F)_2$	Orthorhombic **prismatic,** granular	1 perfect	8	variable: white, yellow, blue, brown	yellow var. is gemstone; vitreous luster
Beryl $Be_3Al_2[Si_6O_{18}]$	Hexagonal **hexagonal prisms**	poor	7½–8	variable	blue var. (igneous) is aquamarine; green (metamorphic) is emerald
Quartz SiO_2	Hexagonal 6-sided crystals, irregular	conchoidal fracture	7	Colorless, white, gray, variable	**vitreous luster** purple var. is amethyst
Cryptocrystalline quartz SiO_2		conchoidal fracture	7	variable	agate is banded; opal has waxy luster
Sillimanite Al_2SiO_5	Orthorhombic prismatic	1 good	7	white, colorless	vitreous to silky luster
Andalusite Al_2SiO_5	Orthorhombic prismatic, square cross section	1 good	6½–7½	**pink,** gray, white	**chiastolite shows black cross in cross section**
Kyanite Al_2SiO_5	Triclinic prismatic **bladed crystals**	1 perfect, 1 good	5½–7	white to **blue,** gray	vitreous luster; **hardness is 7 across bladed crystal, 5½ along its length**
Plagioclase $Ca[Al_2Si_2O_8]$ to $Na[AlSi_3O_8]$	Triclinic **tabular**	1 perfect, 1 good at 90°	6	white to dark gray, bluish	**albite twinning;** vitreous luster when fresh, to dull
Potassium feldspar $K[AlSi_3O_8]$	Monoclinic, var. tabular	1 perfect, 1 good at 90°	6	white to pink, tan, salmon	perthite structure; vitreous luster when fresh, to dull
Tremolite– Actinolite $Ca_2(Mg,Fe)_5[Si_8O_{22}](OH)_2$ (an amphibole)	Monoclinic prismatic	1 perfect	5½–6	white, gray, **pale green**	vitreous luster; **clusters of long radiating prismatic crystals**
Nepheline $Na[AlSiO_4]$	Hexagonal irregular masses	2 poor	5½–6	white, gray, colorless	greasy luster

Nonmetallic Luster, Light Color, Softer Than a Knife Blade

Name and Formula	Crystal Group and Habit	Cleavage/ Fracture	Hardness	Color	Other (Streak, Luster)
Fluorite CaF_2	Isometric **cubic crystals**	**4 perfect**	4	light green, blue, yellow, purple	vitreous luster
Dolomite $CaMg(CO_3)_2$	Hexagonal varied	3 perfect at 75°	3½–4	light tan, brown	vitreous to pearly luster; bubbles in HCl only when powdered or heated
Calcite $CaCo_3$	hexagonal varied	3 perfect at 75°	3	white, transparent, variable	vitreous to pearly luster; **bubbles in dilute HCl**
Gibbsite $Al(OH)_3$	Monoclinic	1 perfect	2½–3½	white, variable pale colors	pearly luster; **strong clay odor** when breathed on
Barite $BaSO_4$	orthorhombic platy, massive, granular	3	2½–3½	white, gray, variable pale colors	rosette clusters of platy crystals colored pink by impurities are "desert roses"; **density 4.5**
Halite $NaCl$	Isometric **cubic crystals,** massive	**3 perfect at 90°**	2–2½	clear to white	**salty taste**
Muscovite $KAl_2[AlSi_3O_{10}](OH)_2$ (a mica)	Monoclinic platy	1 perfect	2–2½	white, transparent	vitreous to pearly luster
Gypsum $CaSO_4 \cdot 2H_2O$	Monoclinic varied	1 perfect	2	white, transparent	vitreous, pearly, silky luster; var. alabaster often used for carvings
Clays (complex silicates)	Monoclinic earthy masses	1 perfect, not usually visible	<2	white, gray, variable	dull earthy luster; clay odor when breathed on
Talc	Monoclinic foliated masses, fine grained	1 perfect	1	white, brown, pale to dark green	**greasy feel; pearly luster;** var. soapstone used for carvings

(Continued)

Nonmetallic Luster, Dark Color, Harder Than a Knife Blade

Name and Formula	Crystal Group and Habit	Cleavage/ Fracture	Hardness	Color	Other
Staurolite $FeAl_4Si_2O_{10}(OH)_2$	Monoclinic (pseudo-orthorhombic) prismatic	1 poor	7–7½	dark brown red-brown	vitreous to resinous luster, twins form a cross
Tourmaline (complex boron silicate)	Hexagonal prismatic (rounded triangular cross sections)	poor	7	commonly black, occ. pink or green	vitreous luster
Cryptocrystalline quartz	Massive	conchoidal fracture	7	variable	dark gray → flint rust red → jasper dull red, green → chalcedony tan, red, green → chert
Garnet $Ca, Mg, Fe, Al\,[SiO_4]$	Isometric equant (dodecahedra)	irregular fracture	7–7½	usually **shades of red**, occ. tan, green	resinous to vitreous luster
Olivine $(Fe, Mg)_2\,SiO_4$	Orthorhombic equant, sugary masses	**conchoidal fracture**	6½–7	**olive green**, occ. golden	vitreous luster
Epidote $Ca_2(Al,Fe)_3[Si_3O_{12}](OH)$	Monoclinic tabular, prismatic	1 perfect 1 poor	6½–7	pistachio **green**	vitreous luster
Glaucophane $Na_2(Mg,Fe,Al)_5$ $[Si_8O_{22}](OH)_2$ (an amphibole)	Monoclinic prismatic	2 perfect	6–6½	**blue**, lavender blue	vitreous to silky luster
Hornblende $NaCa_2(Mg,Fe,Al)$ $[(Si,Al_8)O_{22}](OH_2)$ (an amphibole)	Monoclinic prismatic	**2 perfect at 124°**	5–6	black, dark green	vitreous luster; crystals can be long, needlelike
Pyroxene (a common pyroxene is augite, $Ca(Mg,Fe)Si_2O_6$)	Monoclinic orthorhombic prismatic	**2 good at 90°**	5–6	black, dark green	vitreous luster; crystals shorter than hornblende
Ilmenite $FeTiO_3$	Hexagonal thick, tabular	none	5–6	black	submetallic

Nonmetallic Luster, Dark Color, Softer Than a Knife Blade

Name and Formula	Crystal Group and Habit	Cleavage/ Fracture	Hardness	Color	Other
Apatite $Ca_5(PO_4)_3(F,OH)$	Hexagonal prismatic or massive	2 poor	5	variable (green, brown, blue, red)	commonly transparent; vitreous luster; normal teeth are hydroxyapatite
Hematite Fe_2O_3	Hexagonal earthy granular masses	not obvious	5	**reddish-brown**	**red-brown streak**; earthy luster
Serpentine $Mg_6[Si_4O_{10}](OH)_8$ (serpentine is a mixture of two minerals)	Monoclinic massive, fibrous	none	4–5	black, **mottled pale green or aqua**	fibrous → asbestos
Limonite $FeO(OH \cdot nH_2O)$	amorphous	none	< 4	yellow to brown	**earthy luster; yellow-brown streak**
Sphalerite ZnS	Isometric equant	**4 perfect**	3½–4	brown, black, yellow	**resinous luster**; yellowish streak
Biotite $K(Mg,Fe)_3$ $[AlSi_3O_{10}](OH)_2$ (a mica)	Monoclinic **platy**	1 perfect	**2½–3**	**black**, occ. brown	vitreous luster; shinier than hornblende
Chlorite $(Mg,Fe,Al)_6$ $[Al,Si)_4O_{10}](OH)_8$ (a mica)	Monoclinic **platy**	1 perfect	2–2½	**green**	vitreous to pearly luster; white to pale green streak
Glauconite complex silicate (a mica)	Monoclinic **granular**	1 perfect, not obvious	2	**dark blue-green**	occurs as sand-sized grains in marine sand-stones; green streak
Clays (complex silicates)	Monoclinic earthy masses	1 perfect not usually visible	< 2	brown, gray, variable	dull earthy luster; clay odor when breathed on
Graphite C	Hexagonal	1 perfect	1	black	greasy feel; marks paper easily; black streak

Flow Chart for Rock Identification

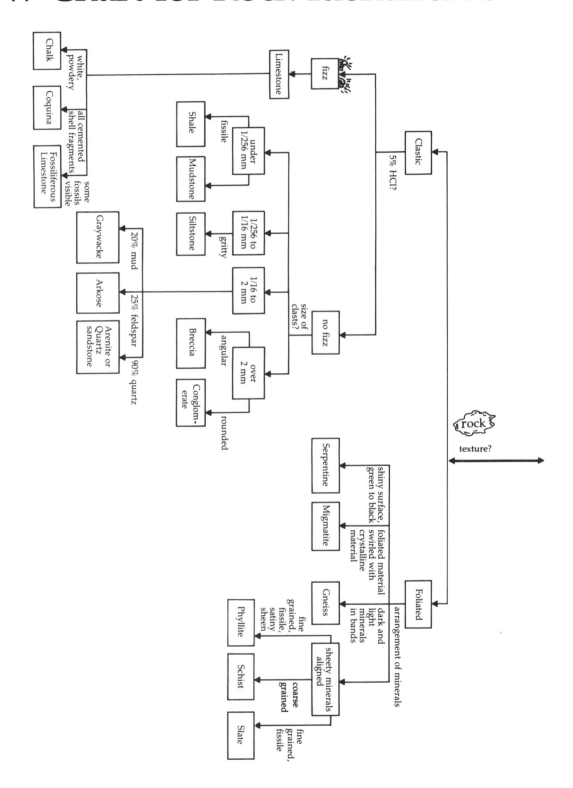

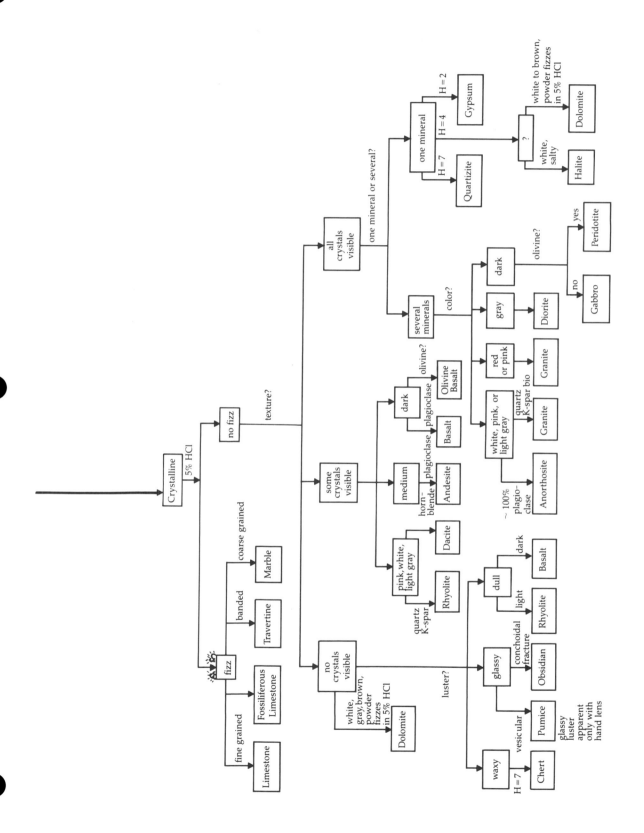

Use of the Hand Lens

Geologists often look at rocks with a small magnifier called a hand lens in order to pick out fine details—the twinning on a plagioclase crystal in gabbro or the shape of quartz grains in a sandstone, for example.

Most hand lenses consist of one or several optical elements protected by a metal or plastic swing-out case. There are $7\times$, $10\times$, $14\times$, and $20\times$ magnifiers. Geologists usually carry a hand lens on a cord around their necks to have it handy.

When looking at a specimen through a hand lens, first bring the lens close to your eye with one hand. Then with the other hand move the sample toward the lens until it comes into focus. You should tilt your head back a bit so that as much light as possible falls on the sample. Natural sunlight is preferable to incandescent or fluorescent lighting.

Stratigraphic Column Symbols

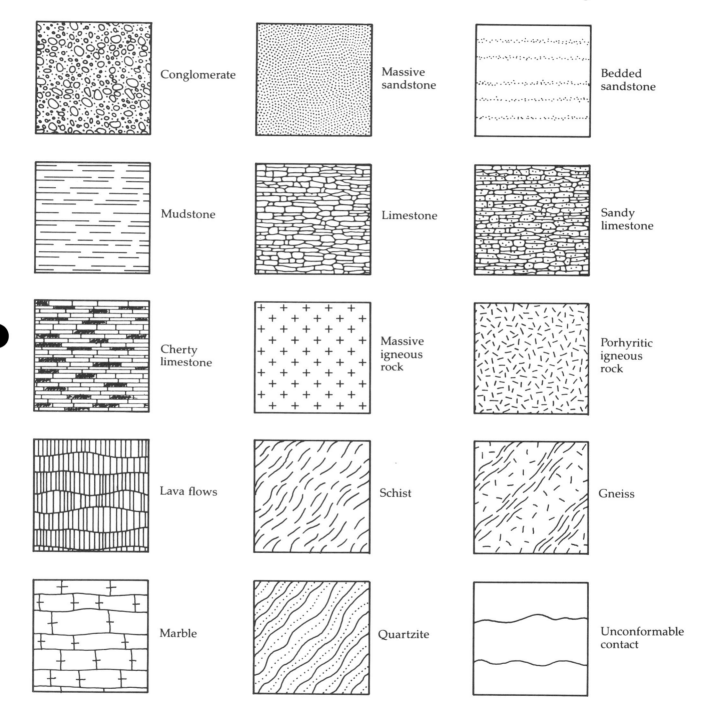

Conglomerate

Massive sandstone

Bedded sandstone

Mudstone

Limestone

Sandy limestone

Cherty limestone

Massive igneous rock

Porhyritic igneous rock

Lava flows

Schist

Gneiss

Marble

Quartzite

Unconformable contact

Topographic Map Symbols

VARIATIONS WILL BE FOUND ON OLDER MAPS

Primary highway, hard surface	
Secondary highway, hard surface	
Light-duty road, hard or improved surface	
Unimproved road	
Road under construction, alinement known	
Proposed road	
Dual highway, dividing strip 25 feet or less	
Dual highway, dividing strip exceeding 25 feet	
Trail	

Railroad: single track and multiple track	
Railroads in juxtaposition	
Narrow gage: single track and multiple track	
Railroad in street and carline	
Bridge: road and railroad	
Drawbridge: road and railroad	
Footbridge	
Tunnel: road and railroad	
Overpass and underpass	
Small masonry or concrete dam	
Dam with lock	
Dam with road	
Canal with lock	

Buildings (dwelling, place of employment, etc.)	
School, church, and cemetery	Cem
Buildings (barn, warehouse, etc.)	
Power transmission line with located metal tower	
Telephone line, pipeline, etc. (labeled as to type)	
Wells other than water (labeled as to type)	Oil Gas
Tanks: oil, water, etc. (labeled only if water)	Water
Located or landmark object; windmill	
Open pit, mine, or quarry; prospect	x
Shaft and tunnel entrance	Y

Horizontal and vertical control station:	
Tablet, spirit level elevation	BM △ 5653
Other recoverable mark, spirit level elevation	△ 5455
Horizontal control station: tablet, vertical angle elevation	VABM △ 95/9
Any recoverable mark, vertical angle or checked elevation	△ 3775
Vertical control station: tablet, spirit level elevation	BM × 957
Other recoverable mark, spirit level elevation	× 954
Spot elevation	× 7369 × 7369
Water elevation	670 670

Boundaries: National	
State	
County, parish, municipio	
Civil township, precinct, town, barrio	
Incorporated city, village, town, hamlet	
Reservation, National or State	
Small park, cemetery, airport, etc.	
Land grant	
Township or range line, United States land survey	
Township or range line, approximate location	
Section line, United States land survey	
Section line, approximate location	
Township line, not United States land survey	
Section line, not United States land survey	
Found corner: section and closing	
Boundary monument: land grant and other	
Fence or field line	

Index contour		Intermediate contour	
Supplementary contour		Depression contours	
Fill		Cut	
Levee		Levee with road	
Mine dump		Wash	
Tailings		Tailings pond	
Shifting sand or dunes		Intricate surface	
Sand area		Gravel beach	

Perennial streams		Intermittent streams	
Elevated aqueduct		Aqueduct tunnel	
Water well and spring		Glacier	
Small rapids		Small falls	
Large rapids		Large falls	
Intermittent lake		Dry lake bed	
Foreshore flat		Rock or coral reef	
Sounding, depth curve	10	Piling or dolphin	
Exposed wreck		Sunken wreck	
Rock, bare or awash; dangerous to navigation			

Marsh (swamp)		Submerged marsh	
Wooded marsh		Mangrove	
Woods or brushwood		Orchard	
Vineyard		Scrub	
Land subject to controlled inundation		Urban area	

Geologic Map Symbols

Symbols for Geologic Maps

Bedding

⊕ Horizontal beds

⊤₂₅ Strike and dip of beds

✕ Strike of vertical beds
90

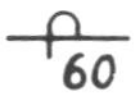

⊕ Strike and dip of
60 overturned beds

Contacts

〜 Definite contact

------ Approximate contact

········ Concealed contact

Folds

Axis of anticline

Axis of syncline

Plunging anticline

Plunging syncline

Faults

⊤50 Fault showing dip

— — Inferred fault

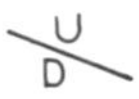

High-angle fault, U on up
side, D on down side

Thrust or reverse fault,
barbs on side of upper
plate

Normal fault, hachures on
down side

Thrust or low-angle
reverse fault, T on upper
plate

Geologic System	Map Symbol	Color
Quaternary	Q	Brownish yellow
Tertiary	T	Yellow
Cretaceous	K	Green
Jurassic	J	Blue green
Triassic	Tr	Peacock blue
Permian	P	Blue
Pennsylvanian	P	Blue
Mississippian	M	Blue
Devonian	D	Blue gray
Silurian	S	Blue purple
Ordovician	O	Blue purple
Cambrian	Є	Brick red
Precambrian	pЄ	Brownish red

Metric Conversion Table

Conversion Factors (c.f.)

1. Length (A × c.f. = B)

A	Centimeters (cm)	Meters (m)	Kilometers (km)	Inches (in)	Feet (ft)	Yards (yd)	Miles (mi)
B							
cm	1	100	100000	2.54	30.4	91.4	1.609×10^5
m	0.01	1	100	0.0254	0.3048	0.9144	1.609
km	1×10^{-5}	0.001	1	2.54×10^{-5}	3.048×10^{-4}	0.144×10^{-4}	1.609
in	0.3937	39.37	3.937×10^4	1	12	36	6.336×10^4
ft	0.03281	3.281	3,281	0.08333	1	3	5,280
yd	0.01094	1.094	1,094	0.02778	0.3333	1	1,760
mi	6.214×10^{-6}	6.214×10^{-4}	0.6214	1.578×10^{-4}	1.894×10^{-4}	5.682×10^{-4}	1

2. Area (A × c.f. = B)

A	Acres	Square Centimeters (cm²)	Square Meters (m²)	Square Kilometers (km²)	Square Inches (in²)	Square Feet (ft²)	Square Miles (mi²)
B							
acres	1	2.471×10^{-8}	2.471×10^{-4}	247.1	6.27×10^{-6}	3.296×10^{-5}	640
cm²	4.047×10^7	1	1000	1×10^{10}	6.452	929	2.59×10^{10}
m²	4.047×10^4	1×10^{-4}	1	1×10^6	6.452×10^{-4}	0.0929	2.59×10^6
km²	4.047×10^{-3}	1×10^{-10}	1×10^{-6}	1	6.452×10^{-10}	9.29×10^{-8}	2.59
in²	6.273×10^6	0.1550	1,550	1.55×10^9	1	144	1.55×10^9
ft²	4.356×10^4	1.076×10^{-3}	10.76	1.076×10^7	6.944×10^{-3}	1	2.788×10^7
mi²	1.562×10^{-3}	3.861×10^{-11}	3.861×10^{-7}	0.3861	4.25×10^{-9}	3.587×10^{-7}	1

3. Volume (A × c.f. = B)

A	Cubic Centimeters (cm³)	Cubic Meters (m³)	Cubic Inches (in³)	Cubic Feet (ft³)	Cubic Yards (yd³)
B					
cm³	1	1×10^6	16.39	2.832×10^4	7.646×10^5
m³	1×10^{-6}	1	1.639×10^{-5}	2.832×10^{-2}	0.7646
in³	6.102×10^{-2}	6.102×10^4	1	1,728	46,656
ft³	3.531×10^{-5}	35.31	5.787×10^{-4}	1	27
yd³	1.308×10^{-6}	1.308	2.143×10^{-5}	3.704×10^{-2}	1

4. Mass (A × c.f. = B)

A	Grams (g)	Kilograms (kg)	Ounces (oz)	Pounds (lb)
B				
g	1	1000	28.35	453.6
kg	0.001	1	2.835×10^{-2}	0.4536
oz	3.527×10^{-2}	35.27	1	16
lb	2.205×10^{-3}	2.205	6.25×10^{-2}	1

5. Temperature

$$°C = \frac{(°F - 32°)}{1.8} \qquad °F = (°C \times 1.8) + 32°$$

Useful Formulas and Partial Table of Tangents

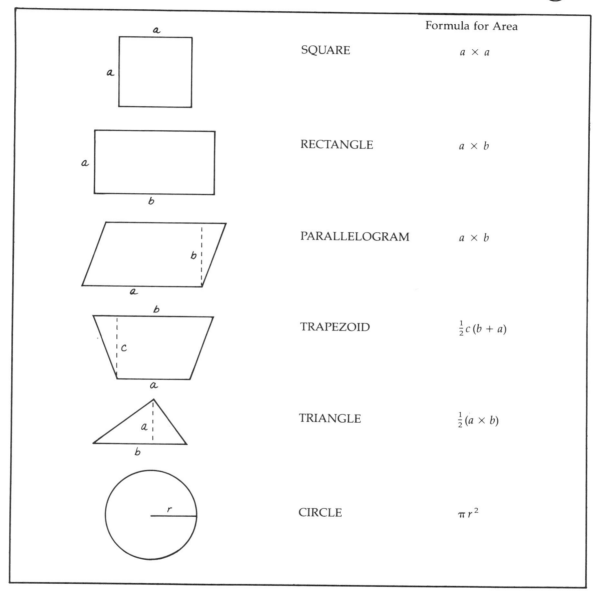

		Formula for Area
SQUARE		$a \times a$
RECTANGLE		$a \times b$
PARALLELOGRAM		$a \times b$
TRAPEZOID		$\frac{1}{2} c (b + a)$
TRIANGLE		$\frac{1}{2}(a \times b)$
CIRCLE		πr^2

The tangent of any angle in a right triangle (one in which one internal angle is 90°) is defined as the length of the side opposite the angle divided by the length of the side adjacent to the angle.

$$\tan \alpha = \frac{a}{b}$$

If sides a and b are known, the tangent of the angle may be calculated. The table of tangents will then give the approximate angle in degrees. The slope, α, of a hill is easily determined in this way.

$$\tan \alpha = \frac{v \;(\text{vertical distance})}{h \;(\text{horizontal distance})}$$

411

Partial Table of Tangents

Angle	Tangent	Angle	Tangent	Angle	Tangent	Angle	Tangent
0°	0.0000	22°30'	0.4142	45°	1.0000	67°30'	2.4142
0°30'	0.0087	23°	0.4245	45°30'	1.0176	68°	2.4751
1°	0.0175	23°30'	0.4348	46°	1.0355	68°30'	2.5386
1°30'	0.0262	24°	0.4452	46°30'	1.0538	60°	2.6051
2°	0.0349	24°30'	0.4557	47°	1.0724	69°30'	2.6746
2°30'	0.0437	25°	0.4663	47°30'	1.0913	70°	2.7475
3°	0.0524	25°30'	0.4770	48°	1.1106	70°30'	2.8239
3°30'	0.0612	26°	0.4877	48°30'	1.1303	71°	2.9042
4°	0.0699	26°30'	0.4986	49°	1.1504	71°30'	2.9887
4°30'	0.0788	27°	0.5095	49°30'	1.1708	72°	3.0777
5°	0.0875	27°30'	0.5206	50°	1.1918	73°30'	3.1716
5°30'	0.0963	28°	0.5318	50°30'	1.2131	73°	3.2708
6°	0.1051	28°30'	0.5430	51°	1.2349	73°30'	3.3759
6°30'	0.1139	29°	0.5543	51°30'	1.2572	74°	3.4874
7°	0.1228	29°30'	0.5658	52°	1.2799	74°30'	3.6059
7°30'	0.1317	30°	0.5773	52°30'	1.3032	75°	3.7321
8°	0.1405	30°30'	0.5890	53°	1.3270	75°30'	3.8667
8°30'	0.1495	31°	0.6009	53°30'	1.3514	76°	4.0108
9°	0.1584	31°30'	0.6128	54°	1.3764	76°30'	4.1653
9°30'	0.1673	32°	0.6249	54°30'	1.4020	77°	4.3315
10°	0.1763	32°30'	0.6371	55°	1.4282	77°30'	4.5107
10°30'	0.1853	33°	0.6494	55°30'	1.4550	78°	4.7046
11°	0.1944	33°30'	0.6619	56°	1.4826	28°30'	4.9152
11°30'	0.2034	34°	0.6745	56°30'	1.5108	79°	5.1446
12°	0.2126	34°30'	0.6873	57°	1.5399	79°30'	5.3955
12°30'	0.2217	35°	0.7002	57°30'	1.5697	80°	5.6713
13°	0.2309	35°30'	0.7133	58°	1.6003	80°30'	5.9758
13°30'	0.2401	36°	0.7265	58°30'	1.6318	81°	6.3138
14°	0.2493	36°30'	0.7400	59°	1.6643	81°30'	6.6912
14°30'	0.2586	37°	0.7536	59°30'	1.6977	82°	7.1154
15°	0.2680	37°30'	0.7673	60°	1.7320	82°30'	7.5958
15°30'	0.2773	38°	0.7813	60°30'	1.7675	83°	8.1444
16°	0.2868	38°30'	0.7954	61°	1.8040	83°30'	8.7769
16°30'	0.2962	39°	0.8098	61°30'	1.8418	84°	9.5144
17°	0.3057	39°30'	0.8243	62°	1.8807	84°30'	10.3854
17°30'	0.3153	40°	0.8391	62°30'	1.9210	85°	11.4301
18°	0.3249	40°30'	0.8541	63°	1.9626	85°30'	12.7062
18°30'	0.3346	41°	0.8693	63°30'	2.0057	86°	14.3007
19°	0.3443	41°30'	0.8847	64°	2.0503	86°30'	16.3499
19°30'	0.3541	42°	0.9004	64°30'	2.0965	87°	19.0811
20°	0.3640	42°30'	0.9163	65°	2.1445	87°30'	22.9038
20°30'	0.3739	43°	0.9325	65°30'	2.1943	88°	28.6363
21°	0.3839	43°30'	0.9490	66°	2.2460	88°30'	38.1885
21°30'	0.3939	44°	0.9657	66°30'	2.2998	89°	57.2900
22°	0.4040	44°30'	0.9827	67°	2.3558	89°30'	114.589
						90°	∞

Glossary

A

aa Hawaiian term for a rough, jagged, clinkery lava flow.

abandoned channel a stream channel that no longer drains water from the land.

ablation formation of residual deposits by the separation and removal of sediment by wind action.

abrasion mechanical erosion by the grinding, scraping, or rubbing of a rock surface by wind, water, or ice, plus other rock particles.

absolute age geologic age expressed in units of time, usually years.

abyssal plain a flat region of the ocean floor with a slope of less than 5 feet per mile (0.001).

acre a unit of land area equal to 1/640 of a square mile.

aftershock an earthquake that follows a larger earthquake and has its focus close to that of the larger earthquake.

aggradation the building up of the earth's surface by deposition, primarily by stream processes.

alluvial resulting from deposition by running water.

alluvial fan a fan-shaped, gently sloping stream deposit at the base of a mountain valley.

alluvium mud, sand, or gravel sediment deposited by running water.

alpine glacier a glacier occurring in mountainous area that originates in a cirque and flows downhill in a valley.

altitude the vertical distance of a point above a surface, generally sea level.

amphibole a group of dark ferromagnesian silicate minerals closely related in crystal structure and composition; includes hornblende.

anaerobic in the absence or near absence of oxygen.

andesite extrusive igneous rock of intermediate composition.

angle of repose the maximum angle of slope at which loose, cohesionless material will rest in a pile.

angular unconformity an unconformity between rocks in which the overlying younger rock has been deposited on older folded or tilted rock.

anticline a concave-downward fold where the older rocks are at the center of the fold.

aqueous of or pertaining to water.

aquiclude rock body that is relatively impermeable, resulting in the entrapment of groundwater.

aquifer a permeable rock body sufficiently saturated with water to yield economical quantities of water.

Archean the earlier of the two Precambrian Eras, lasting from 3800 million years ago (the age of the oldest known rocks on earth) to 2500 million years ago.

arcuate curved or bent.

arenite a general term for sedimentary rock composed of sand-sized particles.

arete a narrow jagged mountain ridge formed by headward glacial erosion.

arid climate characterized by dryness, having less than 10 inches (25 cm) of annual precipitation.

arkose feldspar-rich sandstone.

ash fine pyroclastic material under 4.0 mm in diameter, created by explosive volcanic eruption.

attitude the position of a structural surface relative to the horizontal, expressed by measurements of strike and dip.

B

bankful stage the elevation of the water surface of a stream flowing at the capacity of the channel.

bar a deposit of sediment along a shoreline, usually elongate in shape.

barchan a moving crescent-shaped sand dune with the horns pointing downwind.

barrier island an elongate offshore ridge of sand, generally parallel to the coastline.

basalt an extrusive igneous rock of mafic composition.

base level the level below which erosion cannot take place, especially stream erosion; the level where the gradient is zero.

basic igneous rocks having low silica content; mafic.

basin a topographic low area characterized by sedimentary deposition.

batholith large discordant igneous intrusion with a surface exposure greater than 100 km^2.

bed a subdivision of stratified sedimentary rocks, usually a rock layer 2 cm to 1 m in thickness, that exhibits some degree of homogeneity.

Benioff zone a zone of shallow-to-deep earthquakes beneath the trenches of the Pacific, dipping toward a continent or island chain; where the oceanic crust descends into the mantle.

boulder a large rounded rock fragment greater than 256 mm (10 inches) in diameter.

Bowen's reaction series a series of minerals in which any early formed mineral tends to react with the melt to yield a new mineral further down in the series.

braided stream a stream that separates into two or more channels because of islands or channel bars.

breccia a clastic sedimentary rock composed of angular rock fragments greater than 2 mm in diameter.

C

cactolith a quasihorizontal chonolith composed of anastomosing ductoliths whose distal ends curl like a harpolith, are thin like a sphenolith, or bulge discordantly like an akmolith or ethmolith.

calcareous containing calcium carbonate.

calcite a common rock-forming mineral composed of calcium carbonate.

Cambrian the first period of the Paleozoic Era, between 570 and 500 million years ago, characterized by the first appearance of organisms with hard body parts.

carbonate rock a rock chiefly composed of carbonate minerals; usually limestone or dolomite.

Carboniferous a period of the upper Paleozoic Era between 360 and 285 million years ago, characterized by the abundant growth of land plants.

carbon-14 dating a method of radiometric age dating using the concentration of carbon-14 remaining in organic material. Carbon-14 has a half-life of 5570 ± 30 years.

cement a mineral substance chemically precipitated in the void spaces of clastic rocks that binds the rock together.

Cenozoic an era of geologic time that includes the Tertiary and Quaternary Periods, characterized by the evolution of mammals (from 67 million years ago to the present).

channel the natural depression where surface water flows.

chemical rock a sedimentary rock composed primarily of minerals precipitated from aqueous solutions.

chemical weathering the process of weathering through chemical reaction.

cinder cone a type of volcano formed by the accumulation of ejected material, generally basaltic or andesitic in composition.

cirque an amphitheaterlike depression formed by erosion at the head of an alpine glacier.

clastic referring to a rock or sediment composed primarily of broken fragments of pre-existing rocks or organisms.

clay (1) a sedimentary particle with a diameter of less than 1/256 mm. (2) a group of sedimentary sheet silicate minerals.

cliff any high, nearly vertical face of rock.

closed basin a drainage basin without an outlet.

cobble a rock fragment having a diameter in the range from 64 to 256 mm.

colluvium any rock material transported and deposited by slope processes.

composite volcano *see* **stratovolcano.**

conglomerate a coarse-grained clastic sedimentary rock composed mainly of rounded-to-subrounded fragments larger than 2 mm in diameter.

contact metamorphism metamorphism caused by heat adjacent to an igneous intrusion or flow.

continent a major land mass on the earth's surface.

continental crust that type of the earth's crust that underlies the continents and continental shelves; intermediate to granitic in composition; ranging in thickness from 35 to 60 km.

continental divide a drainage divide separating streams flowing into different oceans.

continental glacier a large moving ice mass of considerable thickness covering over 50,000 km^2 of land.

continental shelf the margin of the continent that is underwater.

contour interval the difference in value between two adjacent contour lines.

contour line a line composed of points having the same value.

core the central zone of the earth's interior, composed primarily of metallic iron and nickel.

correlation the determination of the association of related geologic phenomena, usually rock strata.

craton part of the continental crust that is stable and has been without major deformation over an extended period of geologic time.

creek a small stream channel, generally a tributary of a river.

creep (1) the slow, imperceptible motion of rock material downslope by gravitational forces. (2) the continuous, usually slow deformation of rock resulting from constant stress acting over a long period of time.

Cretaceous the last period of the Mesozoic Era, between 145 and 66 million years ago, characterized by the appearance of flowering plants. The end of the Cretaceous featured the extinction of the dinosaurs.

crevasse a deep, nearly vertical crack in glacial ice.

crossbedding the internal arrangement in sedimentary rock of strata deposited at an angle to the horizontal by moving water or wind.

cross-section a drawing illustrating geologic features perpendicular to the general structural trend.

crust the outer continuous rock shell of the earth, composed primarily of granite or basalt, with a thickness between 10 and 60 km.

crystal   a homogeneous solid body of a chemical element or compound with a definite internal arrangement that is expressed by planar faces.

crystalline   consisting of crystals or crystal fragments.

crystalline-granular   a texture of igneous rocks in which the minerals are intergrown and are visible to the unaided eye.

crystal settling   the sinking of crystals due to density differences between crystals and magma.

current   the movement of a fluid body (air or water) in a definite direction and with a definite velocity.

cutbank   the outer bend in a river meander.

cutoff   the new and relatively short channel formed when a stream cuts through a narrow neck of land between loops of a meander.

D

debris avalanche   the rapid and sudden slide of unconsolidated rock material.

debris flow   the rapid flow of rock material en masse down a slope.

declination   the horizontal angle between true (geographic) north and magnetic north.

deflation   the sorting, lifting, and transport of unconsolidated fine-grained particles by the wind.

deformation   a general term for the process of folding, faulting, shearing, compression, or extension of rocks as a result of various earth forces.

delta   a landform at the mouth of a river resulting from the deposition of alluvium upon entering a standing body of water.

dendritic drainage   the drainage pattern or stream channels that branch in a treelike pattern.

deposit   (1) the dropping of rock or organic material by natural processes. (2) rock or organic material that has accumulated by natural processes.

desert   an area of low precipitation, less than 10 inches (25 cm) annually.

desert pavement   the thin residual concentration of gravel over the surface in an arid area, resulting from the removal of smaller particles.

detrital   pertaining to loose, transported rock material.

detritus   loose fragmental material that has been displaced from its place of origin.

Devonian   a geologic time period between 410 and 360 million years ago, characterized by the development of fishes and the appearance of abundant land plants, insects, and amphibians.

diagenesis   the postdepositional modification of sediments by chemical, physical, and organic processes, resulting in a sedimentary rock.

dike   a tabular, discordant igneous intrusion.

dip   the angle that a bedding plane or fault surface forms with the horizontal, as measured perpendicular to the strike of the surface.

dip-slip fault   a fault along which the movement is parallel to the dip of the fault plane.

discharge   the flow of water in a channel measured in volume per unit time.

dissolved load   that part of the total stream load carried in solution.

distributary   a stream channel that flows away from the main stream channel and does not return to it, as on a delta or alluvial fan.

dome   a circular or elliptical uplift or anticlinal structure.

drainage basin   the entire region that contributes water to a stream channel.

drainage density   the ratio of total stream lengths of all channels in a basin to the area of the basin.

drainage divide   the topographic high separating adjacent drainage basins.

drift   all rock material transported by a glacier.

drowned valley   a valley that is partially submerged by water, either by a rise in water level or by a subsidence of the land.

drumlin   a smooth, elongated, oval hill composed partially or entirely of glacial till.

dune   a low ridge or mound composed of loose granular material, usually deposited by wind but sometimes by water.

dust   dry solid material less than 1/16 mm in size.

E

earthquake   the sudden oscillatory motion of the ground caused by the abrupt release of energy in faulting or volcanic activity.

effluent stream   a stream receiving water from the zone of saturation.

ejecta   material thrown out into the air in a volcanic eruption or meteor impact.

elastic   that property of a body in which deformations are instantly and totally recoverable to the original shape of the body.

elevation   the vertical height of a point on the earth's surface above a certain level, usually sea level.

emergent shoreline   a shoreline where either the land is rising or water level is falling.

end moraine   a moraine formed at the terminus of a glacier.

entrenched meander   a stream meander cut vertically downward below the land surface on which the stream originally formed.

Eocene the epoch of the lower Tertiary Period extending from 58 to 37 million years ago, characterized by the rise of mammals.

eolian having been eroded, transported, or deposited by the wind.

eon the longest geologic time unit.

ephemeral stream a stream that flows only periodically in response to precipitation.

equator the great circle on the earth on which every point is equidistant from the two poles.

erosion the general process or processes by which rock material is transported from a land area.

erratic a generally large rock fragment deposited by glacial ice.

eruption the ejection of rock or magma from a volcano.

escarpment a long, generally continuous cliff.

esker a long narrow ridge composed of stratified drift deposited by a subglacial stream.

estuary an extension of the ocean inland where fresh and salt water mix.

euhedral said of a mineral grain bounded by its own regular crystal faces.

evaporite a chemical sedimentary rock formed by the precipitation of minerals from an evaporating water body.

evolution the theory that life has developed gradually from simple forms to more complex forms.

exposure an area on the surface where a geologic feature is visible.

extinction the total disappearance of a species of living organisms from the earth.

extrusive pertaining to an igneous rock that cooled at the earth's surface.

F

facies the sum of all the lithologic and biologic characteristics exhibited by a sedimentary rock from which the environment of deposition can be inferred; also, in metamorphism, a set of mineral assemblages produced under certain conditions of temperature and pressure.

fault the surface or zone of rock fracture along which displacement has taken place.

fault scarp a cliff or slope formed directly by fault movement.

fault trace the line of intersection between a fault and the surface.

felsic pertaining to a light-colored silica-rich igneous rock.

field the outdoors, where geological observations and data are collected.

fine grained a rock texture in which the minerals are relatively small and difficult to observe with the unaided eye.

firn granular ice that is transitional between snow and glacial ice.

fissility the splitting of rocks into thin, generally parallel sheets.

fissure a crack or fracture in the earth's crust.

floodplain the area of land adjacent to and formed by a stream that is covered by water in times of flood.

fluvial related to streams.

focus the point within the earth at which the initial motion occurs in an earthquake.

fold a curve or bend in rock.

foothills a region of relatively low hills adjacent to a mountain range.

footwall the underside of a dipping fault surface.

formation the fundamental rock stratigraphic unit characterized by some degree of internal homogeneity and with a sufficient thickness to be mapable.

fossil any remains, trace, or imprint of a plant or animal that has been preserved by natural processes in rock or rock material through geological time.

fracture (1) breakage of a mineral in a nonordered fashion and not along cleavage planes. (2) a general term for a break in a rock.

fresh water water containing very little dissolved material.

friable crumbly, easily pulverized.

frost wedging or riving the mechanical breakdown of rock by the great pressure exerted by water freezing in pores and cracks.

fumarole an opening on the surface from which gases and vapors are emitted.

G

gabbro a dark-colored basic intrusive igneous rock composed principally of plagioclase and pyroxene.

gem a cut and polished stone that has intrinsic value and possesses characteristics of beauty, durability, rarity, and size for use as jewelry.

geochronology the study of time in relationship to the history of the earth, especially absolute time.

geologic history the history of the earth, its products, and its processes throughout geologic time for a particular area or duration of time.

geologic map a map that records geologic information, such as the distribution and nature of rock units and the occurrence of structural features.

geologic time the age of a particular event or material with respect to the age of the earth.

geology the study of the origin, nature, and development of the planet earth.

geomorphology the study of the configuration and features of the earth's surface.

geophysics the study of the earth using the applications of physics.

glacier a large mass of ice present throughout the year on the earth's surface formed by the recrystallization of snow that moves under its own weight.

glassy an extrusive igneous rock texture that is similar to that of broken glass and develops by the rapid cooling of lava.

gneiss a foliated metamorphic rock where the granular minerals and platy minerals are segregated into distinct bands or lenses.

goniometer an instrument in mineralogy to measure the angle between crystal faces.

graded bedding a type of sedimentary bedding in which the clast size decreases gradually from the base to the top.

graded stream a stream in equilibrium, with a velocity just sufficient to carry the sediment load provided by its drainage basin.

gradient the degree of inclination of the earth's surface; the steepness of slope.

granite a felsic plutonic rock composed primarily of potassium feldspar, plagioclase, and quartz, with biotite and hornblende.

gravel an unconsolidated natural accumulation of rounded rock fragments consisting predominantly of particles larger than sand.

gravity the gravitational attraction at or near the earth's surface resulting from the mutual attraction between the earth and other masses.

graywacke a sandstone containing over 15 percent mud.

grid a network of uniformly spaced imaginary lines superimposed on a map for reference of locations.

groundmass the interstitial fine-grained material in a porphyritic igneous rock.

groundwater all subsurface water, especially that which comprises the zone of saturation beneath the water table.

group a major rock stratigraphic unit consisting of two or more associated rock formations.

H

habit the characteristic form of a mineral.

hackly a rock or mineral fracture having a jagged surface.

halite salt; a mineral or rock composed of sodium chloride.

hanging valley a tributary valley whose floor at its mouth is significantly higher than the floor of the main valley, usually due to glacial action.

hardness the resistance of a mineral to scratching.

headland a stretch of land that extends from the coast into a standing body of water.

headwater the upper portion of a drainage basin.

Holocene the epoch of the upper Quaternary Period extending from 10,000 years ago to the present.

hornfels a fine-grained metamorphic rock lacking preferred orientation, typically formed by contact metamorphism.

hummocky a type of topography that is uneven and characterized by rounded mounds or small hills.

hydrology the science that deals with global water, its properties, circulation, and distribution on and under the surface.

hydrosphere the waters of the earth, as distinguished from the rocks, air, and living organisms.

hydrothermal related to the products and action of hot waters.

hypothesis a concept that is tested for validity by experimentation and observation.

I

ice the mineral form of water.

igneous a rock or mineral that formed from molten material.

impermeability the condition of a rock or rock material through which fluids are not able to travel.

inclination (1) the slope of any geological surface. (2) the angle at which magnetic field lines dip.

infiltration the movement of water or other fluids into rock or rock material.

influent stream a stream that contributes water to the zone of saturation.

intensity a measure of the effects of an earthquake, described usually by the Modified Mercalli Scale.

interbedded rock layers that alternate with others of a different nature; "interbedded sandstone and shale."

isotope various species of atoms of a chemical element with the same atomic number but different atomic weights because the number of neutrons in the nucleus is different.

J

joint a crack in a rock without displacement along it.

Jurassic the second period of the Mesozoic Era, occurring approximately from 210 to 145 million years ago, characterized by proliferation of ferns and ammonites and the appearance of primitive birds.

K

kame an irregular hill or mound consisting of stratified drift deposited by a subglacial stream.

karst an erosional topography characterized by sinkholes, caves, and disappearing streams formed by groundwater in limestone, dolomite, and evaporite bedrock.

kettle a shallow basin or depression in stratified drift thought to have formed from the melting of stagnant ice remaining from the retreat of a glacier; frequently occupied by a kettle lake.

L

laccolith a mushroom-shaped concordant igneous intrusion.

lacustrine pertaining to, produced by, or formed in a lake.

lagoon a shallow stretch of seawater partly separated from the sea by a low narrow strip of land such as a barrier beach or reef.

lake an inland body of standing water formed in a depression of appreciable size on the surface.

landform a recognizable characteristic feature of the earth's surface produced by natural processes.

landscape a set of closely related landforms within a given area.

landslide the downslope movement or rock and rock material along a discrete surface by gravitational forces.

latitude the angular distance north or south of the equator.

lithic related to or composed of stone.

lithification the process of solidifying unconsolidated sediment into a coherent rock.

lithology the description of rocks on the basis of color, mineralogic composition, grain size, and other important characteristics.

lithosphere the solid rock-bearing portion of the earth.

load the total material transported by a sedimentary agent, such as a stream, glacier, wind, or waves.

loess wind-blown dust deposited as a blanket of homogeneous nonstratified fine-grained material downwind from desert or glaciated areas.

longitude the angular distance east or west of a specified north–south line through Greenwich, England.

longitudinal dune a long, narrow accumulation of sand oriented parallel to the prevailing wind direction.

longshore current a coastal current created by the approach of waves at an angle to the shoreline.

low-energy environment a sedimentary environment characterized by a lack of current or water movement that results in the deposition of some sediment.

M

mafic an igneous rock composed of iron and magnesian minerals and having relatively low silica content.

magma molten rock material.

magma chamber a reservoir of magma in the upper portion of the earth's crust.

magnetic anomaly regions on the earth's surface where the measured magnetic field is greater or less than the average magnetic field of the earth.

magnitude a measurement of the strength of an earthquake or the energy released by it as determined by seismographic measurements expressed by the Richter magnitude scale.

map a diagram illustrating the physical features of part or all of the surface of the earth, moon, or other planetary body.

map scale the relationship between distance on a map and the surface it represents.

marble a metamorphic rock composed of calcite and/or dolomite.

marker bed an easily recognizable rock unit that is used as a reference in stratigraphic correlation.

massive a sedimantary rock layer that is internally homogeneous and lacking recognizable stratification.

mass wasting the downslope movement of rock material by gravitational forces.

matrix the smaller or finer grained material filling the interstices between the larger grains of a sedimentary rock or between the phenocrysts of a porphyritic igneous rock.

meander a sinuous bend in a stream channel.

meandering stream a stream channel having successive meander bends.

mechanical weathering the process of rock weathering whereby rocks are broken into smaller fragments.

Mercalli scale an arbitrary scale of earthquake intensity that rates the effects of an earthquake from I to XII.

meridian an imaginary line of equal longitude.

Mesozoic an era of geologic time between 245 and 66 million years ago, beginning at the end of the Paleozoic and ending at the beginning of the Cenozoic, characterized by predominance of reptiles.

metamorphic grade the intensity of metamorphism, implying the pressure and temperature conditions that caused metamorphism.

metamorphic rock any rock derived from preexisting rock that has been altered by pressure, temperature, or chemically active fluids while in the solid state.

microcrystalline a rock texture having crystals too small to be identified without the aid of a microscope.

migmatite a rock of high metamorphic grade, perhaps where partial melting has occurred.

mineral a naturally occurring inorganic solid with a definite chemical composition or range of compositions and an orderly internal arrangement of atoms and molecules.

mineralogy the study of minerals.

Miocene an epoch of the Tertiary Period between 24 and 5 million years ago, after the Oligocene and before the Pliocene, characterized by the appearance of primative apes, whales, and grazing animals.

Mississippian   a period of the Paleozoic Era between 360 and 320 million years ago, after the Devonian and before the Pennsylvanian; it is an American classification corresponding to the Lower Carboniferous of European usage.

Modified Mercalli scale   an arbitrary scale of earthquake intensity, revised in 1931 from the Mercalli scale.

Mohorovicic discontinuity   the boundary surface between the earth's crust and mantle.

Mohs hardness scale   a standard classification of mineral hardness based on ten separate minerals.

monocline   a type of fold consisting of a local steepening in an otherwise uniformly dipping layer.

moon   a natural satellite of a planet.

moraine   an accumulation of till deposited directly by a glacier, generally forming a mound, ridge, or undulating plain.

mountain   an elevated area of the earth's surface of significant relief that is higher than a hill.

mouth   the location where a stream channel discharges water into a larger body of water, where a tributary channel enters the main stream, or where a channel enters a standing body of water.

mud   unconsolidated sediment, clay and silt in size.

mudstone   a clastic sedimentary rock composed of clay- and silt-sized materials.

mylonite   a granulated sheared metamorphic derived from extreme pressure, especially along fault zone.

N

natural levee   a long, broad, low ridge or embankment adjacent to a stream channel composed of sand and silt and deposited during floods.

neap tide   an unusually small tide occurring during the first and third quarters of the moon.

normal fault   a dip-slip fault where the block above a dipping fault surface (hanging wall) moves down relative to the block below the fault surface (footwall).

nucleation   the beginning of crystal growth at one or more points.

O

oblique fault   a fault that has moved both horizontally and vertically.

obsidian   a black or dark glass of rhyolitic composition.

ocean   the continuous salt water body that surrounds the continents.

oceanography   the study of the oceans.

offset   a measurement of displacement along a fault.

oil   petroleum.

Oligocene   an epoch of the lower Tertiary Period approximately 37 to 24 million years ago.

Ordovician   the second period in the Paleozoic Era, about 500 to 440 million years ago.

ore   a naturally occurring economic concentration of a mineral or minerals.

organic   related to biologically derived compounds containing carbon.

orogeny   the formation of mountains.

outcrop   part of a geologic formation or structure that is exposed on the surface.

outwash   drift that is transported and deposited by meltwater in front of a glacier.

overland flow   that part of the water runoff that flows on the surface as a sheet toward a stream channel.

P

pahoehoe   a type of lava flow, generally basaltic in composition, characterized by a smooth, glassy, "ropy" texture.

paleo-   a prefix indicating an ancient time or ancient conditions.

Paleocene   the earliest epoch of the Tertiary Period extending from 66 to 58 million years ago, characterized by the appearance of placental mammals.

peat   an unconsolidated deposit of organic material formed in a water-saturated environment.

pebble   any rounded rock fragment having a diameter between 4 and 64 mm.

pegmatite   a coarse-grained igneous rock generally occurring as veins or dikes in or adjacent to stocks or batholiths; pegmatites represent the last stage of crystallization in felsic igneous intrusions, are rich in quartz and potassium feldspar, and may contain many rare minerals.

pelagic   pertaining to the deep open ocean environment.

Pennsylvanian   a period of the Paleozoic Era after the Mississippian and before the Permian, occurring between 320 and 285 million years ago; a period of proliferation of land plants; in European usage, the Upper Carboniferous.

perched groundwater   groundwater located above the main body of groundwater by an impermeable layer (aquiclude).

perennial   pertaining to a feature, generally a water body, that is present throughout the year.

peridotite   a plutonic igneous rock composed of olivine and pyroxene.

period   the fundamental geologic time unit, a subdivision of an era and longer than an epoch.

permeability   the ability of a porous rock or rock material to transmit fluids.

Permian   the last period of the Paleozoic Era, occurring after the Pennsylvanian Period, approximately 285 to 245 million years ago. The end of the Permian featured the extinction of major groups of marine invertebrates, especially tribolites.

petroleum a complex, naturally occurring liquid composed primarily of hydrocarbons.

petrology the branch of geology that studies the origin, occurrence, structure, and history of rocks.

phaneritic an igneous rock texture in which all the individual mineral crystals are large enough to be seen with the unaided eye.

Phanerozoic an eon of geologic time incorporating the Paleozoic, Mesozoic, and Cenozoic Eras; generally defined as geologic time in which the evidence of life is abundant.

phenocryst a relatively large crystal in a porphyritic igneous rock.

plain a flat to gently sloping land surface.

planet any of the nine celestial bodies that revolves around the sun in elliptical orbits.

plate the basic feature in plate tectonic theory; the upper 50 to 250 km of the earth is divided into large, mobile, but somewhat rigid rock slabs called plates.

plateau a predominantly flat to gently sloping surface at relatively high elevations.

plate tectonics a geologic theory that envisions the earth's outer layer (50–250 km) as being composed of from 10 to 25 large rigid slabs that "float" and move on a more mobile underlayer in the mantle.

Pleistocene the first epoch of the Quaternary Period, occurring after the Pliocene and before the Holocene between 1.6 and 0.01 million years ago; characterized by glacial periods and the rapid evolution of hominids.

Pliocene the epoch of the Tertiary Period after the Miocene, approximately 5 to 1.6 million years ago, characterized by the appearance of distinctly modern plants and animals.

plucking a process of glacial erosion where blocks of rock are loosened, detached, and moved by the freezing and flowing of ice.

plunging fold a fold where the fold axis is inclined from the horizontal.

pluton an igneous intrusion that has cooled at depth.

point bar in a stream channel, the arcuate body of alluvium on the inside of a meander bend.

polymorph minerals having the same chemical composition but varying in atomic arrangement; calcite and aragonite, for example, are both $CaCO_3$.

porosity the property of rock or rock material containing void or open spaces; porosity is the percentage of void space in the overall volume of the solid.

porphyritic an igneous rock texture having two or more distinct crystal sizes.

potassium–argon dating a type of radiometric age dating using the decay of potassium-40 to argon-40.

Precambrian that portion of geologic time prior to the Pha-

nerozoic Eon characterized by the lack of hard-shelled organisms.

precipitation the water (rain, snow, etc.) that falls to the earth from the atmosphere.

pressure the force exerted on a surface divided by the area.

profile the outline produced where a vertical plane intersects the surface of the ground.

Proterozoic the more recent of the two Precambrian Eras, lasting from 2500 to 570 million years ago, characterized by the appearance of very primitive soft-bodied organisms.

pumice a light-colored vesicular volcanic glass of rhyolitic composition.

P-wave a seismic body wave that travels in the direction of propagation by compression and extension of the solid; the fastest moving seismic wave.

pyroclastic pertaining to an accumulation of rock material broken by volcanic activity.

Q

quadrangle a rectangular area bounded by parallels of latitude and meridians of longitude, used as a unit in mapping.

quarter section an area of land in the township and range grid system containing 160 acres and equal to a fourth of a section.

quartzite a nonfoliated metamorphic rock composed primarily of quartz.

Quaternary the second period of the Cenozoic Era, beginning approximately 1.6 million years ago and extending to the present.

R

radial drainage pattern a drainage pattern in which streams diverge outward from a high central area.

radioactivity the emission of energetic particles or radiation during the decay of an unstable nucleus.

range the north–south columns of townships in the township and range grid system of the midwestern and western United States; also, a group of mountains.

Recent the present geologic age; Holocene.

recessional moraine an end moraine behind the terminal moraine, formed during the retreat of a glacier.

recrystallization the formation of new crystalline mineral grains in a rock by metamorphic processes.

recurrence interval the average length of time between natural events of a certain type, such as earthquakes or floods.

reef (1) a ridge or moundlike sedimentary rock formed by and composed of organisms, usually corals. (2) a landform composed of a jagged ridge of upturned rock, common in the western states.

regional metamorphism a category of metamorphic environment where extensive volumes of the earth's crust have been altered by elevated temperature and temperature.

relative age the chronological order of geologic events with respect to each other, not to the absolute time scale.

relief the distance between maximum and minimum elevations in an area.

reservoir (1) an accumulation of petroleum in the subsurface. (2) a storage area for water.

retreat a decrease in the length of a glacier.

rhyolite a group of volcanic rocks of felsic composition characterized by abundant potassium feldspar and quartz.

Richter scale an arbitrary scale measuring the magnitude of an earthquake, developed by C.F. Richter in 1935.

ridge a long, narrow, elevated part of the surface.

rift zone an elevated area characterized by normal faulting and mafic volcanic activity.

ripple a small ridge formed by the accumulation of sand in transport by water or wind.

river a channel of water on the land with a constant or seasonal flow of water of considerable volume.

roche moutonee a small elongate knob of bedrock, sculpted by a glacier; its long axis is parallel to the direction of ice flow, and its steep end is downstream.

rock (1) any naturally formed accumulation of minerals. (2) popularly, any hard consolidated material from the earth.

rock cycle the sequence by which rocks change to different rock types in the different environments in the crust.

rock type one of the three major groups of rocks, igneous, sedimentary, or metamorphic.

rounded the characteristic of a rock particle having no angular surfaces.

rule of V's five rules describing the pattern of a bed eroded by a stream.

runoff water moving on the earth's surface in channels or sheets.

S

sag pond a small depression in the earth's crust along and within fault zones, filled with water by the surrounding drainage.

saline a natural accumulation of salts.

saltation the process in which particles are transported by bouncing off the surface; common in water and wind transport.

salt dome a dome-shaped accumulation of salt that has risen above its source bed due to density differences with the surrounding sediments.

sand any sedimentary fragment with a diameter between 1/16 and 2 mm.

sand dune an accumulation of sand formed by the wind.

sandstone a sedimentary rock consisting primarily of sand-sized clasts.

saturated a condition where fluids occupy the void spaces in a rock.

scale the proportion between linear distance on a map and the true distance on the surface it represents.

scarp a linear arrangement of hills or cliffs formed by erosion or faulting.

schist a medium- to coarse-grained foliated metamorphic rock.

schistosity the parallel planar arrangement of platy, tabular, or prismatic minerals.

scouring the process of erosion by flowing ice, water, or wind.

seafloor spreading the hypothesis that the basaltic ocean floor is created at the ocean ridges and spreads laterally away from them by further volcanism.

seamount a sharp mountain-shaped rise in the surface of the seafloor; seamounts rise approximately 1000 m above the ocean floor.

section one of the 36 one-mile-square areas in a township in the township and range grid system.

sediment the unconsolidated fragmental material in the process of being transported on the earth's surface.

sedimentary rock all rock formed at or near the surface of the earth through a nonmagmatic process.

seismic pertaining to the movement of the earth's crust, especially during an earthquake.

seismogram the record made by a seismograph.

seismograph the instrument that records vibrations of the earth.

seismology the study of the earth's internal structure and movement.

semiarid a type of climate where precipitation is between 10 and 20 inches (25–50 cm) per year; grasses are the predominant semiarid vegetation.

shale a claystone having parallel laminations.

shear (1) the crushing and scratching of rocks sliding past each other. (2) deformation in which two parts of a body slide past each other in direction parallel to their plane of contact.

sheet any tabular body of rock, especially an intrusive igneous body.

shield volcano a volcano characterized by low slopes and basaltic lava flows.

silicate a compound whose crystal lattice contains the SiO_4 tetrahedra.

sill a tabular igneous body parallel to the structure of the surrounding rock.

silt rock fragments between 1/256 and 1/16 mm in diameter; silt combined with clay forms mud.

Silurian a period of the lower Paleozoic Era, between 440 and 410 million years ago.

sinkhole a depression resulting from groundwater erosion beneath the surface.

sinuosity the ratio of stream channel length to down-valley length.

slate a compact, clay-rich, foliated metamorphic rock.

slide shortened form of *landslide*.

slope the angle or surface of incline of the earth's surface; the gradient.

slump a landslide characterized by a curved slip surface such that a rotarylike motion occurs.

stratovolcano a large, more or less symmetric andesite cone built from a central eruptive vent consisting of lava deposits; also called a composite volcano.

submarine beneath the ocean surface.

submergence the change in relative level of a standing body of water and the adjacent land in which the water level appears to rise.

suspension the process of transport where rock materials are held within the surrounding fluid.

S-wave a seismic body wave in which the motion is perpendicular to the direction of propagation; it does not travel through liquids.

syncline a fold that is concave-up in shape; the youngest layers are exposed at the center.

T

tableland a type of topography where flat slopes occur above steeper slopes.

tabular a planar shape in which two dimensions are much larger than the third.

talus the accumulation of broken rock fragments, generally at the base of a cliff or steep slope.

tarn a small lake or pond within a cirque; a cirque lake.

tectonic pertaining to the forces involved in or the structures related to crustal motions.

tension the state of stress on a body being pulled apart.

terminal moraine the end moraine that marks the farthest advance of a glacier.

terrace a flat narrow area adjacent to a stream, composed of alluvium.

Tertiary the first period in the Cenozoic Era, covering the time span between 66 and 1.6 million years ago.

texture the physical appearance of a rock, including such characteristics as size, shape and arrangement of minerals, crystallinity, and granularity.

thalweg the imaginary line connecting the lowest points in a stream bed.

theory a hypothesis that is well supported by factual evidence.

thermal pertaining to or caused by heat.

thermal metamorphism a type of metamorphism in which heat is the primary or sole agent of change.

thickness the perpendicular distance between the upper and lower boundaries of a rock unit.

thin section a polished thin (0.03 mm thick) slab of rock that is observed optically by a geologist.

thrust fault a dip-slip fault where the block above the fault moves up the dip of the fault relative to the lower block; suggests compression of the crust.

tidal bulge the large undulation in sea level created primarily by the gravitational and centrifugal forces generated by the earth, moon, and sun.

tide the repeating oscillation in water level along shorelines of large standing water bodies.

till unstratified drift deposited directly by a glacier.

tombolo a neck of land composed of wave-deposited sediment that connects an offshore island to the mainland.

topographic map a map illustrating the shapes, dimensions, and arrangements of surface features.

topography the general configuration of the land surface of the earth.

topsoil surface soil.

township a unit of land 6 miles on a side in the grid system developed by the U.S. Public Land Survey, popularly known as the township and range system.

township line one of the imaginary east–west reference lines in the township and range grid system.

trace the line of intersection between the earth's surface and a rock unit.

transform fault a strike-slip type fault present between active plate margins.

transgression the advance of a standing water body over the land surface, resulting from a relative rise in water level with respect to the land.

transpiration the process of water passing into the atmosphere by biological processes.

transportation movement of rock material to lower elevations on the surface.

travel time the time required for an earthquake wave to travel from the focus to a given location.

trellis drainage pattern a stream drainage pattern characterized by parallel main streams intersected at right angles by tributaries.

trend general direction or bearing of a geologic surface.

Triassic the first period in the Mesozoic Era, covering the time between 245 and 210 million years ago.

tributary a stream whose mouth intersects another stream.

tributary glacier a glacier that flows into a larger glacier.

trough with respect to folds, the line connecting the lowest points of a fold; the axis of a syncline.

true north the direction to the geographic north pole.

trunk the principal channel into which smaller channels flow.

tsunami a sea wave produced by earthquakes, landslides, or volcanic eruptions.

tuff a volcanic rock composed of solidified pyroclastic material.

turbidite the deposit from a submarine density flow.

twin an intergrowth of two or more crystals of the same mineral in a certain repeating pattern.

U

ultramafic said of an igneous rock composed primarily of mafic minerals (pyroxene, olivine) with little or no feldspar.

unconformity the physical break in strata representing the nondeposition of erosion of rock material in geologic time.

undulating having a wavy outline or appearance.

uniformitarianism the geologic principle that the processes and natural laws acting at the present time have acted in a similar manner and with the same general intensity throughout geologic time.

uplift an area that has been rising relative to the surface around it.

upstream in the direction of higher elevation along a stream.

U-shaped valley a valley having a wide valley floor and steep side.

V

vadose zone the zone of aeration; that volume of the crust occupied by gases.

valley an elongate area bordered by areas of higher elevation.

valley glacier an alpine glacier.

varve a thin layer of fine sediment deposited in a still body of water within a year's time, usually in groups representing many years.

vein a thin filling or intrusion in a rock.

velocity the rate and direction of movement, a vector quantity.

ventifact a stone eroded or polished from the impact of windblown particles.

vertical exaggeration the technique of making a profile in which the vertical scale is increased relative to the horizontal to accentuate the vertical dimension of the surface.

vesicle a hole or cavity in volcanic rocks from the entrapment of gas in the cooling lava.

viscosity the property of internal resistance to flow of a substance.

vitreous having a glasslike luster.

volatile component the gaseous material derived from a magma, generally water or carbon dioxide.

volcano an opening in the surface of the earth from which magma, gas, and ash erupt.

volcanology the branch of geology that deals with the materials, activity, and causes of magma rising, erupting, and solidifying on the earth's surface.

V-shaped valley a valley characterized by a narrow floor and steeply inclined sides.

W

wall the rock mass on a given side of a fault; the hanging wall or footwall.

wallrock the rock mass enclosing a vein or intrusion.

water table the boundary between the zone of saturation and the zone of aeration in the groundwater system.

wave-cut any landform eroded by waves.

wave refraction the process by which waves are bent laterally toward shallower areas near the shoreline because of friction of the wave against the bottom.

waxy a type of luster having the soft quality of wax.

weathering the alteration of rock due to exposure to the atmosphere.

wildcat well an exploratory well, usually for oil.

windward the side facing the wind.

X

xenolith a foreign rock occurring as an inclusion in an igneous rock.

x-ray a type of electromagnetic radiation of very short wavelengths used by geologists to identify internal structures of rocks and minerals.

Z

zenith the point in the sky directly overhead.

zone of aeration the generally tabular area beneath the surface where the void spaces are filled with gas and water.

zone of saturation the area beneath the surface where the void spaces within the rock are filled entirely with water.

Color Plates

MINERALS

Color Plate 1. Well-formed crystals of halite (bottom), quartz (upper left), and pyrite (upper right). Pyrite has a distinctive metallic luster as a result of its chemical composition and bonding. Notice the angles between crystal faces of these minerals.

Color Plate 2. Albite twinning in a large crystal of plagioclase, the white mineral; the gray translucent mineral is quartz. The straight parallel lines of dark and light running along the cleavage face of this plagioclase crystal result from reflections of the structure in a single crystal as it crystallizes.

Color Plate 3. Perthitic structure in feldspar. Notice the difference between perthitic structure and albite twinning. Twin lines are straight and parallel, but perthitic structure consists of light and dark (or translucent and opaque) wavy subparallel intergrowths of two feldspars.

Color Plate 4. Cleavage in calcite. Calcite has excellent cleavage in three directions at 120°, resulting in three intersecting sets of smooth faces.

PLUTONIC ROCKS

Color Plate 5. Biotite granite, 55 percent pinkish-tan potassium feldspar, 15 percent white plagioclase, 10 percent black biotite, 20 percent light gray translucent quartz with typical vitreous luster. Thin section (crossed polarizers) shows brightly colored biotite grains and white to dark gray grains of quartz, plagioclase, and potassium feldspar. The "plaid" pattern in a few grains is typical of microcline, a mixed potassium–sodium feldspar.

Color Plate 6. Granite. Red potassium feldspar is about 60 percent of the rock. Note the vitreous luster of light gray translucent quartz.

Color Plate 7. Porphyritic biotite granodiorite. The large rectangular pinkish crystals are potassium feldspar phenocrysts that show the typical tabular habit of feldspars. Note the larger proportion of dark minerals (15 percent to 20 percent) as compared with granite. The thin section (crossed polarizers) shows white to dark gray quartz and potassium feldspar (not

a phenocryst), albite twinning in plagioclase at lower left, and brown biotite.

Color Plate 8. Diorite. The percentage of dark minerals (primarily hornblende) is 30 to 40 percent; the white mineral is plagioclase (about 60 percent); quartz is absent. The thin section in plane light (top left) shows green to brown hornblende crossed by multiple closely spaced lines; these are cleavages running the length of the crystals. Two small olive-green grains at the extreme right center are views of the ends of hornblende crystals, showing the two amphibole cleavages at 120 degrees. The large brown grain at the upper left is biotite; the white area is plagioclase. The thin section with crossed polarizers reveals the typical bright colors of mafic minerals; the grains with black and white "stripes" are plagioclase with albite twinning.

Color Plate 9. Olivine gabbro. Dark gray plagioclase (note albite twinning in large tabular grain) constitutes from 50 to 60 percent of this rock. Some olivine and pyroxene are also present. The thin section (crossed polarizers) shows dramatic albite twinning and the tabular habit of plagioclase that crystallized early. The bright-colored grains with no consistent cleavage direction are olivine crystals. A few colored grains have parallel cleavage lines; these are pyroxene. The red pyroxene grain at the bottom right and the brown pyroxene at top center (below the orange olivine) both show cleavage traces at 90 degrees to each other, the typical cleavage of pyroxene.

Color Plate 10. Gabbro. Here the mineral composition is pure plagioclase in the type of gabbro called anorthosite. Note the albite twinning in one of the grains. The thin section (crossed polarizers) shows very coarse albite twinning.

Color Plate 11. Peridotite. Green olivine and green-black pyroxene make up this ultramafic rock.

VOLCANIC ROCKS

Color Plate 12. Biotite rhyolite. Note the well-formed flakes of biotite (compare with Figure 1.4D) suspended in an aphanitic matrix. The thin section (plane light) reveals thin plates of biotite (brown) and well-formed six-sided quartz phenocrysts (white) suspended in a very fine crystalline matrix.

Color Plate 13. Dacite. Phenocrysts of plagioclase, quartz, potassium feldspar, and hornblende.

Color Plate 14. Andesite. Like many andesites, about half of this rock is phenocrysts. Here the tabular crystal habit of plagioclase is well expressed. Dark phenocrysts are hornblende.

Color Plate 15. Hornblende andesite. The typical prismatic habit of hornblende is particularly well developed in this rock. Small phenocrysts of plagioclase (white) are also present. The thin section (crossed polarizers) shows a coarse needle of hornblende. Near the middle of the needle, at its left edge, is an end view of another hornblende crystal that shows two cleavage traces at 120 degrees. The matrix consists largely of plagioclase. Small phenocrysts of plagioclase are also present.

Color Plate 16. Vesicular olivine basalt. Equant olivine crystals (green) and a few small plagioclase phenocrysts are visible in a vesicular aphanitic matrix. The thin section (crossed polarizers) shows brightly colored olivine crystals and small plagioclase phenocrysts in a fine crystalline matrix of plagioclase and olivine.

Color Plate 17. Limburgite. This rock is a rather unusual mafic rock, not included on your rock chart. The purpose for including it here is to show the stubby prismatic crystal habit of pyroxene as it appears in a matrix of mafic glass.

Color Plate 18. Pumice. Notice the highly vesicular nature of this felsic glass. Some pumice is so full of vesicles it is light enough to float. The thin section (plane light) shows rounded vesicles.

Color Plate 19. Scoria. Vesicular mafic glass.

Color Plate 20. Obsidian. This felsic glass is dark because of submicroscopic particles of magnetite uniformly dispersed in the glass. Note the conchoidal fracture.

Color Plate 21. Rhyolite tuff. The dark patches in this fragmental textured rock are not vesicles but fragments of pumice and other rocks. Some small phenocrysts of quartz, potassium feldspar, and biotite are also present. This rock, the Bishop tuff, formed as an ash-fall deposit from a catastrophic volcanic explosion in east-central California. The ash fall covered a wide area of the western and central United States and is widely used as a stratigraphic marker bed to correlate rock layers from place to place.

Color Plate 22. Tuff. This rock is called a welded tuff because the material was pressed together while still hot, resulting in extreme compression of dark fragments of volcanic glass. The thin section (plane light) shows these compressional "flame" structures. Phenocrysts are white areas.

VOLCANIC STRUCTURES

Color Plate 23. Columnar basalt. Thick flows of basalt often form prismatic joints perpendicular to the flow surface as they cool and contract. In some cases well-defined six-sided columns result. Devil's Postpile in California and Devil's Tower in Wyoming display spectacular columnar jointing.

Color Plate 24. Aa surface. *Aa* is one of the few Hawaiian words in the scientific vocabulary (along with *pahoehoe*); it refers to a jagged, clinkery, blocky surface on a basalt flow.

Color Plate 25. Pahoehoe. Less viscous basalt forms this shiny, ropy, wrinkled surface.

Color Plate 26. Basaltic bombs and spatter. Volcanic bombs are streamlined forms created when blobs of magma (usually basaltic) are ejected from a volcano and cool in mid-air.

SEDIMENTARY ROCKS

Color Plates 27 through 33 show rocks with clastic texture whose origin is clearly detrital. Color Plates 37, 38, 39, 40, and 42 show rocks of biogenic origin with both clastic and interlocking textures. Color plates 41, 43, and 44 are chemical sedimentary rocks with interlocking textures formed by direct precipitation from water. Since they are very fine grained, chalk and chert might also be considered as chemical rocks; their biogenic origin is apparent only with a microscope.

Detrital Rocks

Color Plate 27. Conglomerate. Most of the identifiable clasts are lithic fragments of granule size (2 to 4 mm) and pebble size (4 to 64 mm). There is abundant fine matrix and calcite cement. The thin section (crossed polarizers) shows the rounded clasts clearly as lithic fragments. The pale pink and green color between the grains is the calcite cement.

Color Plate 28. Breccia. The angularity of the granules and pebbles distinguishes this breccia from a conglomerate.

Color Plate 29. Quartz sandstone. The clasts are well-rounded, well-sorted quartz grains in the fine sand size range (1/16 to 1/4 mm). The cement is silica with some iron oxides. The thin section (crossed polarizers) shows the excellent rounding and sorting.

Color Plate 30. Arkosic sandstone (arkose). Sorting is only moderate, and the grains are subangular in this feldspar-rich (pink-grained) sandstone.

Color Plate 31. Immature lithic sandstone. Note the typical dark color and poor sorting of this rock, which may be called graywacke. The thin section (crossed polarizers) shows the angularity of the clasts and the abundant clay matrix. A large number of the clasts are fragments of other rocks, as shown by smaller crystals within the clasts.

Color Plate 32. Mudstone. Very fine particles (<1/16 mm) make up this massive (nonfissile) mudstone. The thin section (plane light) shows abundant clay with some larger grains (mostly quartz).

Color Plate 33. Shale. Contrast this fissile mudstone (shale) with the massive mudstone in Color Plate 32. The particle size is approximately the same; only their arrangement differs.

Sediment

Color Plate 34. Pure quartz sand. The extraordinary degree of sorting and rounding of grains is typical of wind-worked sediments.

Color Plate 35. Fine to coarse sand. Sorting is only moderate, and the grains are angular in this sediment. Composition is mixed; quartz, mica flakes, and light-colored opaque feldspar predominate.

Color Plate 36. Poorly sorted coarse sand to pebble-sized sediment. Many lithic fragments are visible.

Varieties of Limestone

Color Plate 37. Coquina. In this clastic textured limestone, coarse shell debris is cemented by calcite. Clearly this rock has a detrital origin as well as a biogenic origin. The thin section (plane light) shows faint grains of calcite cement between shells as well as many voids.

Color Plate 38. Fossiliferous limestone. Fossils are still clearly visible in this rock, but sufficient recrystallization has occurred that the rock's texture is interlocking rather than clastic. The photograph shows a drop of hydrochloric acid fizzing on the surface of the calcite. The thin section (crossed polarizers) shows many shells in a recrystallized calcite matrix.

Color Plate 39. Chalk. Microscopic organisms form a porous, very fine-grained limestone. Note the earthy luster. The thin section (plane light) shows various foraminifera pressed together.

Color Plate 40. Oolitic limestone. Spherical calcite concretions (cemented with calcite) give this rock the appearance of a well-sorted, well-rounded coarse sandstone. Oolites form when tiny particles of silt or shell are rolled back and forth by tides and waves in a calcite-rich ooze in shallow seawater. The thin sec-

tion (plane light) shows oolites as concentric rings of calcite (with radial structure as well) around a central core.

Color Plate 41. Travertine. The banded appearance of this chemical limestone is caused by thin layers of calcite precipitating from water flowing over the surface.

Chemical Rocks

Color Plate 42. Chert. Note the waxy luster of this siliceous rock. Chert may form as a chemical precipitate from silica-saturated seawater or from the accumulation of radiolaria (or, less commonly, diatoms), microscopic marine organisms.

Color Plate 43. Rock salt. Coarsely crystalline halite shows cubic crystal form and cleavage in three directions at right angles.

Color Plate 44. Rock gypsum. The variety of gypsum called selenite is translucent and displays clearly the oblique cleavage of gypsum.

SEDIMENTARY STRUCTURES

Color Plate 45. Crossbedding. Layers of sediment that meet at an angle indicate a change in the wind or water current that deposited the sediment. Wind crossbeds (inset) are typically several feet high, whereas water-laid crossbeds are typically several inches high.

Color Plate 46. Graded bedding. This outcrop of graded bedding shows granules and fine sand.

Color Plate 47. Ripple marks. These are symmetric ripple marks, formed by oscillating water direction, typically in a wave zone.

Color Plate 48. Mud cracks. This specimen consists of fine sediments that washed into mud cracks and subsequently lithified.

METAMORPHIC ROCKS

Color Plates 49 through 57 show foliated metamorphic rocks; Color Plates 58 and 59 show nonfoliated rocks; Color Plate 60 shows a variety of serpentine rocks, some foliated and others nonfoliated.

Color Plate 49. Slate. This rock represents the first stage of metamorphism. Some of the clay minerals have begun to recrystallize into larger flakes, and there is a definite alignment of these sheetlike minerals in response to directed pressure. Compare the thin section of slate (plane light) with that of mudstone, its nonfoliated precursor (Color Plate 32).

Color Plate 50.   Phyllite. The recrystallization of clay minerals is more extensive, and the rock has acquired a shinier luster. The thin section (plane light) shows that some mica has formed. See also Color Plate 62.

Color Plate 51.   Muscovite garnet schist. Recrystallization from clay into mica is complete, and some quartz has begun to form. Most of this rock is muscovite mica with small garnets (dark equant grains about 5 mm across); minor quartz is present. The thin section (crossed polarizers) shows the increase in grain size from phyllite. The large black crystal is garnet.

Color Plate 52.   Garnet schist. This schist contains spectacular large garnets displaying well-formed crystal faces.

Color Plate 53.   Amphibole biotite gneiss. Light feldspar and quartz and dark amphibole and biotite have segregated into different areas in this banded rock. The thin section (crossed polarizers) shows this segregation well. Note the larger crystal size in the gneiss as compared with the schist.

Color Plate 54.   Augen gneiss. *Augen*, the German word for eye, refers to the shape of the large pink clots of feldspar. Some augen gneiss, formed in very high-pressure shear zones, is considered to be a cataclasite or mylonite.

Color Plate 55.   Chlorite schist thin section (plane light) shows the alignment of sheety chlorite grains and the folding by directed pressures. Since this section was photographed with uncrossed polarizers, it suggests the true color of chlorite.

Color Plate 56.   Hornblende amphibolite. This lineated (foliated) rock is 90 percent hornblende because the original parent rock was mafic, probably a basalt. The thin section (crossed polarizers) shows aligned hornblende needles. The gray crystals are calcic plagioclase; note the rare albite twinning in some grains.

Color Plate 57.   Garnet glaucophane schist. This schist is produced in high-pressure regional metamorphism of basalt. The section at top left shows glaucophane in plane light, suggesting the true color of the mineral. A well-formed six-sided crystal of garnet is pale red-brown. On the right is the same view with crossed polarizers, where garnets now appear black.

Color Plate 58.   Quartzite. Virtually pure quartz grains have been sutured together. Compare the thin section (crossed polarizers) of this quartzite with that of the pure quartz sandstone, Color Plate 29.

Color Plate 59.   Marble. The black streaks in this white marble are carbon impurities. The pale pink to green colors in the thin section with crossed polarizers are typical of calcite (see Color Plates 27 and 38). Notice the twin striations cutting obliquely across the crystals.

Color Plate 60.   Serpentine varieties. Dark crystalline serpentine is a metamorphosed mafic material. The pale aqua and pale green serpentines were emplaced in the upper crust along fault zones by high pressure. The yellow-green serpentine is cut by streaks of asbestos, another variety of serpentine. Note the silky luster of asbestos.

MISCELLANEOUS SAMPLES

Color Plate 61.   This slab of banded Dakota sandstone shows small "left lateral" offsets in the colored layers due to faulting.

Color Plate 62.   This phyllite is deformed into tight "chevron" folds by compression.

Color Plate 63.   A freshly broken cobble of diorite shows the depth to which weathering has proceeded.

Color Plate 64.   Bands of tan sandstone and gray mudstone show different susceptibilities to weathering.

Color Plate 65.   These soil samples were collected in the western foothills of the Sierra Nevada, northern California.

Color Plate 66.   Limestone bedrock has been gouged in several different directions by rock particles carried along the bottoms of glaciers that moved in slightly different directions.

Color Plate 67.   Granodiorite has been scratched by coarse particles and polished by very fine rock dust carried at the bottom of an alpine glacier.

Color Plate 68.   The polished facets of these ventifacts are produced when rocks are abraded by fine sand carried by the wind.

Color Plate 1. Halite (bottom); quartz (upper left); pyrite (upper right).

Color Plate 2. Albite twinning in plagioclase (white mineral); the gray mineral (vitreous luster) is quartz.

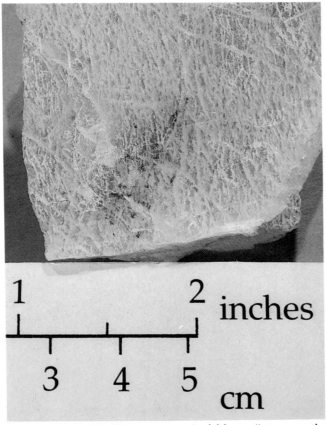

Color Plate 3. Perthitic structure in feldspar (intergrowth of sodium-rich and potassium-rich phases).

Color Plate 4. Cleavage in calcite, three directions at 120 degrees to each other.

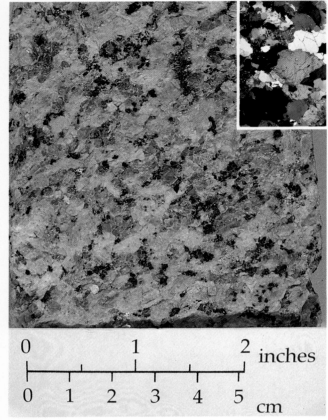

0   1   2 inches

0   1   2   3   4   5   cm

Color Plate 5. Biotite granite (potassium feldspar, plagioclase, quartz, biotite). Thin section, 20X, crossed.

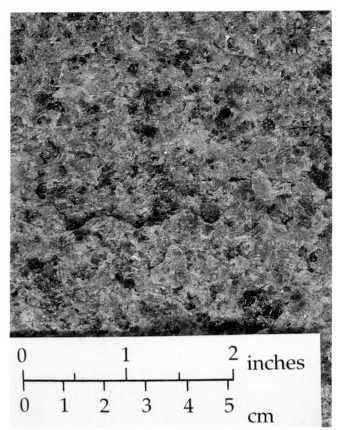

0   1   2 inches

0   1   2   3   4   5   cm

Color Plate 6. Granite (red potassium feldspar = 60 percent, gray translucent quartz = 20 percent).

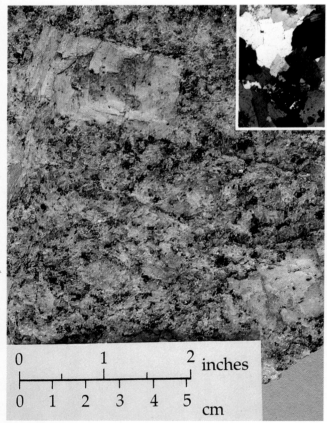

0   1   2 inches

0   1   2   3   4   5   cm

Color Plate 7. Porphyritic biotite granodiorite (potassium feldspar phenocrysts). Thin section, 20X, crossed.

0   1   2 inches

0   1   2   3   4   5   cm

Color Plate 8. Hornblende diorite. Thin sections, 20X, plane light (top left) and crossed polarizers (top right).

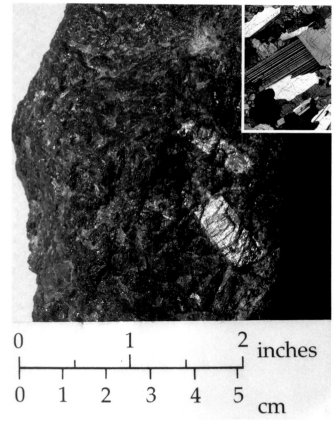

0   1   2   inches

0   1   2   3   4   5   cm

Color Plate 9.   Olivine gabbro. Thin section 20X, crossed polarizers.

0   1   2   inches

0   1   2   3   4   5   cm

Color Plate 10.   Gabbro (anorthosite; albite twinning in plagioclase crystal). Thin section, 20X, crossed polarizers.

0   1   2   inches

0   1   2   3   4   5   cm

Color Plate 11.   Peridotite (olivine and pyroxene).

0   1   2   inches

0   1   2   3   4   5   cm

Color Plate 12.   Biotite rhyolite. Thin section, 20X, plane light.

0   1   2   inches

0   1   2   3   4   5   cm

Color Plate 13.   Dacite.

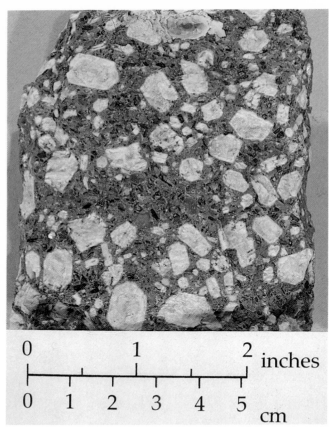

0   1   2   inches

0   1   2   3   4   5   cm

Color Plate 14.   Andesite (tabular plagioclase phenocrysts).

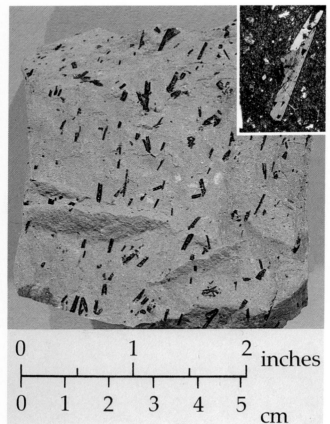

0   1   2   inches

0   1   2   3   4   5   cm

Color Plate 15.   Hornblende andesite (prismatic hornblende phenocrysts). Thin section, 20X, crossed polarizers.

0   1   2

0   1   2   3   4   5

Color Plate 16.   Vesicular olivine basalt (green olivine phenocrysts). Thin section, 20X, crossed polarizers.

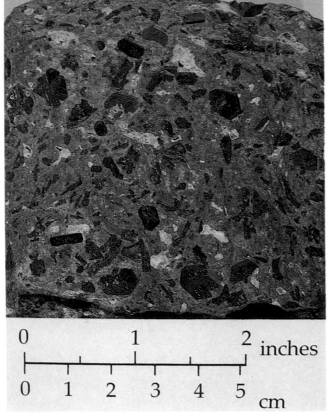

Color Plate 17. Limburgite displaying stubby prismatic habit of pyroxene.

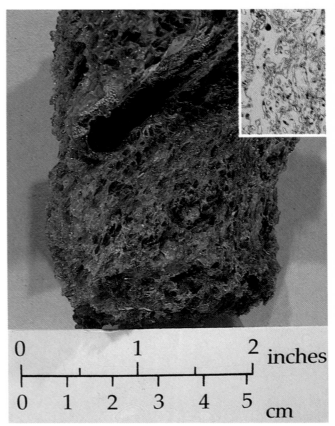

Color Plate 18. Pumice (highly vesicular felsic glass). Thin section, 100X.

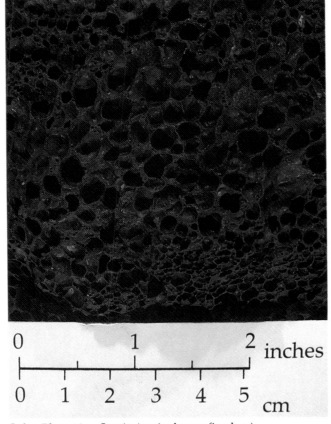

Color Plate 19. Scoria (vesicular mafic glass).

Color Plate 20. Obsidian (felsic glass).

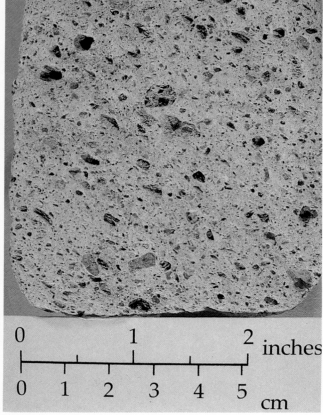

Color Plate 21. Rhyolite tuff (fragmental texture; phenocrysts of quartz, potassium feldspar, biotite).

Color Plate 22. Welded tuff. Thin section, 20X, plane light.

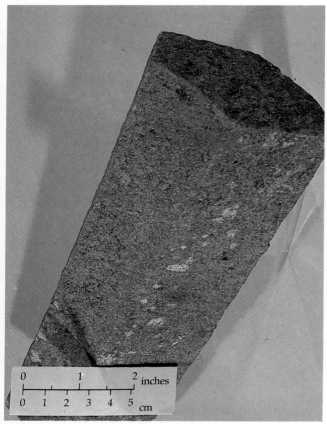

Color Plate 23. Columnar basalt. Prismatic joints have formed perpendicular to flow surface.

Color Plate 24. Aa (jagged, clinkery surface on basalt flow).

Color Plate 25. Pahoehoe (shiny, ropy, wrinkled surface on basalt flow).

Color Plate 26. Bombs and spatter, streamlined forms created when blobs of magma solidify in mid-air.

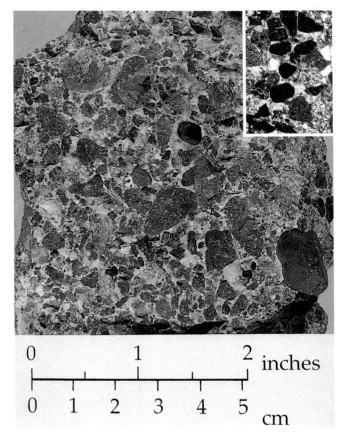

Color Plate 27. Conglomerate (lithic fragments in fine matrix). Thin section, 20**X**, crossed polarizers.

Color Plate 28. Breccia.

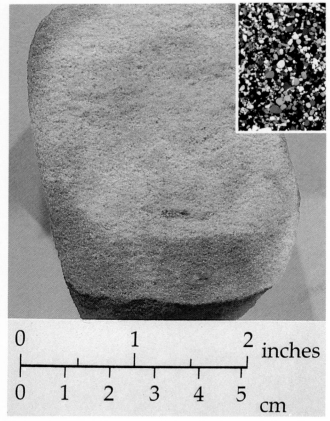

Color Plate 29. Quartz sandstone (well-rounded, well-sorted quartz grains). Thin section, 20**X**.

Color Plate 30. Arkosic sandstone (subangular grains, moderate sorting, feldspar-rich).

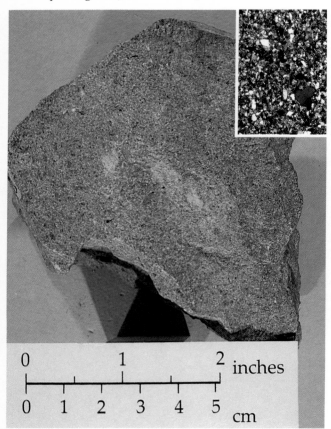

Color Plate 31. Immature lithic sandstone, or graywacke (angular, poorly sorted grains). T.s., 20**X**, crossed polarizers.

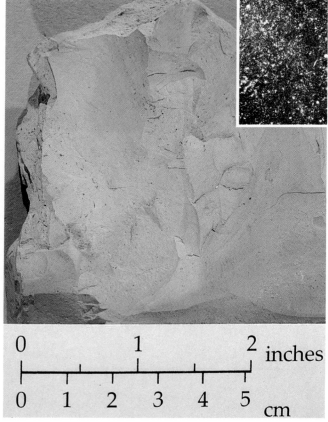

Color Plate 32. Mudstone (clay- and silt-sized grains). Thin section, 100**X**, plane light.

Color Plate 33.   Shale (fissile mudstone).

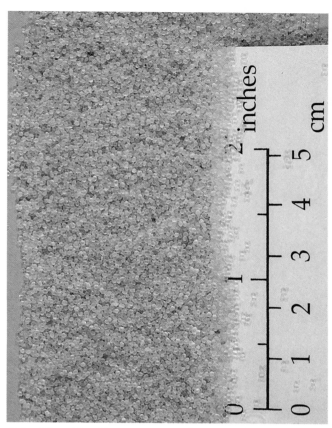

Color Plate 34.   Pure quartz sand (extremely well-sorted, well-rounded grains of quartz).

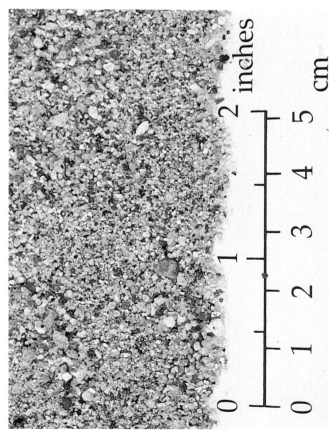

Color Plate 35.   Fine to coarse sand, moderate sorting, angular grains.

Color Plate 36.   Coarse sand- to pebble-sized sediment.

Color Plate 37. Coquina (clastic limestone). Thin section, 20**X**, plane light.

Color Plate 38. Fossiliferous limestone (interlocking texture). Thin section, 20**X**, crossed polarizers.

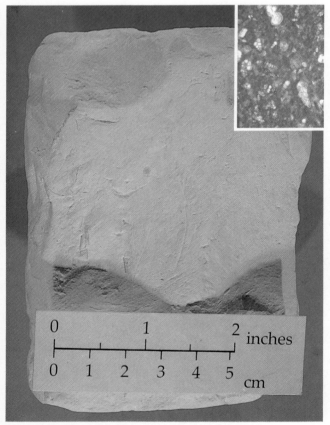

Color Plate 39. Chalk (foraminiferous limestone). Thin section, 100**X**, plane light.

Color Plate 40. Oolitic limestone. Thin section, 20**X**, plane light.

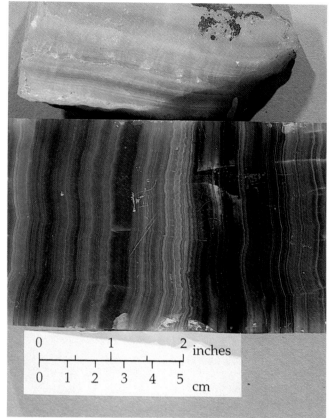

Color Plate 41. Travertine (interlocking texture, chemical origin).

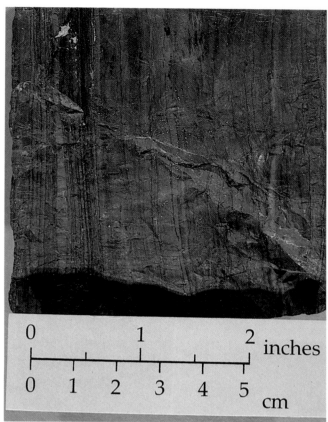

Color Plate 42. Chert (siliceous rock, may be either chemical or biogenic).

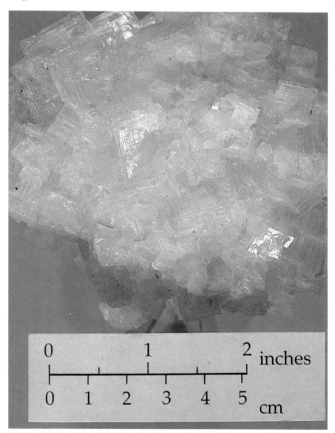

Color Plate 43. Rock salt (note the cubic crystal form of halite).

Color Plate 44. Rock gypsum (note the oblique angles between cleavages).

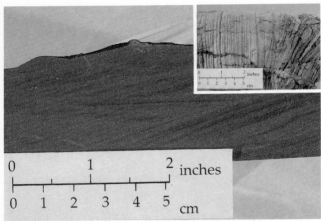

Color Plate 45. Cross bedding, water (dark color) and wind (inset, light color).

Color Plate 46. Graded bedding—granules and fine sand.

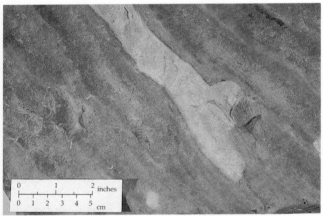

Color Plate 47. Ripple marks—symmetric ripples formed in wave zone.

Color Plate 48. Lithified sediments from mud cracks.

Color Plate 49. Slate (fine-grained foliated rock). Thin section, 100X, plane light.

Color Plate 50. Phyllite. Thin section, 20X, plane light.

Color Plate 51. Muscovite garnet schist (muscovite mica, quartz, and garnet). Thin section, 20X.

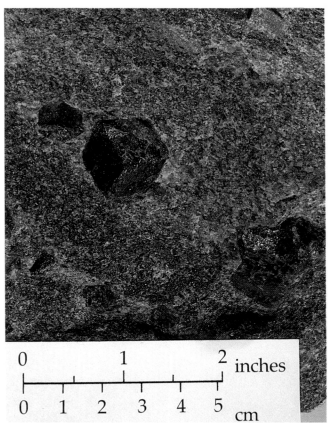

Color Plate 52. Garnet schist (very large, well-formed garnets).

Color Plate 53. Amphibole biotite gneiss (coarse grained, foliated). Thin section, 20X.

Color Plate 54. Augen gneiss—"eyes" of feldspar in foliated matrix.

Color Plate 55. Chlorite schist. Thin section with plane light, 20X.

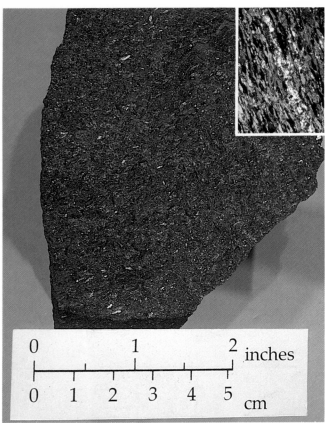

Color Plate 56. Hornblende amphibolite (90 percent hornblende). Thin section, 20X, crossed polarizers.

Color Plate 57. Garnet glaucophane schist. Thin section with plane light (top left) and crossed polarizers (right).

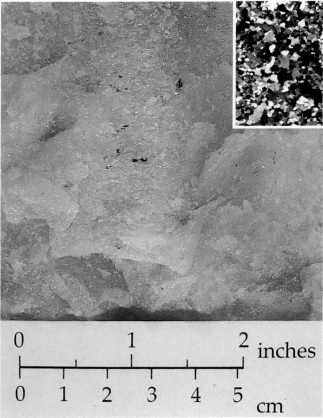

Color Plate 58. Quartzite (metamorphosed quartz sandstone). Thin section, 20X, crossed polarizers.

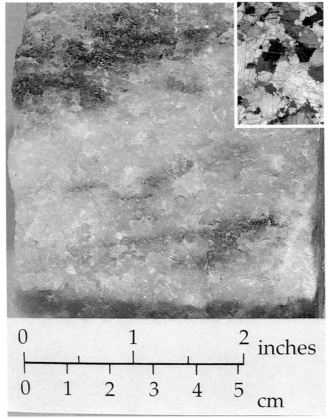

Color Plate 59. Marble (metamorphosed limestone). Thin section, crossed polarizers, 20X, shows twinning in calcite.

Color Plate 60. Serpentine varieties.

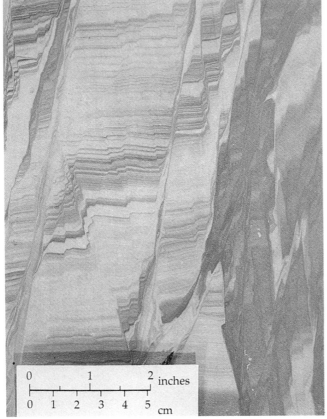

Color Plate 61. Small faults in banded sandstone.

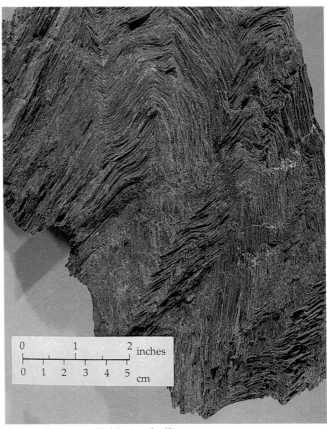

Color Plate 62. Folds in phyllite.

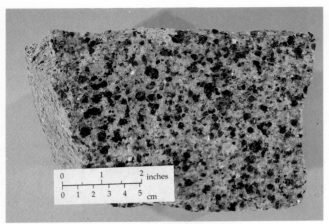

Color Plate 63. Weathering rind in diorite.

Color Plate 64. Differential weathering in layered sandstone and shale .

Color Plate 65. Soil samples taken in the Sierra Nevada (California).

Color Plate 66. Glacial gouges in limestone from glaciers moving in different directions.

Color Plate 67. Glacial polish and scratches in granodiorite.

Color Plate 68. Ventifacts, polished by fine sand.

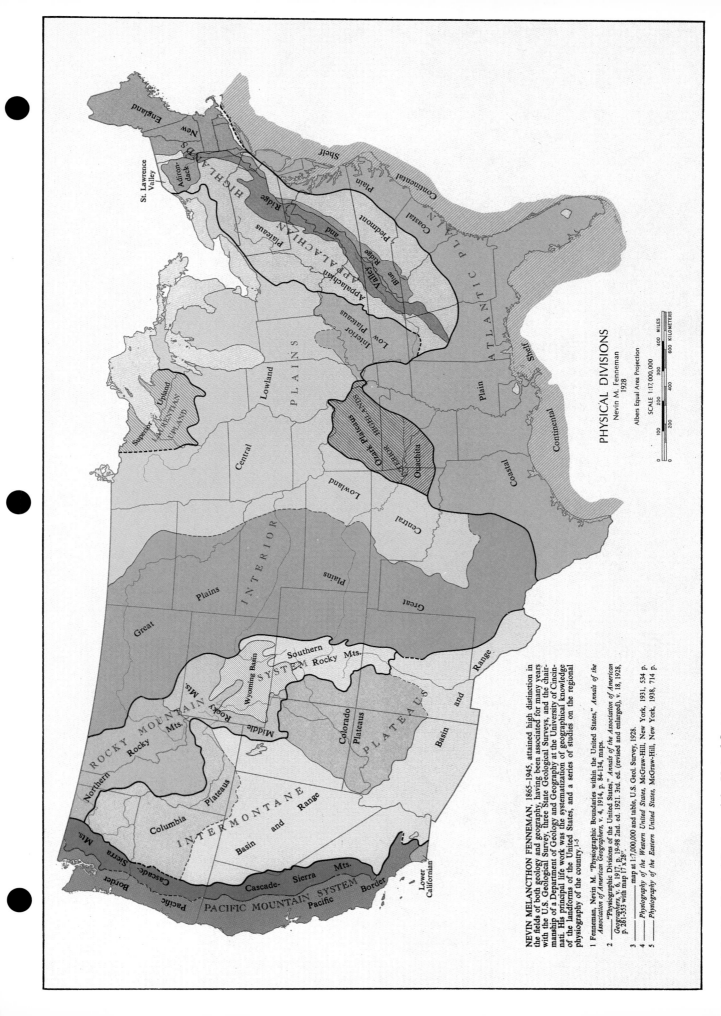

PHYSICAL DIVISIONS

Nevin M. Fenneman

1928

Albers Equal Area Projection

SCALE 1:17,000,000

NEVIN MELANCTHON FENNEMAN, 1865–1945, attained high distinction in the fields of both geology and geography, having been associated for many years with the U.S. Geological Survey, three State Geological Surveys, and the chairmanship of a Department of Geology and Geography at the University of Cincinnati. His principal life work was the systematization of geographical knowledge of the landforms of the United States, and a series of studies on the regional physiography of the country.[1-5]

1 Fenneman, Nevin M. "Physiographic Boundaries within the United States," *Annals of the Association of American Geographers*, v. 4, 1914, p. 84-134, maps.

2 "Physiographic Divisions of the United States," *Annals of the Association of American Geographers*, v. 6, 1917, p. 19-98 2nd. ed. 1921. 3rd. ed. (revised and enlarged), v. 18, 1928, p. 261-353 with map 17 x 28".

3   map at 1:7,000,000 and table, U.S. Geol. Survey, 1928.

4 ——— *Physiography of the Western United States*, McGraw-Hill, New York, 1931, 534 p.

5 ——— *Physiography of the Eastern United States*, McGraw-Hill, New York, 1938, 714 p.

Color Map 1.   Physical Regions of the United States.

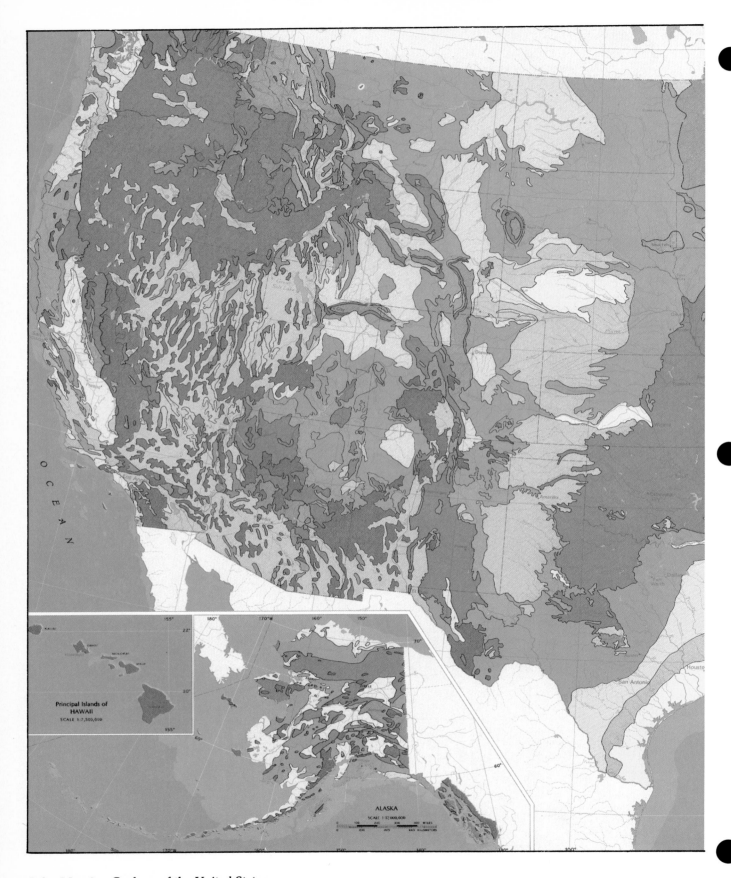

Color Map 2. Geology of the United States.

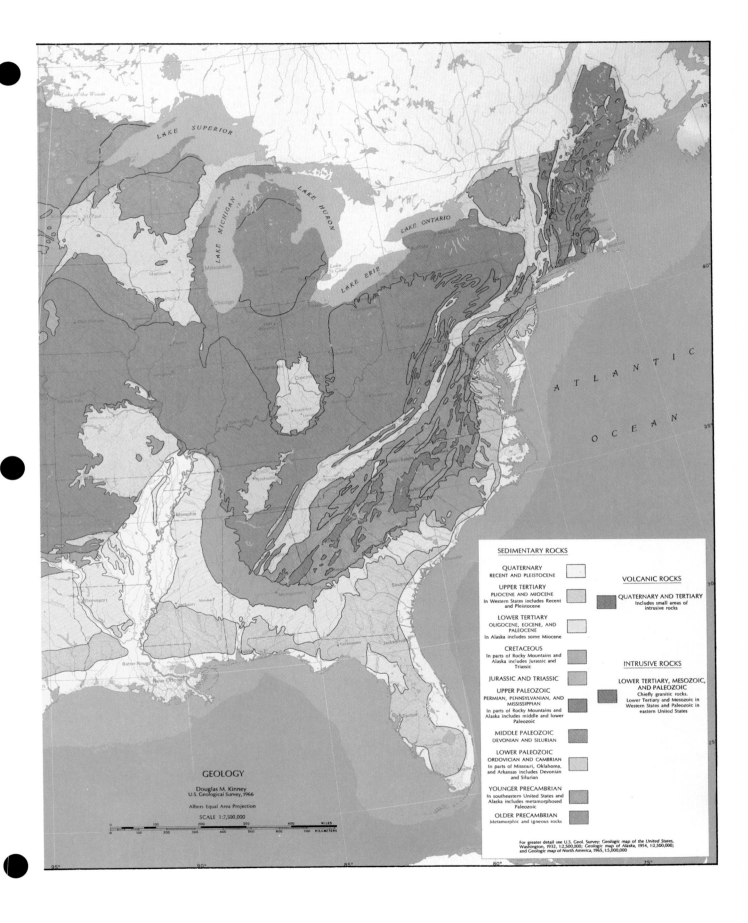

SEDIMENTARY ROCKS

QUATERNARY
RECENT AND PLEISTOCENE

UPPER TERTIARY
PLIOCENE AND MIOCENE
In Western States includes Recent
and Pleistocene

LOWER TERTIARY
OLIGOCENE, EOCENE, AND
PALEOCENE
In Alaska includes some Miocene

CRETACEOUS
In parts of Rocky Mountains and
Alaska includes Jurassic and
Triassic

JURASSIC AND TRIASSIC

UPPER PALEOZOIC
PERMIAN, PENNSYLVANIAN, AND
MISSISSIPPIAN
In parts of Rocky Mountains and
Alaska includes middle and lower
Paleozoic

MIDDLE PALEOZOIC
DEVONIAN AND SILURIAN

LOWER PALEOZOIC
ORDOVICIAN AND CAMBRIAN
In parts of Missouri, Oklahoma,
and Arkansas includes Devonian
and Silurian

YOUNGER PRECAMBRIAN
In southeastern United States and
Alaska includes metamorphosed
Paleozoic

OLDER PRECAMBRIAN
Metamorphic and igneous rocks

VOLCANIC ROCKS

QUATERNARY AND TERTIARY
Includes small areas of
intrusive rocks

INTRUSIVE ROCKS

LOWER TERTIARY, MESOZOIC,
AND PALEOZOIC
Chiefly granitic rocks.
Lower Tertiary and Mesozoic in
Western States and Paleozoic in
eastern United States

GEOLOGY

Douglas M. Kinney
U.S. Geological Survey, 1966

Albers Equal Area Projection

SCALE 1:7,500,000

MILES

KILOMETERS

For greater detail use U.S. Geol. Survey: Geologic map of the United States,
Washington, 1932, 1:2,500,000; Geologic map of Alaska, 1954, 1:2,500,000;
and Geologic map of North America, 1965, 1:5,000,000

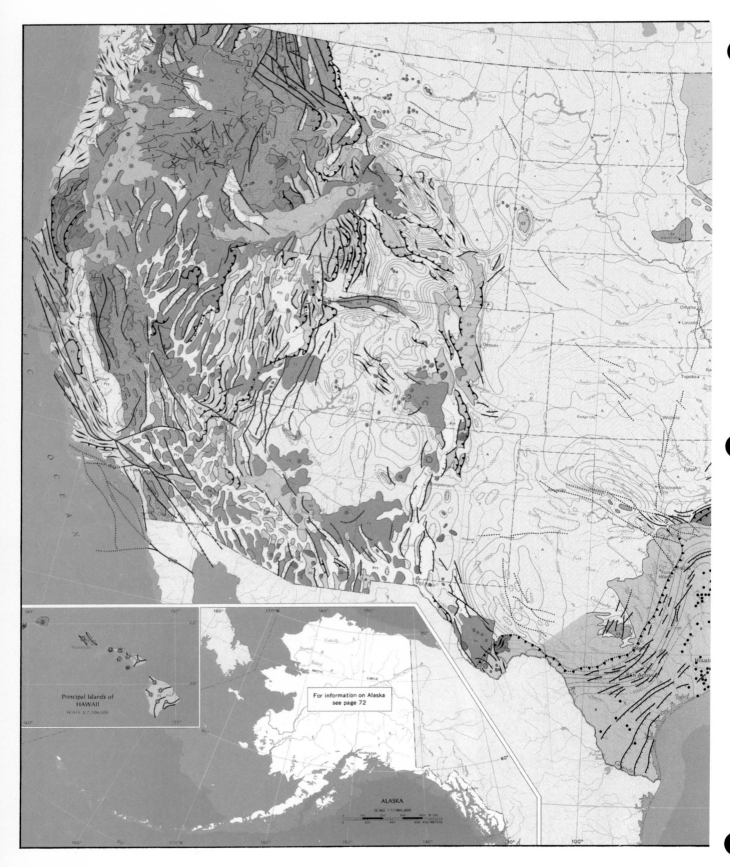

Color Map 3. Structure Map of the United States.

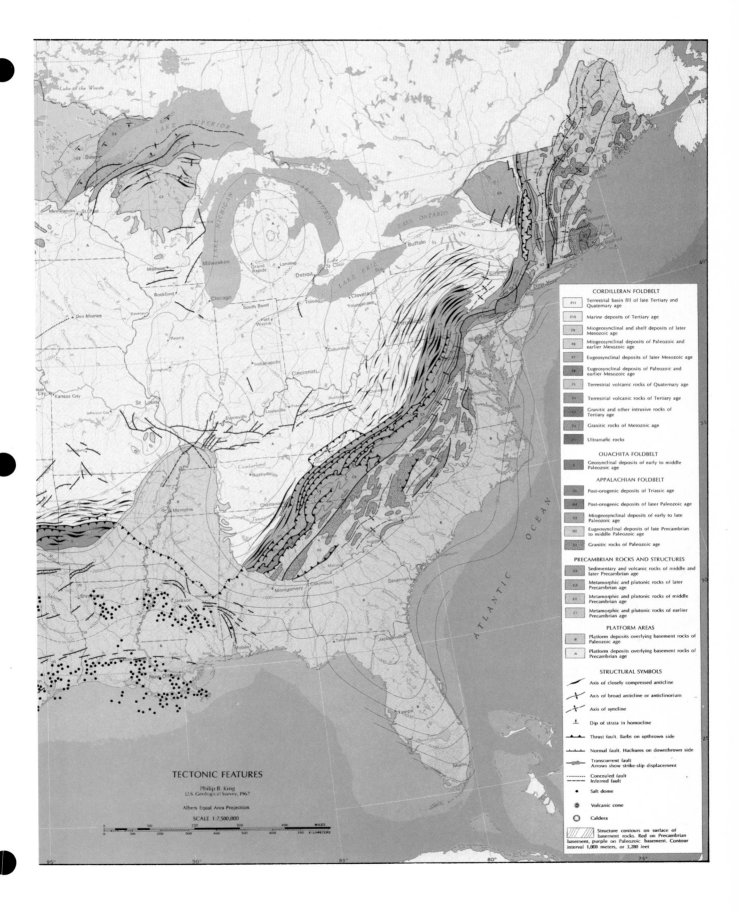

TECTONIC FEATURES

Philip B. King
U.S. Geological Survey, 1967

Albers Equal Area Projection

SCALE 1:7,500,000

CORDILLERAN FOLDBELT

- F11 Terrestrial basin fill of late Tertiary and Quaternary age
- F10 Marine deposits of Tertiary age
- F9 Miogeosynclinal and shelf deposits of later Mesozoic age
- F8 Miogeosynclinal deposits of Paleozoic and earlier Mesozoic age
- F7 Eugeosynclinal deposits of later Mesozoic age
- F6 Eugeosynclinal deposits of Paleozoic and earlier Mesozoic age
- F5 Terrestrial volcanic rocks of Quaternary age
- F4 Terrestrial volcanic rocks of Tertiary age
- F3 Granitic and other intrusive rocks of Tertiary age
- F2 Granitic rocks of Mesozoic age
- F1 Ultramafic rocks

OUACHITA FOLDBELT

- E Geosynclinal deposits of early to middle Paleozoic age

APPALACHIAN FOLDBELT

- D5 Post-orogenic deposits of Triassic age
- D4 Post-orogenic deposits of later Paleozoic age
- D3 Miogeosynclinal deposits of early to late Paleozoic age
- D2 Eugeosynclinal deposits of late Precambrian to middle Paleozoic age
- D1 Granitic rocks of Paleozoic age

PRECAMBRIAN ROCKS AND STRUCTURES

- C4 Sedimentary and volcanic rocks of middle and later Precambrian age
- C3 Metamorphic and plutonic rocks of later Precambrian age
- C2 Metamorphic and plutonic rocks of middle Precambrian age
- C1 Metamorphic and plutonic rocks of earlier Precambrian age

PLATFORM AREAS

- B Platform deposits overlying basement rocks of Paleozoic age
- A Platform deposits overlying basement rocks of Precambrian age

STRUCTURAL SYMBOLS

- Axis of closely compressed anticline
- Axis of broad anticline or anticlinorium
- Axis of syncline
- Dip of strata in homocline
- Thrust fault. Barbs on upthrown side
- Normal fault. Hachures on downthrown side
- Transcurrent fault. Arrows show strike-slip displacement
- Concealed fault
- Inferred fault
- Salt dome
- Volcanic cone
- Caldera

Structure contours on surface of basement rocks. Red on Precambrian basement, purple on Paleozoic basement. Contour interval 1,000 meters, or 3,280 feet.

EXPLANATION

CENOZOIC

 s   Slumps, landslides and rockfalls

PALEOZOIC

PERMIAN

Pt   Toroweap Formation

Pc   Coconino Sandstone; sharp level contacts with Toroweap Formation and Hermit Shale

Ph   Hermit Shale

PENNSYLVANIAN   SUPAI GROUP

Pe   Esplanade Sandstone

MISSISSIPPIAN

Mr   Redwall Limestone

DEVONIAN

Dtb   Temple Butte Limestone; local channel filling only in the Kwagunt, Nankoweap and Marble Canyon areas

CAMBRIAN

Єm   Muav Limestone; includes overlying undifferentiated rocks

Єba   Bright Angel Shale; gradational contacts with Muav Limestone and Tapeats Sandstone

Єt   Tapeats Sandstone

YOUNGER PRECAMBRIAN

UNKAR GROUP

P€i   Predominantly Diabase intrusives; sills and dikes

P€d   Dox Formation; sandstone

P€s   Shinumo Formation; quartzite

P€h   Hakatai Formation; shale

P€b   Bass Formation; includes Hotauta Conglomerate Member

OLDER PRECAMBRIAN

ZOROASTER PLUTONIC COMPLEX

P€gr₁   Granite to granodiorite, relatively poor in mafic minerals

P€gr₂   Granodiorite to quartz diorite and rarely diorite, relatively rich in mafic minerals

VISHNU GROUP

P€vs   Predominantly mica schist and quartzo-feldspathic schist, with minor units of para-gneiss, amphibolite and calc-silicate

P€va   Predominantly amphibolite

P€vc   Predominantly calc-silicate rock

Color Map 4. Part of the Bright Angel quadrangle, Arizona.

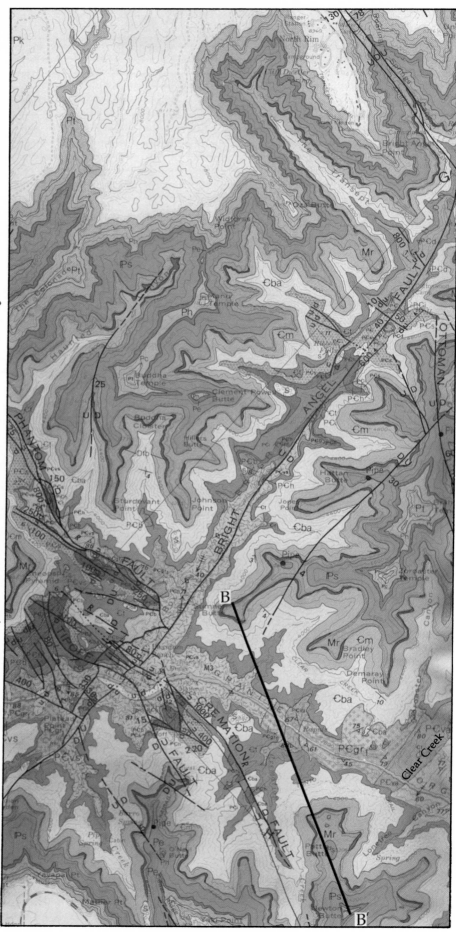

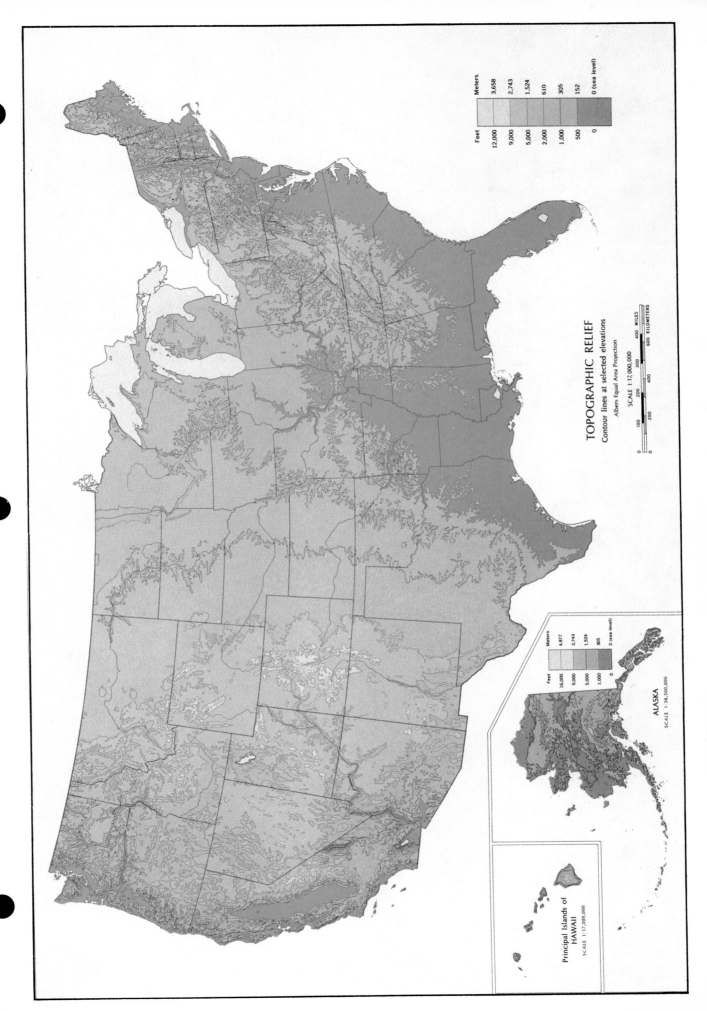

Color Map 5. Topographic Relief of the United States.

Color Map 6. Classes of Land Surface Forms, United States.